FOOD JOURNAL

Breakfast	Servings	Calories	
		Subtotal	

Snack			
		Subtotal	

Lunch			
		Subtotal	

Snack			
		Subtotal	

Dinner			
		Subtotal	

Snack			
		Subtotal	

Total Calories From Food

FITNESS ACTIVITY JOURNAL

	Duration	Calories

Total Calories From Fitness

NOTES

FOOD JOURNAL

Breakfast	Servings	Calories
	Subtotal	

Snack		
	Subtotal	

Lunch		
	Subtotal	

Snack		
	Subtotal	

Dinner		
	Subtotal	

Snack		
	Subtotal	

Total Calories From Food

FITNESS ACTIVITY JOURNAL

	Duration	Calories

Total Calories From Fitness

NOTES

Made in the USA
Monee, IL
03 November 2026

67376786R00056

Southern Space Studies

The Southern Space Studies series presents analyses of space trends, market evolutions, policies, strategies and regulations, as well as the related social, economic and political challenges of space-related activities in the Global South, with a particular focus on developing countries in Africa and Latin America. Obtaining inside information from emerging space-faring countries in these regions is pivotal to establish and strengthen efficient and beneficial cooperation mechanisms in the space arena, and to gain a deeper understanding of their rapidly evolving space activities. To this end, the series provides transdisciplinary information for a fruitful development of space activities in relevant countries and cooperation with established space-faring nations. It is, therefore, a reference compilation for space activities in these areas.

More information about this series at http://www.springer.com/series/16025

Annette Froehlich
Editor

Space Fostering African Societies

Developing the African Continent through Space, Part 1

Springer

Editor
Annette Froehlich
SpaceLab
University of Cape Town
Rondebosch, South Africa

ISSN 2523-3718 ISSN 2523-3726 (electronic)
Southern Space Studies
ISBN 978-3-030-32929-7 ISBN 978-3-030-32930-3 (eBook)
https://doi.org/10.1007/978-3-030-32930-3

This Springer imprint is published by the registered company Springer Nature Switzerland AG
The registered company address is: Gewerbestrasse 11, 6330 Cham, Switzerland

Dedication

Dedicated to Voortrekker—Namibia's Old Man of the Desert

I dream of our vast deserts, of our forests, of all our great wildernesses. We must never forget that it is our duty to protect this environment.
—Nelson Mandela[1]

"Voortrekker", credits: EHRA, http://www.desertelephant.org/

This publication is dedicated to the iconic Namibian desert elephant known worldwide as Voortrekker ("Pioneer"/"Leader"), who was killed by a trophy hunter in June 2019. Voortrekker's plight in many ways exemplifies the plight of African wildlife in general, and of African elephant populations in particular, which are

[1]Scott Ramsay, "Wild Mandela: conservationist and lover of the land", *News24*, January 29, 2014, https://www.news24.com/Travel/South-Africa/Wild-Mandela-conservationist-and-lover-of-the-land-20140128 (accessed July 30, 2019).

facing increasing pressure from poaching, habitat loss and human-wildlife conflict.[2] The World Wide Fund For Nature estimates that there are around 4,15,000 African elephants remaining, down from 3 to 5 million in the early twentieth century.[3] Accompanying this dramatic decline in population numbers is the loss of over 50% of the African elephant range since 1979.

Namibia is well known for its desert elephant population (one of only two such populations, the other being in Mali's Sahara Desert).[4] Voortrekker was considered to be the most important (and best-known) bull elephant of the remaining herd of 120 Namibian desert elephants.[5] He was a pioneer for this elephant population in the Ugab and Huab rivers region of Namibia and was one of the first to return to the region following the years of poaching, hunting and conflict which resulted in a decimation of the elephant population.[6] Historically, all elephants had been wiped out of the region by the early 1980s, "shot out by poachers and for sport by former apartheid-era Cabinet Ministers—and of course cattle farmers intent on driving them off their land and back into the Etosha National Park", but with Voortrekker's hopeful return to the area by the late 1980s (along with a group he led from Etosha National Park), "they had become a permanent feature and unique tourist attraction".[7]

Despite this hopeful return, the small population of Namibian desert elephants in the region (reportedly numbering 62 individuals in 2016[8]) is nevertheless declining yearly, and Voortrekker was one of only two breeding bulls left, while all nine calves born since 2014 had died within a week—a "sign of a distressed population".[9] With Voortrekker's killing, "the writing appears to be on the wall for this

[2]World Wide Fund for Nature, "African Elephants", 2019, http://wwf.panda.org/knowledge_hub/endangered_species/elephants/african_elephants/ (accessed July 30, 2019).

[3]Ibid.

[4]Louzel Lombard Steyn, "Namibia's desert elephants back on the butcher's block", *Conservation Action Trust*, August 30, 2017, https://conservationaction.co.za/media-articles/namibias-desert-elephants-back-butchers-block/ (accessed July 30, 2019).

[5]Megan Carr, "Who killed 'Voortrekker' the dominant iconic desert elephant in Namibia", *Global March for Elephants and Rhinos*, June 30, 2019, https://www.change.org/p/us-interior-secretary-global-march-for-elephants-and-rhino-demands-ban-on-imports-of-trophies-of-endangered-species-from-africa/u/24767796?recruiter=398200968&utm_source=share_update&utm_medium=facebook&utm_campaign=facebook (accessed July 30, 2019).

[6]Jasmine Stone, "The Terrible Truth Behind The Shooting of An Iconic Namibian Elephant", *2oceansvibe*, July 1, 2019, https://www.2oceansvibe.com/2019/07/01/the-terrible-truth-behind-the-shooting-of-an-iconic-namibian-elephant/ (accessed July 30, 2019).

[7]John Grobler, "Iconic Namibian elephant 'Voortrekker' killed by trophy hunter", *IOL Independent Media*, July 1, 2019, https://www.iol.co.za/news/africa/iconic-namibian-elephant-voortrekker-killed-by-trophy-hunter-28317901 (accessed July 30, 2019).

[8]Louzel Lombard, "Namibian Tourism Ministry brushes off questionable killing of desert elephants", *Conservation Action Trust*, October 24, 2017, https://conservationaction.co.za/media-articles/namibian-tourism-ministry-brushes-off-questionable-killing-desert-elephants/ (accessed July 30, 2019).

[9]John Grobler, "Iconic Namibian elephant 'Voortrekker' killed by trophy hunter", *IOL Independent Media*, July 1, 2019, https://www.iol.co.za/news/africa/iconic-namibian-elephant-voortrekker-killed-by-trophy-hunter-28317901 (accessed July 30, 2019).

small group of hardy survivors".[10] This is especially tragic since the population cannot be replaced by savannah elephants, since they do not possess the behaviour or skills needed for desert survival (for example, storing water in a pouch in their throat, using their tusks to dig wells, or finding water where there seems to be none on the surface).[11] A 2016 study found that Namibian desert-dwelling elephants have adapted to survive in very harsh conditions by "covering their bodies with sand wetted by their urine or regurgitated water from a specialized pouch beneath their tongue that holds many gallons of water", that they "remember the location of scarce water and food resources across their home ranges, which are unusually large compared to those of other elephants", and that these elephants "play a critical role in this arid ecosystem by creating paths and digging watering holes".[12] In particular, these elephants possess "unique knowledge and survival skills", such as how to survive in the desert—knowledge which is "crucial to the survival of future generations of elephants in the arid habitat, and pressure from hunting and climate change may only increase in the coming decades".[13] The adaptations of this desert-dwelling elephant population are "not genetically transferred to the next generation, rather through the passing on of knowledge by mature individuals", while "[m]orphological differences, like the adapted elephants' thinner bodies and wider feet, also distinguish them from typical savannah elephants".[14]

Local communities of Otjimboyo, Sorris Sorris and Tsiseb conservancies voiced their concern at the killing of Voortrekker (for a hunting license cost of N $1,20,000–about 7,600 Euro or US$ 8,500), noting that:

> These elephants are our resources, and we object to them being hunted for problems caused by different populations of elephants. (…) It is our belief that the shooting of elephants does not solve the problem. In fact, this only makes it worse. We want to keep our communities safe and to do this we need to ensure that our elephants are calm and relaxed when entering villages. It is our belief that the shooting of elephants or scaring them off with gunshots, screaming or chasing them off results in aggressive animals and this cannot be tolerated.[15]

The government of Namibia had claimed that Voortrekker was a "problem elephant" causing property and infrastructure damages in the area, while others

---

[10]Ibid.

[11]C. J. Carrington, "Namibia And The Sacrifice of Rare Desert Elephants", *The Dodo*, July 17, 2014, https://www.thedodo.com/namibia-and-the-sacrifice-of-r-625424514.html (accessed July 30, 2019).

[12]Carl R. Woese Institute for Genomic Biology, University of Illinois at Urbana-Champaign, "Desert elephants pass on knowledge—not mutations—to survive", *ScienceDaily*, August 3, 2016, www.sciencedaily.com/releases/2016/08/160803161607.htm (accessed July 30, 2019).

[13]Ibid.

[14]Louzel Lombard Steyn, "Questionable killing of Namibia's desert elephants", *Africa Geographic*, October 25, 2017, https://africageographic.com/blog/questionable-killing-namibias-desert-elephants/ (accessed July 30, 2019).

[15]Jasmine Stone, "The Terrible Truth Behind The Shooting of An Iconic Namibian Elephant", *2oceansvibe*, July 1, 2019, https://www.2oceansvibe.com/2019/07/01/the-terrible-truth-behind-the-shooting-of-an-iconic-namibian-elephant/ (accessed July 30, 2019).

disputed this by pointing to a "smaller herd, aggressive and frightened by farmers shooting at them, [which] may be the real cause of the Omatjete constituency's complaints [of damages] that led to Voortrekker's death warrant being issued".[16] Voortrekker was remembered as an "incredibly gentle, peaceful and magnificent elephant. His presence has often calmed other inexperienced elephants around him. He was known locally as the 'Old Man', that was always welcome because he never caused any problems or induced fear."[17] The Namibian government had previously issued a hunting permit for Voortrekker in 2008.[18] At that time, it took the actions of 10 dedicated women who:

> took up Voortrekker's cause, and walked 140 kilometers (about 87 miles) through the desert in order to raise the funds needed to buy the bull elephant's permit. His hunting tag was successfully purchased from the Government for a total of $12,000 USD, as a live trophy. The other five elephants had lost their lives, but Voortrekker was now a living legend.[19]

Voortrekker's killing earlier this year (2019), despite this previous effort, elicited a heartfelt outcry, including a statement by Johannes Haasbroek, whose organisation (Elephant Human Relations Aid) is devoted to minimising human-wildlife conflict, particularly in terms of the desert-adapted elephants:

> The iconic bull "Voortrekker" has been murdered by a trophy hunter. He was the last large dominant bull amongst the 120 desert dwelling elephant left in the North West deserts of Namibia. He was targeted not for anything but his fame. We bought a license to hunt him in 2008 and for 10 years the hunting outfitters and their sick clients conspired to get this gentle giant declared a problem to justify a hunt. He never stepped out of line. I lived and fought and cried for that gentleman. I have no words anymore. Let the planet die now. With him. All that is left for me is to watch and weep. See you in a better world my friend. This one was not meant for us. I failed you. Rest in peace "Voortrekker".[20]

Instead of allaying these concerns, the response to this outcry by the Namibian Ministry of Environment and Tourism (MET) raised further questions:

> It was shot to generate funds for the affected communities. We had the elephant hunted as a trophy and we do not entertain the naming of wild animals. That is one of the characteristics that separates wild from domestic animals. Naming animals also triggers emotional attachment to a certain or specific animal which may overshadow our judgement in wildlife conservation. It should be noted further that the MET is not here for a popularity contest.

---

[16]Ibid.

[17]Johannes Haasbroek of Elephant Human Relations Aid, quoted in Africa Geographic Editorial, "Iconic desert-adapted elephant 'Voortrekker' killed by trophy hunter in Namibia", African Geographic, June 30, 2019, https://africageographic.com/blog/iconic-desert-adapted-elephant-voortrekker-killed-by-trophy-hunter-in-namibia/ (accessed July 30, 2019).

[18]C. J. Carrington, "Voortrekker, Legendary Elephant, Under Threat in Namibia", *The Dodo*, May 8, 2014, https://www.thedodo.com/living-legend-of-the-namib-voo-659043534.html (accessed July 30, 2019).

[19]Ibid.

[20]Johannes Haasbroek of Elephant Human Relations Aid, quoted in Global March for Elephants and Rhinos, "Voortrekker", July 1, 2019, https://www.facebook.com/March4Elephants/posts/3069517189733078?comment_id=3070219739662823 (accessed July 30, 2019).

> We make decisions based on what is good for our conservation based on the existing principles, policies and legislation. It's unfortunate that the elephant was put down but we were left with no other alternative after this specific animal continued to cause damages to property in the area.[21]

Despite the MET's rejection of the naming of Voortrekker, he was an ambassador for Namibian tourism and conservation, and was beloved by tourists, as a statement issued by German tourism operators following the killing clearly echoes:

> Many people very much enjoy watching elephants in Namibia. They love elephants. Especially the desert elephants of Namibia are fascinating creatures. Elephant bulls with big tusks are not only important genetically, they are also critical for the education of younger bulls. And of course those bulls are the highlights of every photo safari. Our guests love nature and leave a lot of money in Namibia. The pictures of Voortrekker are up on many walls in Germany. We as nature lovers and tourists are shocked about the decision of MET to have Voortrekker killed. (…) We cannot bring Voortrekker back to life. But we are very much in favour of not issuing any permits for hunting of desert elephants in future. The elephants of Namibia are worth a lot more than those dollars of hunters.[22]

It remains to be seen whether Namibia's iconic desert elephants will endure, or whether photographs on walls will be the only trace left of them for the enjoyment of future generations. Nelson Mandela's words ring louder than ever before, but it is up to us whether we will conserve this great African legacy or sell it off to the highest bidder.

[21]Ibid., Statement by MET Spokesperson Romeo Muyunda four days after the hunt of Voortrekker.

[22]Rainer Stoll, "Voortrekker", *travel-to-nature*, July, 2019, https://www.facebook.com/VoortrekkerTheDesertElephant/photos/a.433136714205248/433137520871834/?type=3&theater (accessed July 30, 2019). Translated from German.

Foreword

The publication on African Countries activities in Space demonstrates the ever-growing interest of not only State actors, but individuals of African descent in Space exploration and its con-commitment benefits. The UN Committee of the Peaceful Uses of Outer Space has provided the platform for such interests to be actualized through partnerships and collaboration with other State Parties quite advanced in knowledge about space exploration and activities. The Committee has over the years helped to enhance international corporation in the peaceful uses of Outer Space and its benefit to all members.

International cooperation in Space exploration and the use of Space technology application should be a priority to the African Continent because of its benefits in agriculture, food security, water availability, disaster management and health, which would help to advance the African Agenda 2063.

To this end, the recent establishment of the African Space Agency at the 32nd Ordinary Session of the African Union is a welcome development. Africa needs to build up its capacity in the use of Space technology applications and this publication which aims to share knowledge and best practices is a step in the right direction. It is equally important for Space actors with proven capabilities, to make available the information and know-how, through the writings in this publication for the benefit of its readers.

I commend this publication for the important role it continues to play in raising awareness on the place of Space science and technology in Africa's socio-economic development.

Vienna, Austria

Ambassador Vivian N. R. Okeke
Permanent Representative
of Nigeria to UNOOSA

Contents

Estimation of the Impact of Alien Trees in the Cape Town Water Crisis Using Satellite Data

James Wilson and Maureen Tanner

Abstract
This section investigates a proposed space-based application for South Africa: using remote sensing data to identify alien trees located in catchment areas in the Western Cape. The various satellite datasets used for the analysis are discussed and then the identification process is defined. Once a forest of alien trees is identified, calculations are performed to quantify how much water is being lost to these trees. The result of a sample calculation is that a small section of alien forest consumes approximately ten million litres a day, comparable to the amount of fresh water a medium-sized desalination plant can consume.

1 Introduction

Tall alien trees, such as pines, eucalypts and wattles can have a negative impact on the supply of water to dams when situated in catchment areas. Studies in South Africa have demonstrated that, in contrast to indigenous vegetation, such alien trees tend to reduce the annual and low-season streamflows while increasing the amount

J. Wilson (✉) · M. Tanner
University of Cape Town, Rondebosch, South Africa
e-mail: WLSJAM015@myuct.ac.za

M. Tanner
e-mail: mc.tanner@uct.ac.za

A. Froehlich (ed.), *Space Fostering African Societies*, Southern Space Studies,
https://doi.org/10.1007/978-3-030-32930-3_1

Table 1 Water reduction estimation for eucalypts, pines and wattle

Taxon	Common names	Relative water-use	Growth rate	Growing conditions	Annual flow reduction factor (%)
Eucalyptus spp.	Eucalypts	High	Fast	Optimal	90
				Sub-optimal	72
Pinus spp.	Pines	High	Moderate	Optimal	87
				Sub-optimal	57
Wattle species	Black, silver, green wattle	High	Fast	Optimal	90

of water lost to evapotranspiration due to their high biomass.[1] Some studies have even estimated the reduction in annual run-off to be in the order of 100–300 mm per year.[2] A closer examination of the water usage rate of these alien trees reveals even more striking figures as shown in Table 1.[3]

Invasive alien tree species have high water usage needs and exhibit moderate to fast growth rates thus resulting in high annual flow reductions. For instance, a eucalypt tree has an average consumption of 15–20 L of water daily,[4] while a pine tree has an average daily consumption of 25–45 L.[5] Altogether, such invasive alien trees are estimated to use more than 28 mm^3 of water yearly (see footnote 3). On a daily basis, this amounts to 104 MLD (million litres of water per day) that is being consumed by invasive alien trees located in catchment areas. Without the presence of these trees in the catchment areas, a large portion of this water would runoff the catchment area and be collected in the dams.

The City of Cape Town has an augmentation target of 150 MLD. It is therefore argued that the identification, removal and monitoring of alien trees around catchment areas could assist the City in meeting its augmentation target and are thus viable solutions to the water crisis. This is in line with past studies that indicate that such interventions could have positive impacts on the economics of hydrological planning.[6]

[1]Dye P. and Jarmain C. (2004). Water use by black wattle (Acacia mearnsii): implications for the link between removal of invading trees and catchment streamflow response. S. Afr. J. Sci. 100, 40–44.

[2]Gorgens, A.H.M. and Van Wilgen, B.W., 2004. Invasive alien plants and water resources in South Africa: current understanding, predictive ability and research challenges: Working for Water. *South African Journal of Science*, *100*(1–2), pp. 27–33.

[3]Le Maitre, D.C., Forsyth, G.G., Dzikiti, S. and Gush, M.B., 2016. Estimates of the impacts of invasive alien plants on water flows in South Africa. *Water Sa*, *42*(4), pp. 659–672.

[4]Water Requirements for Eucalyptus Plantation, nd, Available at www.eucalyptus.com.br/eucaexpert/Pergunta%20705.doc.

[5]Forestry Facts, n.d. available at https://www.sabie.co.za/about/forestry/.

[6]Gorgens and Van Wilgen (2004).

It is therefore proposed that the use of GIS mapping of infested areas around dams and water catchment areas (focusing on Eucalypts, Pines and Wattles) could assist in the clearing of these invasive species with the aim of addressing the water crisis. Remote sensing data can be used to automate the detection, classification and density of alien trees in target areas to assist the City with removal programmes. Previous research also stated that remotely sensed imagery can be used to sense invasive plants.[7] The results can be integrated into a GIS system, which prioritises areas based on ease of access, site conditions, density of alien trees, potential water savings and other factors to develop effective and efficient removal activities. The application can also be used to ensure that tree removals are satisfactory and have indeed been done, as well as monitor the regrowth in areas where alien trees have already been removed.

2 Existing Approaches

Various solutions have been proposed in past literature to address the invasive alien trees identification issue through remote sensing. Specifically, Moderate Resolution Imaging Spectroradiometer (MODIS) and Hyperspectral Remote Sensing (Infrared) are known to be useful in this regard. In South Africa, there is growing interest in the use of remote sensing across a range of spatial scales, for operational forestry applications.[8]

MODIS offers low spatial resolution and high temporal resolution and has been found to be useful for tracking of landscape and vegetation changes over time.[9] Hyperspectral remote sensing (Infrared) has been found to be useful in the detection of invasive alien plants since it allows for differentiation between plant groups and species using high spectral resolution.[10] Invasive alien plants might also have unique structural and chemical characteristics that can be detected using hyperspectral measurements.[11] Other approaches use statistical models based on the growth characteristics of a species to determine the likelihood of its presence in a specific area.

[7]Bradley, B.A., 2014. Remote detection of invasive plants: a review of spectral, textural and phenological approaches. *Biological invasions*, *16*(7), pp. 1411–1425.
[8]Albaugh, J.M., Dye, P.J. and King, J.S., 2013. Eucalyptus and water use in South Africa. *International Journal of Forestry Research* .
[9]Lu, L., Kuenzer, C., Wang, C., Guo, H. and Li, Q., 2015. Evaluation of three MODIS-derived vegetation index time series for dryland vegetation dynamics monitoring. *Remote Sensing*, *7*(6), pp. 7597.
[10]Schmidt, K.S. and Skidmore, A.K., 2003. Spectral discrimination of vegetation types in a coastal wetland. Remote Sens. Environ, 85, 92–108.
[11]Asner, G.P.; Martin, R.E.; Anderson, C.B.; Knapp, D.E., 2015 Quantifying forest canopy traits: Imaging spectroscopy versus field survey. Remote Sens. Environ, 158, 15–27.

3 Platform Comparison

There exist several platforms and mapping applications that could be used for this project. Remote sensing data from Landsat, Sentinel, Planet and MODIS was considered for this purpose and QGIS was used to process the various datasets. It was found that the resolution of Landsat data was too low for the given purpose and as such analysis focused on the other datasets.

Planet offer a high-end, API driven platform with extensive functionalities that could assist with the mapping and projection of alien plant species around dams and catchment areas. They offer a comprehensive suite of products that could be highly useful to the project (e.g. planet monitoring, planetbasemaps, planet imagery and archives, planet testing, planet explorer and platform).[12] However, the platform is not free of charge and would require some form of investment from the City of Cape Town, not only to obtain the data, but also to develop the software platform based on Planet's API. It is possible that the City could approach Planet to negotiate data exchange to help with the water crisis.

In contrast, QGIS is an open-source mapping application that does not require any funding. In the past, QGIS has successfully been used for the mapping of plant species[13] as well as the tracking of plant invasions.[14] The tool was therefore considered to be ideal for the project.

4 Proposed Solution

A comprehensive solution using remote sensing should include distribution mapping as well as projection of the invasive alien plants. Distribution mapping helps with early detection of new invasions, provide information to support the formulation of management strategies and the assessment of what impact these invasions might have on the environment.[15] Furthermore, the projection of future distribution of the invasive alien plants (e.g. identifying risk areas) can also assist in the management of these invasions (e.g. monitoring).[16] The proposed solution can also

[12]https://www.planet.com/.

[13]Stropp, J., Ladle, R.J., Malhado, M., Ana, C., Hortal, J., Gaffuri, J., H Temperley, W., Olav Skøien, J. and Mayaux, P., 2016. Mapping ignorance: 300 years of collecting flowering plants in Africa. *Global Ecology and Biogeography*, *25*(9), pp. 1085–1096.

[14]Hellmann, C., Werner, C. and Oldeland, J., 2016. A spatially explicit dual-isotope approach to map regions of plant-plant interaction after exotic plant invasion. *PloS one*, *11*(7), p. e0159403.

[15]Guo, Y., Graves, S., Flory, S. L., & Bohlman, S. (2018). Hyperspectral Measurement of Seasonal Variation in the Coverage and Impacts of an Invasive Grass in an Experimental Setting. *Remote Sensing*, *10*(5), 784.

[16]Garzon-Lopez, C. X., Hattab, T., Skowronek, S., Aerts, R., Ewald, M., Feilhauer, H., … & Schmidtlein, S. (2018). The DIARS toolbox: a spatially explicit approach to monitor alien plant invasions through remote sensing. *Research Ideas and Outcomes*, *4*, e25301.

be used to ensure clearances are satisfactorily performed and also to monitor regrowth post-clearance.

The method used to identify the alien species was developed in correspondence with industry experts from CSIR, SANSA and SAEON. Due to the time limits of the project, a statistical approach was ruled out. Furthermore, automated differentiation between the three separate alien species was also eliminated. The envisioned approach was to use Sentinel and Landsat data (in the visual and infrared bands), along with SAR data to identify areas with alien trees. In order to automate the identification process, the following differentiating factors were identified: tree height, leaf colour, forest age and growth rate. The forest age helps identify alien trees as natural fynbos forests reach maturity after three years and then their growth plateaus. After analysing the data, it was determined that automation of the identification process would not be possible in the time constraints. However, considerable work was done on determining forest age using MODIS data, which is discussed briefly below. The Theewaterskloof western catchment area was decided as the target area for the project after discussions with SAEON.

Fynbos tends to grow slower than most alien trees and plateaus in height at around 1–2 m off the ground, this occurs after a period of 3–5 years depending on recovery conditions.[17] This fact about fynbos allows us to easily distinguish it from alien trees which often grow much taller and much faster.[18] The primary way of identifying the approximate age of natural vegetation (not plantations) is to look at when last a fire broke out in the area. Veld fires are very much a part of the fynbos regions natural cycle, periodically burning through areas of dry older bush making way for new plants to grow in their place.[19]

In the last few years there have been more fires than would be expected, coupled with the drought conditions, fires have been wider spread and far more devastating to the landscape and the people living around it. In some of these cases there have been people arrested for arson,[20] however the presence of alien trees has also exacerbated the extent and intensity of these fires as they grow tall and suck up much of the water around them leaving dry vegetation that is not fire resistant. The natural fynbos is naturally drought resistant and there are many succulents that reduce the effects of a fire, however they do not survive under the shade and in the acid soils of the alien trees.[21] Alien tree fires are far more sever and uncontrollable than fynbos fires.

[17]Cape Nature. 2016, Field Guide to Wild Flower Harvesting.
[18]Morris, T.L. and Cramer M.D. (2011). Ecophysiological traits associated with the competitive ability of invasive alien plants. Diversity and Distributions, 17, 898–910.
[19]Kruger, F.J. and Bigalke R.C. (1984). Fire in Fynbos, Ecological effects of fire in South African ecosystems. Ecological studies, 48, 67–114. Springer Verlag, Berlin.
[20]https://www.iol.co.za/capeargus/news/suspected-arsonist-arrested-for-starting-tablemountain-fires-14217949.
[21]Mamders, P.T. and Richardson D.M. (1992). Colonization of Cape Fynbos species by Forest species. Forest Ecology and Management, 48, 277–293.

Downloading the MODIS and VIIRS data, which is freely available on NASA's Fire Information for Resource Management Systems (FIRMS) website through the Active Fire data porthole,[22] one can overlay this data onto a map of the Western Cape and its catchments and then identify areas of forest that have experienced fires in the last 3–5 years—the time frame expected for indigenous and alien vegetation to be distinguishable. This information becomes useful as the areas can then be further compared with active radar imagery and optical imagery to tell if the vegetation is fynbos or alien trees.

The active radar data will indicate the average height of the vegetation by selecting the appropriate bands and analysing the reflection off the vegetation. The optical data should substantiate the data of the SAR by indicating the density of the vegetation by its regularity and colour. Fynbos is usually a mix of greens, ranging from dark to light, whereas alien tree forests like the ones we are looking for are a dark green colour with little variation. These two analysing techniques provide conclusive evidence that the area is in fact fynbos or alien trees.

5 A Simple Approach to Identifying Alien Forests Around Catchments Areas

At present, there is a huge amount of freely available Earth observation data, however the challenge is in finding and interpreting what is useful for a given objective.

For the task of identifying alien trees, many sources were generally too challenging to use. For example, free SAR data is available from the Sentinel-1 satellites, however it was discounted as the resolution is too low to identify trees (at 15 m[23]) without coherence and polarimetric decomposition analysis requiring many hours of computing time and data time. Furthermore, foreshortening and shadows in the mountainous regions around catchments results in the complete loss of applicable data. MODIS data is useful in showing where fires have occurred and also evapotranspiration measurements but again its resolution is much too low to identify smaller clusters of trees.[24] Sample Planet data was obtained as well, with a resolution of 3 m. It is easier to identify vegetation using this data but it is again difficult to identify specific plant species without a statistical approach involving large amounts of processing, most likely involving a computer cluster.

[22] https://earthdata.nasa.gov/earth-observation-data/near-real-time/firms/active-fire-data.

[23] Sentinel Online. 2018. User Guides—Sentinel-1 SAR—Resolution. [ONLINE] Available at: https://sentinel.esa.int/web/sentinel/user-guides/sentinel-1-sar/resolutions. [Accessed 13 June 2018].

[24] DAAC. 2018. Moderate Resolution Imaging Spectroradiometer (MODIS). [ONLINE] Available at: https://ladsweb.modaps.eosdis.nasa.gov/missions-and-measurements/modis/. [Accessed 13 June 2018].

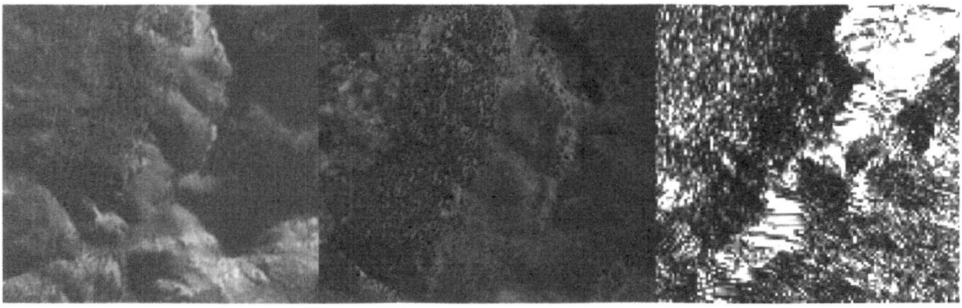

Fig. 1 (From left to right) Planet 3 m—11 June 2018, digital globe 0.5 m—30 November 2017, Sentinel 1 SAR VH—26 May 2018

Within the time constraints of the given project, Google Earth appeared to be the only single useful way to identify trees, albeit without automation and classification. Google Earth uses multiple data sources but in this case Digital Globe 0.5 m imagery is used. Google Earth may be more out-of-date than faster updating sources such as Planet, however pine trees (one of the principle alien threats) take on average 20 years to mature fully,[25] and therefore at least for identifying existing trees, the temporal resolution of imagery is not too important. A comparison of different datasets is shown in Fig. 1, clearly showing the benefit of older yet higher resolution imagery. These images are of a pine forest on a slope just north of Theewaterskloof Dam.

5.1 Identification Procedure

The manual identification of trees using Google Earth—Digital Eye imagery is straightforward. First trees/shrubs were identified visually, thereafter, the diameters of the plants were measured using Google Earth's ruler. This is shown in Fig. 2.

The knowledge of this, combined with the fact that there is no fynbos plant of this size,[26] and that virtually no natural trees occur outside of the riparian zone (this was according to the interview with Dr. Jasper Slingsby of SAEON), means that these plants are definitively alien. They can be identified as pine trees due to the elimination of other species. They are larger and darker green than black wattles, and eucalyptus does not grow on large-gradient slopes. (Ideally, the specific species would have to be verified by field observation, but the fact that they can be identified as alien is sufficient for this objective.)

[25]Forestry. 2006. Sabie. [ONLINE] Available at: https://www.sabie.co.za/about/forestry/. [Accessed 13 June 2018].

[26]Cape Nature. 2016. Field Guide to Wild Flower Harvesting.

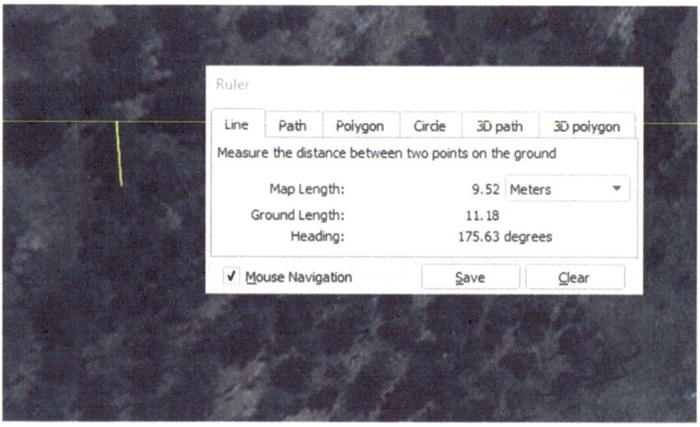

Fig. 2 Measurement of plants—image is taken from same region as Fig. 1

Fig. 3 Identified forests and affected river

Once the trees have been identified, they can be manually highlighted as vector polygons in Google Earth. Highlighted pine forests around the mountain directly north of Theewaterskloof Dam are shown in Fig. 3, along with the affected river that has had its flow reduced by these trees.

5.2 Estimation of Water Use

The water use of a forest depends on many factors, notably season, density, soil type and rainfall amount. It is difficult to be highly accurate in estimation, and therefore the method will be conservative. The forests identified in Fig. 3 will be used for these calculations. Note that this area is in a reserve and therefore is not industrial forestry.

The first step is to calculate the area of the forests. The drawn polygons were converted from KML to a shapefile and imported into QGIS (Fig. 4).

The area function in QGIS was then used to calculate the area of the forests, shown in Table 2.

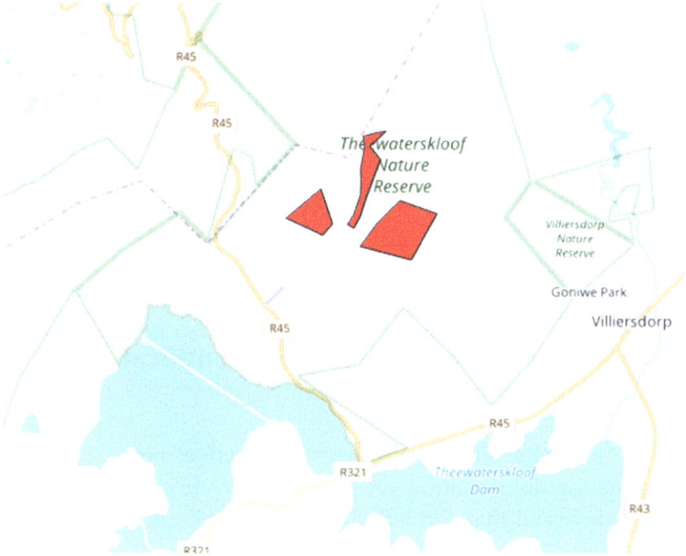

Fig. 4 Forests in QGIS over open street maps

Table 2 Area of forests

Name	Area (m^2)
Pine forest 1	767 287
Pine forest 2	815 665
Pine forest 3	1 904 249
Total	3 487 201

The density of trees then needs to be known. By visual count, the density of trees ranged from 1 per 10 m^2 to about 1 per 25 m^2 or (0.1 to 0.025 per m^2 averaged to 0.075 per m^2). A paper by the CSIR identifies two major species of pine to have a weighted total water use of 129.58 litres per year, or 35.5 litres per day.[27] Therefore, the total water use can be estimated by:

$$Water\ Use = Area \times Density\ of\ trees \times water\ use\ per\ tree$$

$$Water\ Use = 3\,487\,201 \times 0.075 \times 35.5$$

$$Water\ use = 9\,284\,673\ \mathrm{L/day}$$

This number is staggering. A single mountain slope north of Theewaterskloof results in the use of almost 10 million litres of water a day, which constitutes a significant portion of the City's deficit.

People may argue that the transpiration of trees results in the formation of clouds which results in more rain anyway, however this argument is invalid. Again, according to the interview with Dr. Slingsby, the forests are not large enough to transpire enough moisture to form large clouds, and due to the high amount of wind in the Cape, any moisture is quickly blown to a different area. Many people do not realise the impact of pines because they look aesthetically pleasing, but it is massive, and satellite imagery helps to visualise and quantify it.

5.3 Further Work with Regards to This Identification and Estimation Method

This method of manually highlighting trees is effective, however it is slow, and does not account for deforestation measures, or the removal of forests by fire. Therefore, it is suggested that further work be done to automate this process and account for changes.

The easiest way would be to first manually identify the forests around catchment areas as done in the previous section using Google Earth. The catchment regions in the Western Cape are very small, and hardly any natural forests exist outside of the riparian zone. Once these polygons have been drawn they can be extracted to a GIS platform such as QGIS or even Google Earth Engine. Once this has been done, Planet data could be used. Planet has a fully built API with a coherence analysis (visual representation in Fig. 5). With each Planet image of the area becoming available, this coherence analysis could be automatically run within the polygons to determine if there has been pixel change, thereby deducing if the forest has grown more or been removed. This can all be fully automated once the initial polygons

[27]Gush, M.B., Dye, P.J., Geldenhuys, C.J. and Bulcock, H.H., 2011. Volumes and efficiencies of water-use within selected indigenous and introduced tree species in South Africa: Current results and potential applications. In: Proceedings of the 5th Natural Forests and Woodlands Symposium, Richards Bay, 11–14 April.

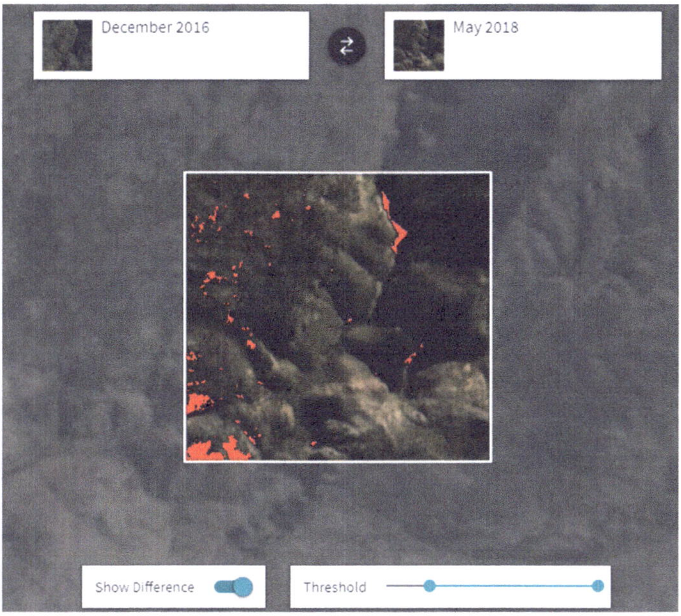

Fig. 5 Coherence analysis in planet explorer (note forest has not changed)

have been manually identified. This is useful in that removal of alien species can be accurately monitored rather than the government or NGOs spending money for removal but not being able to track progress without performing actual site visits.

James Wilson graduated from Electrical Engineering at the University of the Witwatersrand in 2017. He has a large interest in Space technology and hence pursued a Master of Philosophy in Space Science at the University of Cape Town in 2018. He is also a guitar player in an alternative rock band. He aims to work with rocketry as he completes his Master's dissertation designing a control system for a liquid rocket test stand.

A/Prof. Maureen Tanner is in charge of the Honours Programme in the Department of Information Systems of the University of Cape Town. Her research interests lie in Big Data, Space Traffic Management, Agile software development related issues, software engineering and social aspects of social engineering. She is now undertaking a M.Phil. in Space Studies at the University of Cape Town.

Olive Tree Classification and Inventory with Medium Resolution Multi-spectral Satellite Imagery

Ranya Mezzi, Mitchel Alioscha-Perez, Mohamed Allani, Fatma Guedri, Adel Zouabi, Ridha Beji, Hichem Sahli and Ali Sahli

Abstract
Remote sensing is the most widely useful tool for land use planning and decision support system. It gives the accurate information of agricultural activities such as crop identification and classification, crop area, crop growth condition monitoring and yield estimation in a concise and recurring manner for large areas and over long periods of time. This chapter presents a two-step approach for the inventory of olive tree plantations using medium-resolution optical satellite image. First, a multi-scale Conditional Random Fields (CRF) is applied for olive plantations labeling. Second, an algorithm based on a combination of three procedures, mean-sift segmentation, spectral thresholding of Red Band and NDVI and spatial thresholding is carried out to detect and count olive trees. The results of the approach are evaluated on four test sites situated in the Sbikha region, Kairouan-Tunisia and the accuracies are analyzed. Experimental results

R. Mezzi · M. Allani · F. Guedri · A. Sahli (✉)
Institut National Agronomique de Tunisie (INAT), Université de Carthage (UC), 43, Avenue Charles Nicolle, 1082 Tunis-Mahrajène, Tunisia
e-mail: sahli_inat_tn@yahoo.fr

R. Mezzi
e-mail: ranyamezzi@gmail.com

R. Mezzi · M. Alioscha-Perez · H. Sahli
Electronics and Informatics Department (ETRO), Vrije Universiteit Brussel (VUB), Pleinlaan 2, 1050 Brussels, Belgium
e-mail: maperezg@etrovub.be

H. Sahli
e-mail: hsahli@etrovub.be

A. Zouabi · R. Beji
Commissariat Régional au Développement Agricole de Kairouan (CRDA), Cité Sidi Layoun - Kairouan, 3100 El Kef, Tunisia

A. Froehlich (ed.), *Space Fostering African Societies*, Southern Space Studies, https://doi.org/10.1007/978-3-030-32930-3_2

show that the proposed classification and detection schemes achieve good performances with a mean total accuracy about 87%.

1 Introduction

Tunisia is the most important olive-growing country of the southern Mediterranean region. The olive (*Olea europaea* L.) is being cultivated throughout the country. Olive farms cover more than one-third of arable land (1.68 million hectares) producing 6.0% of the world's olive production and contributing 45.0% of food export income, 4.5% of total exports, and 10.0% of the total agricultural production value. The Tunisian olive is dominated by two major varieties, 'Chemlali' in the South and the Center and 'Chetoui' in the North, is rich in cultivars.[1] Olive trees number are estimated at approximately 60 million (30% in the north, 45% in the center and 25% in the south). They are mostly found in single-crop plantations, although they can also be found in combination with other fruit trees and annual crops. There are some 100 olive trees/ha in the north, 60 trees/ha in the center and 20 trees/ha in the south.[2]

Inventories on olive tree plantations are performed with the objective of providing support to the management and conservation activities in rural area or even in tree plantations. Traditional methods for obtaining information on olive tree is to use systematic or random sampling or by sampling stands, so that the final parameters for the plantations are obtained on the basis of statistical extrapolation. This renders the field survey techniques for olive tree inventories expensive, time consuming, and unsuited for large areas. In order to be able to implement agricultural policies effectively, an inventory of olive groves must be geographically known and recorded.

Remote sensing with medium spatial resolution is a cost-effective and reliable way to obtain information about trees. It may be the only practical way to insure sustainable management of olive farms with the necessary information, such as number, density, age, aboveground biomass, disease, etc. in a concise and recurring manner for large areas and over long periods of time.

This study has two fundamental objectives: olive orchards classification and individual olive tree identification. The first objective involves making use of the Conditional Random Fields (CRF) formulation to classify agricultural area at species level, from medium-resolution, single-date optical satellite image. Specifically, the main objective of this work is to explore the performance of this context-based classifier for rural agricultural areas, which has to our knowledge not been investigated in the literature with medium-resolution satellite images. The second objective

[1]Rekik, I.; Salimonti, A.; Grati Kamoun, N.; Muzzalupo, I.; Lepais, O.; Gerber, S.; Perri, E.; Rebai, A. 2008. Characterization and Identification of Tunisian Olive Tree Varieties by Microsatellite Markers. Hortscience, 43(5):1371–1376.

[2]International Olive Council, 2017. The olive sector in Tunisia, Olivae—Official Journal of the International Olive Council, 24, pp 38. International Olive Council, Madrid, Spain.

of this study is to illustrate the potential for individual tree identification and tree crown delineation using a combination of three procedures, mean-sift segmentation, spectral thresholding of Red Band and NDVI and spatial thresholding.

The rest of this chapter is organized as follows: In Sect. 2, the adopted methods and proposed algorithms are introduced and explained and in Sect. 3 the used dataset and experiments are described. Finally, in Sect. 4 some concluding remarks and future work are given.

2 Adopted Methods

2.1 Image Classification by Multi-scale Conditional Random Fields

In this section, we briefly summarize the standard Conditional Random Fields (CRF)-based labeling process. CRF is a flexible framework for contextual classification, introduced by Kumar and Hebert (2006)[3] for image classification. CRFs are undirected graphical models, consisting of nodes n and edges e. The nodes represent the image sites, e.g. pixels or segments. The edges link adjacent nodes and model statistical dependencies between class labels and data at neighbouring image sites.[4] The overall form of a general CRF model is as follows:

$$P(x|y,\theta) = \frac{1}{Z(y,\theta)} \exp\left(\sum_{c \in C} \psi_c(x_c, y, \theta) \right)$$

where Z (y, θ) is a normalization function named the partition function.

The above model is known to follow an exponential distribution, while the potentials ψ_c are functions defined over cliques $c \in C$, usually assumed to be linear and expressed as an inner product between features y and model parameters θ; x_c are the labels limited to factors of order $c \in C$.

This general CRF model can be defined on the multi-scale region-based image labeling context. This includes an appropriate potential definition and features for the energy expression, among others. For region-based image labeling, there are some other stages involved in the overall process, such as the image pre-processing (filtering), segmentation, parameter learning and classification. The general framework is illustrated in Fig. 1.

In order to avoid blurring and the delocalization of the image features, an image adaptive scale-space filtering is used. In this work, we opted for a method that guides the filtering process in such a way that intra-region smoothing is preferred

[3]Kumar, S., Hebert, M., 2006. Discriminative Random Fields. International Journal of Computer Vision 68(2):179–201.

[4]Albert, L.; Rottensteiner, F.; Heipke, C. 2014. Land use classification using conditional random fields for the verification of geospatial databases. ISPRS Annals of Photogrammetry, Remote Sensing & Spatial Information Sciences, 2(4):1–7.

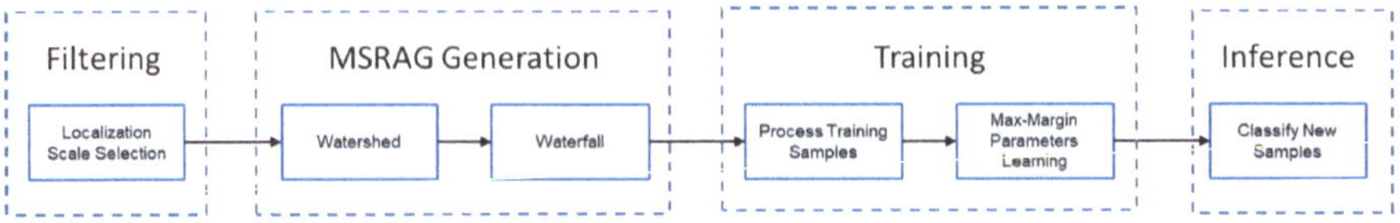

Fig. 1 Stages involved in the proposed multi-scale region-based image labeling process

over inter-region smoothing and edges are gradually enhanced. The employed filter, belongs to the class of nonlinear anisotropic diffusion filters where the multi-scale tower of the vector-valued image is governed by a system of coupled parabolic partial differential equations (PDEs). Thus, a discrete version of the multi-scale tower is obtained by applying the natural scale-space sampling method and a scale, referred to as the localization scale s_0, at which the diffusion filter has removed the noise without affecting or dislocating the edges of the salient features in the image is selected. The localization scale is determined using a maximum correlation criterion which selects the scale that maximize the correlation between the original image and the diffused image.

Starting from the localization scale s_0, the multi-spectral image, is first segmented using the watershed transform and then, the waterfall algorithm is used for producing a nested hierarchy of partitions: the multi-scale region adjacency graph (MSRAG); as illustrated in Fig. 2. To apply the watershed, the gradient of the multi-spectral image is obtained by combining the gradients of the texture (orientations) and the multi-spectral gradient. The waterfall algorithm removes from the current partition (hierarchical level) all of the boundaries completely surrounded by higher boundaries. Thus, the saliency of a boundary is measured with respect to its neighborhood and consider both single segment properties (area, convexity,

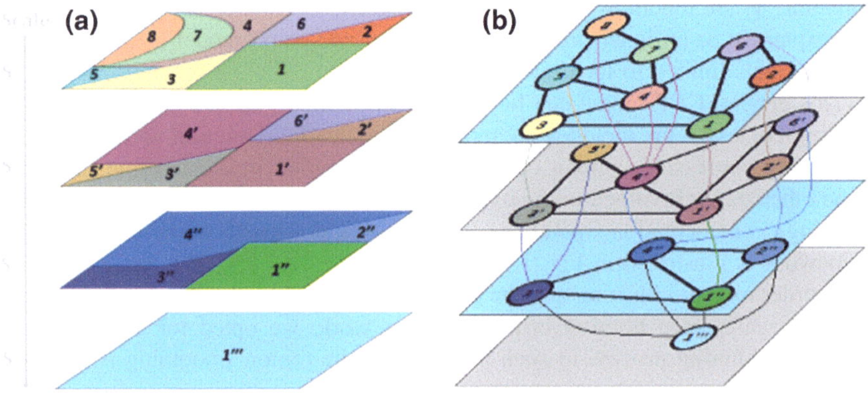

Fig. 2 The hierarchical structure of an image **a** Nested hierarchy of partitions; **b** multi-scale region adjacency graph (MSRAG), defined by the multi-scale region adjacency graph (MSRAG) (Alioscha-Perez, M.; Sahli, H. 2014. Efficient Learning of Spatial Patterns with Multi-Scale Conditional Random Fields for Region-Based Classification. Remote Sens., 6: 6727–6764)

compactness and color variances within the segment) and pair-wise properties (color mean differences between two segments and edge strength). The iteration of the waterfall algorithm ends with a partition of only one region. Such a hierarchy preserves the topology of the initial watershed lines and extracts homogeneous objects of a larger scale.

The general form for determining the optimal labeling of the proposed multi-scale CRF is defined as follows:

$$P(x|y,\theta) = \frac{1}{Z(y,\theta)} \exp\left(\sum_{n=1}^{N} \sum_{s \subset S_n} d_1 \psi_A(x_s^n, y, \omega) + \sum_{n=1}^{N-1} \sum_{s \subset S_n} d_2 \psi_L(x_s^n, x_s^{n+1}, y, \beta^L) + \sum_{n=1}^{N} \sum_{i \subset S_n} \sum_{j \in \eta(i)} d_3 \psi I(x_i^n, x_j^n, y, \beta^I) \right)$$

with s the parent of a node s, η(i) the set of neighbors of site i at the considered level and the parameters $\theta = (\omega, \beta^{I,} \beta^{L})^T$ defined in order to differentiate intra-scale and inter-scale spatial patterns; the weighting parameters dm $\in R^+$ regulate each potential individual contribution to the overall energy expression.

The association potential ψ_A (·; ω), which provides the unary local features, is defined as the probability of a site to acquire a certain label considering the site observations independently. Spectral (mean, standard deviation, kurtosis and skewness for each channel), textural, morphological (shape information, as elongation, area to length ratio and extent) and scale (actual scale) features of the region in question have been considered in the current implementation.

The inter-scale pairwise potential ψ_L (·; β^{L}), which encodes contextual information between parent/child regions at any two consecutive scales aims at modeling the probability distribution of a good region merging having taken place (detecting errors on segmentation) during the segmentation process. This potential function is based on the observations of the region in question (named the actual region) and other additional information, such as, its parent region, all of its siblings, the number of siblings, the scale and the lifetime of the actual region.

The intra-scale pairwise potentials ψ_I (·; β^I), encodes contextual information between neighbors within the same scale, aims at modeling the probability distribution of two regions sharing the same label. This potential function is based on the observations of the two regions in question, but also benefits from additional information, such as, the corresponding neighbor, the lowest and the highest neighbor distances, the lowest and the scale of the actual region.

For parameter learning, a novel strategy combining a max-margin parameter learning approach within the piecewise framework has been used.

The inference process is performed using the loopy believe propagation (LBP) method, which approximate the marginalized posterior by a value named believes, computed using a message passing algorithm.

For further details, please refer to the original publication of Alioscha-Perez and Sahli (2014).[5]

[5]Alioscha-Perez, M.; Sahli, H. 2014. Efficient Learning of Spatial Patterns with Multi-Scale Conditional Random Fields for Region-Based Classification. Remote Sens., 6:6727–6764.

3 Olive Tree Detection Combining Segmentation, Spectral and Spatial Thresholding

The proposed method includes several steps. Firstly, the mean-shift algorithm is employed to segment the agricultural area. Mean-shift is a non-parametric feature-space approach, based on kernel density estimation and has a strong statistical foundation. More detailed explanation of the Mean-Shift can be found in Comaniciu and Meer (2002).[6] Many combinations between the three main parameters of mean-shift algorithm Spatial radius (hs), Range radius (hr) and Minimum Region size (MR) have been tested. The optimum segmentation results of a pansharpened SPOT-7 image is obtained using the parameter combination hs = 5; hr = 5 and MR = 5 and the bands combination: Green, Red, and NIR. Secondly, the Red band and the NDVI images were built from the pansharpened image and converted to a grayscale images in order to exploit the strong differences in plant reflectance in red and NIR bands to determine their spatial distribution. Indeed, live green plants, rich in chlorophyll, absorb solar radiation in the Photosynthetically Active Radiation (PAR) or Visible Red (RED: 630–690 nm) spectral regions. In contrast, leaf cells have evolved to scatter (i.e., reflect and transmit) solar radiation in the near-infrared (NIR: 760–900 nm) spectral regions which carries approximately half of the total incoming solar energy. This happens because the energy level per photon in that domain is not sufficient to be useful to synthesize organic molecules: a strong absorption in NIR would only result in over-heating the plant and possibly damaging the tissues. For these reasons, in a multispectral satellite image the Red band portion appears relatively dark and the NIR band appears relatively bright where green plants are captured.[7] The NDVI (Normalized Difference Vegetation Index) is calculated from these measurements as follows:

$$\mathrm{NDVI} = (\mathrm{NIR} - \mathrm{Red})/(\mathrm{NIR} + \mathrm{Red})$$

Thirdly, from the derived images, the radiometric mean attributes, i.e. mean Red reflectance and mean NDVI, are calculated from the segmented regions. Olive trees were then extracted by using a thresholding technique. Different threshold values have been applied based on an empirical analysis performed on olive trees within the imagery. Following this spectral thresholding, the obtained boundaries of olive trees are divided into 2 regions which represent the crown and the shadow of the tree (Fig. 3a). To improve the obtained results, an automatic fusion of adjacent regions was implemented (Fig. 3b). This made it possible to solve the problem of the shadow and to highlight polygons representing the hedges.

[6]Comaniciu, D.; Meer, P. 2002. Mean shift: a robust approach toward feature space analysis. IEEE Transactions on Pattern Analysis and Machine Intelligence, 24:603–619.

[7]Daliakopoulos, I.N.; Grillakis, E.G.; Koutroulis, A.G.; Tsanis, I.K. 2009. Tree Crown Detection on Multispectral VHR Satellite Imagery. Photogrammetric Engineering and Remote Sensing, 75 (10):1201–1211.

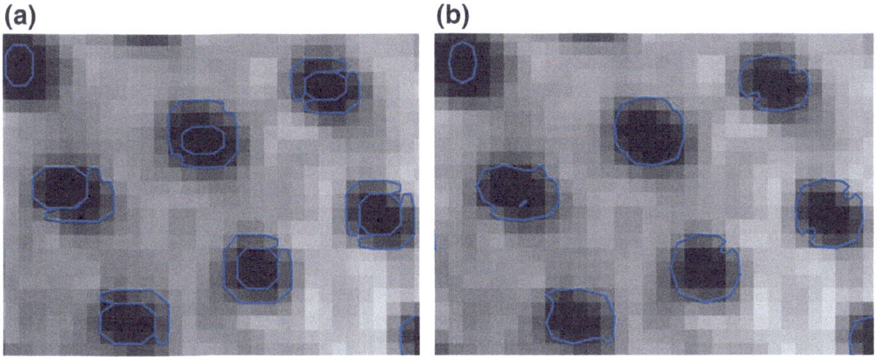

Fig. 3 NDVI thresholding **a** before polygons fusion; **b** after polygons fusion

After polygons (regions) fusion, the mean tree area has been used as spatial attribute for olive recognition. Concerning our test zone, the area parameters seemed well suited to recognize the different boundaries of the objects. The boundary identification differs according to the chosen threshold. The threshold has to be tuned depending on the object surface (i.e. large object, small objects). In summary, the adopted olive tree detection algorithm is based on the combination of the above described three procedures (segmentation, spectral thresholding of Red Band and NDVI and spatial thresholding) as depicted in Fig. 4.

4 Experiments

4.1 Study Area

The performance of the presented approach has been evaluated on an agricultural site, Sbikha region, which is located in Kairouan Governorate, on the center of Tunisia (Fig. 5). It belongs to the upper semi-arid bioclimatic stage, with a cool wet season from November to March and a dry hot season from May to September. Average annual rainfall is 313 mm and average reference evapotranspiration ETo is 1326 mm (Period 1985–2004). The region is covering one of the most productive olive area of Tunisia with more than 70% of the major agriculture area under rainfed conditions.[8] In irrigated areas, the main crops are apricot, citrus, olive and green peas, pepper, barley and wheat while in rainfed areas, there is a predominance of olive crops.[9]

[8]Commissariat Régional au Développement Agricole de Kairouan, 2016. Stratégie d'amélioration du secteur oléicole dans le gouvernorat de Kairouan à l'horizon 2020. Rapport interne, CRDA-Kairouan, pp 8 (in Arabic).

[9]Allani M., R. Mezzi, W. Abdallah, R. Jlassi, N. Boukhalfi, A. Romdhane, F. Stoffner, M.A. Trabelsi, R. Béji, T. Ayoub, M. Kouraichi, E. Chalghaf, A. Sahli, H.W. Müller CRDA Kairouan—Projet CREM-Volet BGR, 2016. Support Technique: Cartes thématiques de l'usage agricole de l'eau dans la zone Nebhana – Sbikha. 141 pp.

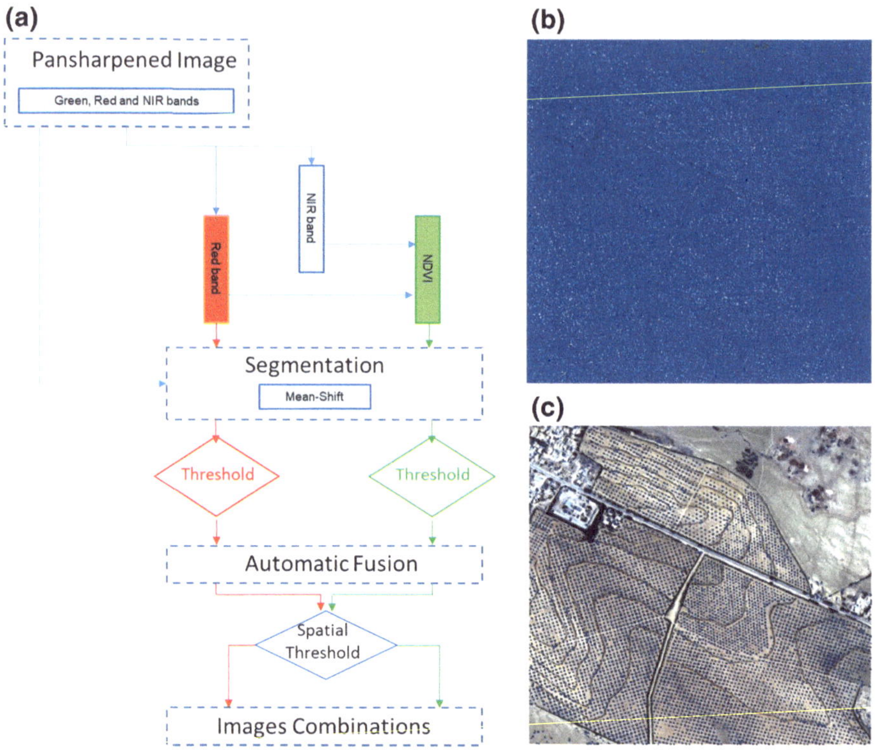

Fig. 4 Olive tree detection **a** algorithm; **b** image after mean-shift segmentation and **c** detected trees after spectral and spatial thresholding

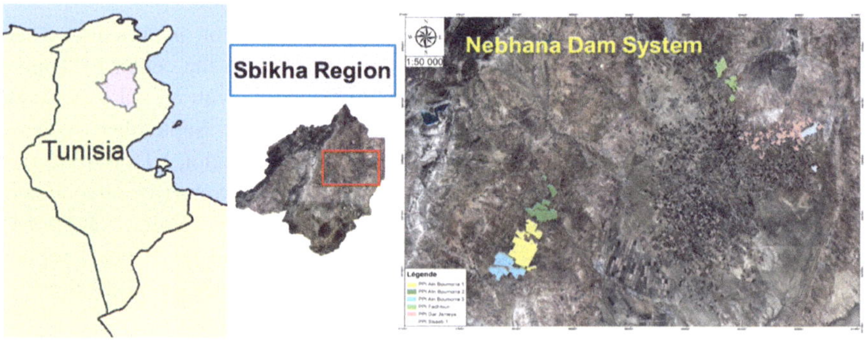

Fig. 5 Geographic localization of the study area: the Sbikha region, Kairouan-Tunisia

4.2 Medium Spatial Resolution Imagery

Two types of images are used in this study. SPOT 7 multispectral (MS) and panchromatic (PAN). The optical images are acquired on August 2015. SPOT 7 MS (5 m) data has four spectral bands: Blue, Green, Red, and Near Infrared (NIR). The spectral ranges of Blue band is between 0.45 and 0.52, Green band is between 0.52 and 0.60, Red band is between 0.63 and 0.69 and NIR band is between 0.76 and 0.90 μm. The spectral range of the PAN image (1.5 m) is between 0.45 and 0.74 μm. A pre-processing is applied to the images; a radiometric correction is used to avoid a radiometric errors of sensor, an atmospheric correction to remove the influence of the atmosphere and a geometric correction is also applied in order to project the images to WGS84. SPOT-7 PAN (1.5 m) and MS (4 m) data are pan sharpened with the Gram Shmidt algorithm since it shows good results with various data sets of different spatial, spectral and temporal resolutions.[10,11]

4.3 Test Sites

We selected four test sites from the region to assess the performance of the proposed approach (Fig. 6).

The sites were chosen according to the landscape considering olive trees distribution and crops homogeneity in the area:

- Test Site 1 with olive trees distributed in a very organized way
- Test Site 2 with olive trees distributed in a moderately organized way
- Test Site 3 with olive trees distributed in an unorganized way with the presence of other types of trees
- Test Site 4 is characterized by fruit trees mixture and olive plots conducted intensively.

Each test site covers an area of 236 hectares.

4.4 Reference Data

Reference data related to fields were collected during field survey performed concurrently with the image acquisitions. For the land use classification, the data are divided into 4 classes: Olive Orchards, Other Fruit Orchards, Bare Soil and

[10]Yuhendra, H.; Joshapat, T.S.; Hiroaki, K. 2011. Performance Analyzing of High Resolution Pan-sharpening Techniques: Increasing Image Quality for Classification using Supervised Kernel Support Vector Machine. Research Journal of Information Technology, 3:12–23.

[11]Jawak, S.; Luis, A. 2013. A Comprehensive Evaluation of PAN-Sharpening Algorithms Coupled with Resampling Methods for Image Synthesis of Very High Resolution Remotely Sensed Satellite Data. Advances in Remote Sensing, 2(4):332–344.

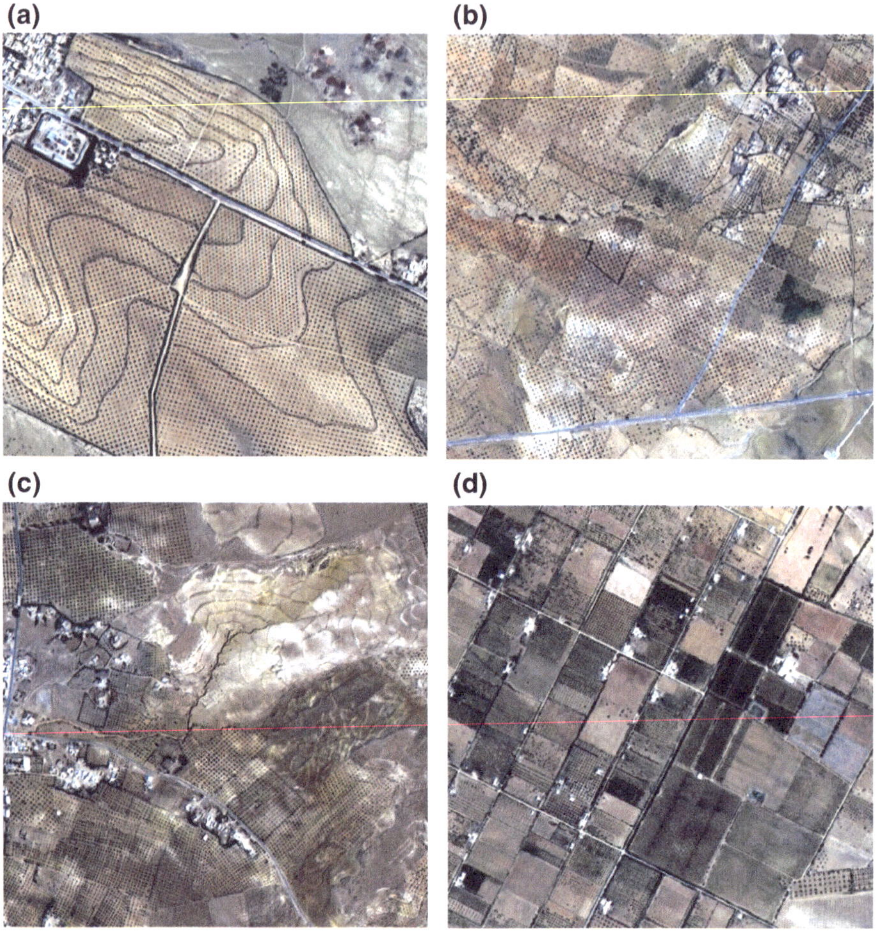

Fig. 6 Pan sharpened SPOT-7 Images of **a** Test Site 1, **b** Test Site 2, **c** Test Site 3 and **d** Test Site 4

Building and Tracks. For analyses, the reference data were separated into two groups: one group to be used for training samples and the rest to be used for assessing the results of the classification. The training samples for each image are selected by an experienced operator. However, note that the samples are collected independently from each site image to achieve the best performance for each dataset. The training and validation samples are detailed in Table 1 and illustrated in Fig. 7.

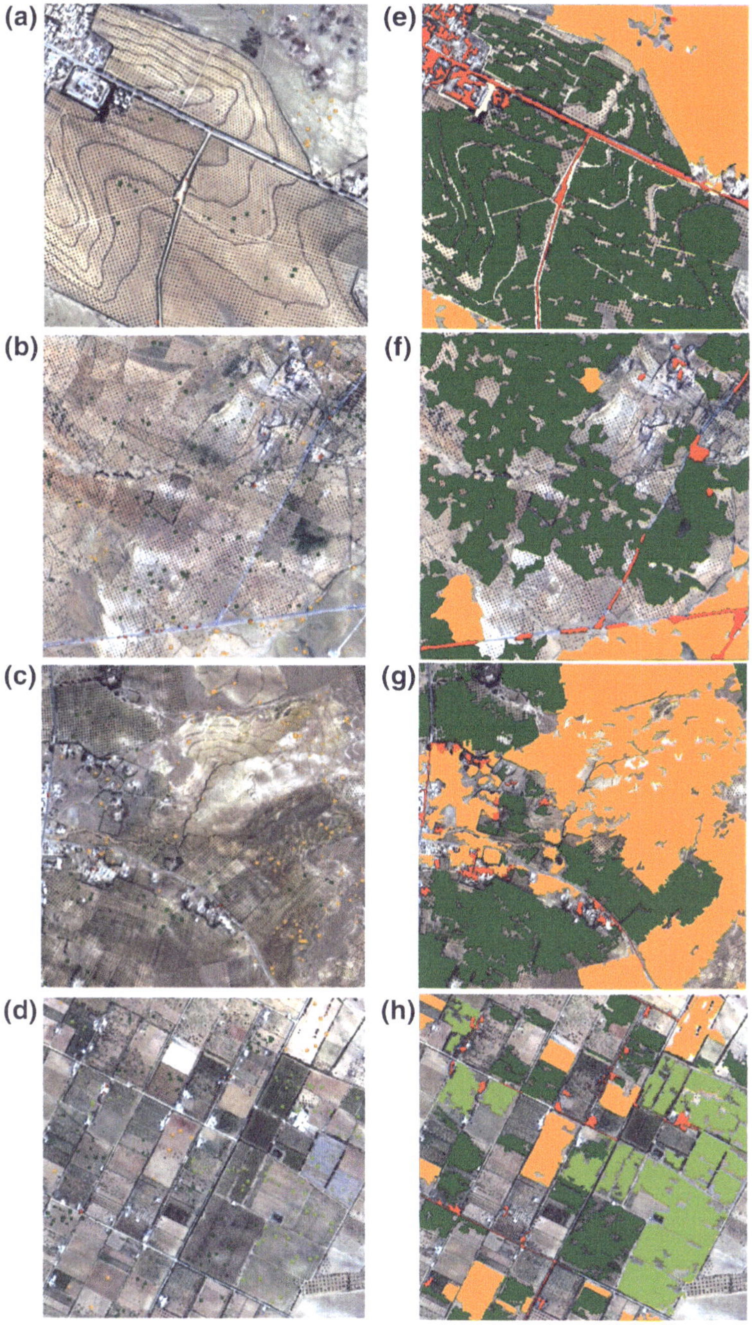

Fig. 7 Training and validation samples for the CRF classification (Green: Olive Orchards, Light Green: Other Fruit Tree Orchards; Orange: Bare Soil and Red: Building and Tracks)

Table 1 Description of selected training and validation samples

Classes/test site	Training area				Validation area			
	Site 1	Site 2	Site 3	Site 4	Site 1	Site 2	Site 3	Site 4
Class 1 Olive orchards	0.47	2.35	0.90	1.62	119.69	20.89	64.68	32.65
Class 2 Other fruit orchards	0	0	0	2.28	0	0	0	43.33
Class 3 Bare soil	0.37	1.48	2.53	0.46	43	97.69	101.86	19.01
Class 4 Building and tracks	0.11	0.36	0.18	0.17	8.99	4.69	3.24	3.75
Selected area (ha)	1.00	4.19	3.61	4.53	179.94	123.27	169.78	98.74
Area ratio[a] (%)	0.40%	1.8%	1.5%	1.9%	72.7%	52.2%	71.9%	41.8%

[a]Ration between selected area and site area

From Table 1, it is apparent that all classes were roughly represented in a balanced way conforming to each site landscape. Note also that training area represents less than 2% of the site area while the validation area represents more than 40% of the site area.

4.5 Land Use Classification

As described above, the MSRAG is based on the creation of a multi-scale tower using the PDE approach. Then, at a manually selected localization scale, s_0, a gradient watershed transformation is performed to detect a set of regions with well-localized contours. To create the hierarchy, the waterfall algorithm is used.

An illustration of the different hierarchical segmentations is provided in Fig. 8 along with the respective classification results. Table 2 shows the corresponding confusion matrix and derived quality measures, more precisely overall accuracy, kappa index, correctness and completeness, also referred to as user's and producer's accuracy.

The CRF approach achieves a mean overall accuracy of 86.8% and a mean kappa index of 0.75. The results for the class olive orchards are quite good for a method using only four broad multispectral channels, with mean PA 90.3% and mean UA 88.3%. Main sources error are confusions between olive orchards and soil. In this case, the bias toward this class is less severe (UA = 87.1%). Mean PA nevertheless stays high (Build./Track 77.2%, Other Fruits Orchards 71.4%).

This is mainly due to the low olive trees density, causing distorted spectral responses. As olives tree are relatively sparse, they are harder to discriminate and they only partially cover the ground, such that bare soil is visible and disturbs the classification. Another more technical factor is the limited number of reference data for two classes, Build./Track and Other Fruits Orchards. Possibly the training set might be insufficient to learn these two classes.

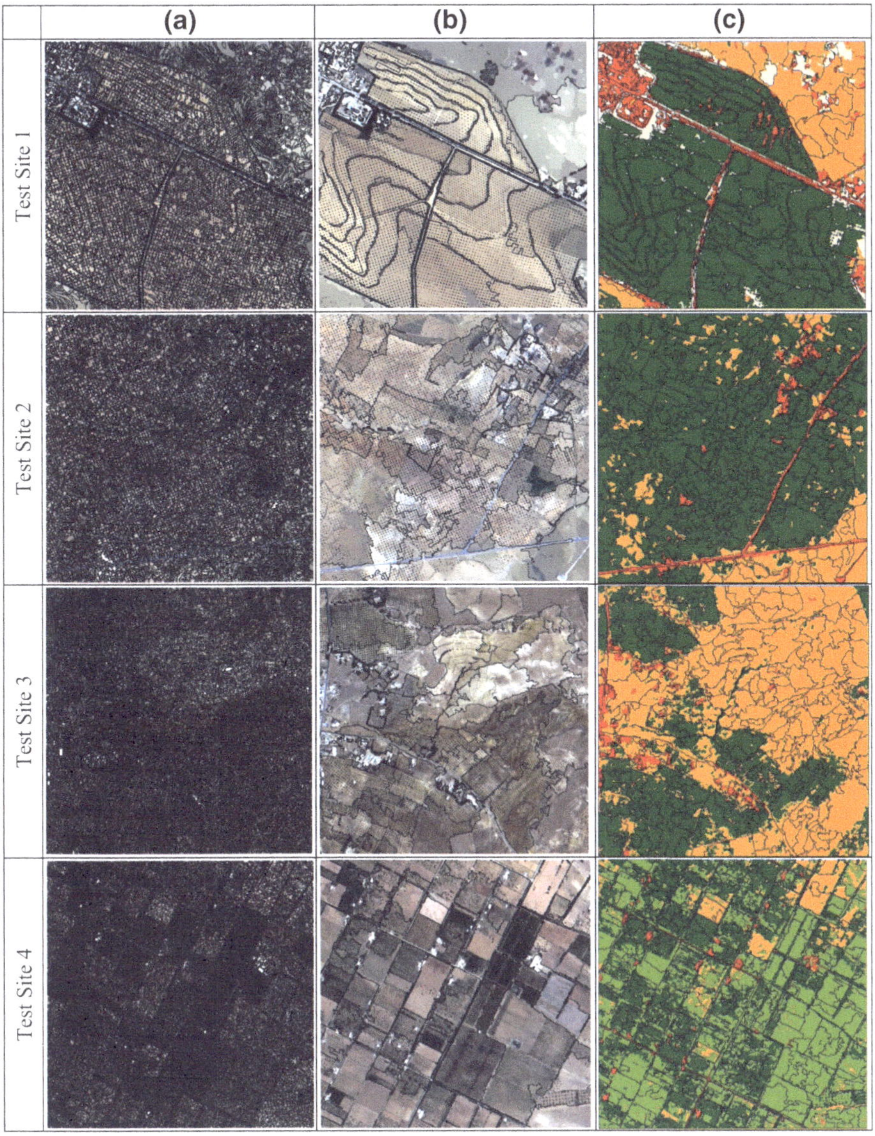

Fig. 8 Scales and classification maps of the four Test Sites. The first two columns provide regions at Scales 1 and 6, respectively, while the last column provides the classification maps

Table 2 Accuracy assessment results for every test data area

Test site	Test circle	Olive tree counted	Ground data	Counting accuracy (%)
1	1	420	436	96%
	2	497	506	98%
2	1	498	513	97%
	2	413	436	95%
3	1	672	789	85%
	2	483	516	94%
4	1	475	1123	42%
	2	87	0	–

4.6 Olive Tree Inventory

The performance of the presented approach is evaluated on the four test sites. We visualize the results in Fig. 9a. As it can be seen, the proposed approach can provide promising results for the detection of olive trees.

An accuracy assessment was applied for the olive tree extraction and compared to the manual counting using on-screen digitization in two circle polygons with 200 m of diameter in each test site area (Fig. 9b).[12,13] The number of olive trees counted manually and the number of extracted ones by the proposed method were selected in the two circle polygons using QGIS; the results of these two counting methods were compared with each other to determine the accuracy of olive tree detection and extraction (Table 3).

The highest mean accuracy obtained is 97% and the lowest is 42%, for test site numbers 1 and 4, respectively. For the sample with 98% accuracy, the difference between the numbers of trees counted using the proposed method and from the ground reference data was 16 trees. The average of accuracy for these 8 datasets was around 87% and the these results show the good potential of the proposed method for olive tree counting using Medium Resolution Multi-Spectral Satellite Imagery.

The discrepancies in the test site 4 occurred for two main reasons. First, errors may be due to the dislocation of the overlapped olive tree crowns, which this method detected as one crown. There were several parts on the image with overlapping crowns, which are expected to be one of the sources of error in tree counting. This situation always occurred in irrigated tree plantations, where the gaps between trees are narrower. However, the effect of the overlapped crowns had been reduced during merging and spatial thresholding process. Second, the error

[12]Shafri, H.Z.M., Hamdan, N., and Saripan, M.I., 2011. Semiautomatic detection and counting of oil palm trees from high spatial resolution airborne imagery, International Journal of Remote Sensing, 32(8):2095–2115.
[13]Ok, A.O., Ozdarici-Ok, A., 2017. Detection of citrus trees from UAV DSMS. ISPRS Annals of the Photogrammetry, Remote Sensing and Spatial Information Sciences, IV-1/W1:27–34.

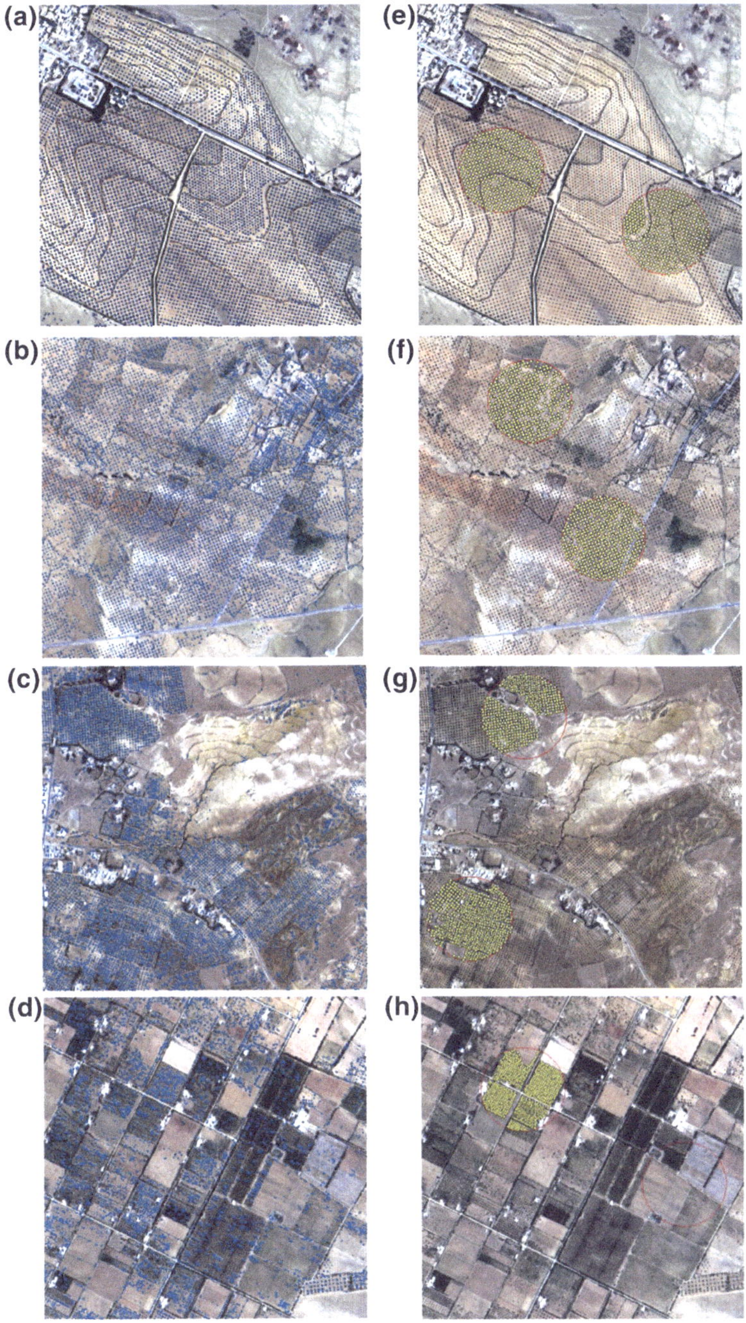

Fig. 9 The results of proposed approach (left column), and the reference data (right column) for the four test sites

Table 3 Confusion matrix of CRF classification for test site images

Test Site 1					
	Olive	Other fruits	Bare soil	Build./tracks	U. A.
Olive	514638	–	3904	1243	99%
Other fruits	–	–	–	–	–
Bare soil	1022	–	155588	1504	98%
Build./tracks	7450	–	11504	34909	65%
P. A.	98%	–	91%	93%	
Overall accuracy (%): 96.4—Kappa index: 0.92					
Test Site 2					
	Olive	Other fruits	Bare soil	Build./tracks	U. A.
Olive	433634	–	41464	2391	91%
Other fruits	–	–	–	–	
Bare soil	32466	–	117582	2176	77%
Build./tracks	4778	–	3270	24632	75%
P. A.	92%	–	72%	84%	–
Overall accuracy (%): 86.9—Kappa index: 0.70					
Test Site 3					
	Olive	Other fruits	Bare soil	Build./tracks	U. A.
Olive	249486	–	12317	1094	95%
Other fruits	–	–	–	–	–
Bare soil	36566	–	437059	3552	92%
Build./tracks	1527	–	3361	9799	67%
P. A.	87%	–	97%	68%	–
Overall accuracy (%): 92.3—Kappa index: 0.84					
Test Site 4					
	Olive	Other fruits	Bare soil	Build./tracks	U. A.
Olive	121812	48422	5303	2105	69%
Other fruits	19138	137528	34543	3686	71%
Bare soil	3771	6346	44471	236	81%
Build./tracks	473	376	211	10682	91%
P. A.	84%	71%	53%	64%	–
Overall accuracy (%): 71.6—Kappa index: 0.56					

may come from the spectral response itself. As in irrigated areas olives tree are relatively dense, they are harder to discriminate from other fruit trees and the exact number of the olive trees provided by spectral thresholding is therefore uncertain.

5 Conclusion and Future Work

The presented work focused on olive orchards classification and individual olive tree identification. The chapter described the main principles of a multi-scale CRF model and illustrated classification results on a set of remote sensing images, which validated the suitability of the method for olive parcels labeling (mean overall accuracy of 86.8% and a mean kappa index of 0.75). A further finding is that moderate training set per class of about 1.5% of the total area is sufficient, in spite of the rather coarse spectral resolution. Note however that it might in practice be easier to collect larger training sets, in order to make sure that the samples are representative of the full class distribution.

Considering the individual classes used in our study, confusions are principally observed between olive orchards/bare soil in rainfed areas and olive Orchards/Other Fruit Trees orchards in irrigated areas. This is due to the spectral similarity of these classes under the limited spectral resolution (B, G, R, NIR) of the images. This, of course, can be mitigated to some extent by collecting more training samples or by integrating additional features into the classification framework. Furthermore, we will be investigating the proposed approaches on more test areas with different characteristics. Moreover, we are planning to compare accuracy and performance with respect to other, more recent, state-of-the-art models.

For individual olive tree identification and inventory, the study concluded that there is a potential in using a combination of three procedures, mean-sift segmentation, spectral thresholding of Red Band and NDVI and spatial thresholding. Results demonstrate that the proposed approach is robust to landscape variability, with an accuracy around 87%. Some problems have been identified in parcels with high density plantations and/or with weeds or highly vegetated grounds, where discriminating olive trees from other fruit trees and/or field background is a difficult task.

In summary, a multi-scale CRF classification method together with segmentation and thresholding techniques are proposed in order to update the olive parcels (area and number) in agricultural regions from medium spatial resolution multispectral image. Good accuracy results are achieved while reducing the time and economic cost required by traditional inventory techniques. Given the importance of the socio-economical and ecological services olive parcels provide, this research brings a contribution towards a more detailed inventories of these plantations by providing valuable data on trees to facilitate their management and monitoring.

Acknowledgements The Study was carried out within the framework of the ETRO-VUB/INAT partnership. We also would like to thank Dr. Hans Werner Müller, Head of the CREM-BGR project (Federal Institute for Geosciences and Natural Resources - BGR) for providing the SPOT-7 data and supports for field works during the growing seasons 2014–2015 and 2015–2016.

Ranya Mezzi is a Master student in Sustainable Management of Water Resources at the National Agronomic Institute of Tunisia (INAT) of Carthage University (UC). Under the guidance of Dr. Ali Sahli, she received an Engineering degree from the National Agronomic Institute of Tunisia (INAT) in 2014. Her research focuses on optical remote sensing and irrigation management.

Mitchel Alioscha-Perez is Post Doc Researcher at the Department of Electronics and Informatics (ETRO), Vrije Universiteit Brussel (VUB). In 2011, he received his Master degree on Signals and Systems, from the Central University of Las Villas, Cuba, and in 2018 a Ph.D. in Engineering from the Vrije Universiteit Brussel. His current work relates to computer vision and machine learning.

Hichem Sahli studied Mathematics and Computer Science at Université Louis Pasteur–Strasbourg (France), and holds a Ph.D. in Computer Science (Informatics/Photonics) from Ecole Nationale Sup. De Physique Strasbourg (France). He is Professor of Computer Vision & Machine learning in the Electronics and Informatics Dept. (ETRO) of the Vrije Universiteit Brussel, and Research Coordinator at the Interuniversity Micro Electronics Center (IMEC). He coordinates the Audio-Visual Signal Processing Laboratory (AVSP) within ETRO. AVSP conducts research on applied and theoretical problems related to machine learning signal and image processing and computer vision. The group explores and capitalizes on the correlation between speech and video data for Computational Information Technology where efficient numerical methods of computational engineering are combined with the problems of information processing. Hichem Sahli has worked in computer vision, mathematical image and signal analysis, and machine learning since 1987. His current work relates to representation learning and scene analysis.

Ali Sahli is Associate Professor and Director of Studies at the National Agronomic Institute of Tunisia (INAT) of Carthage University (UC), where he has been since 1995. From 2003 to 2017 he served as Assistant Professor of agricultural higher education. During the period of 1995–2003 he served as Assistant of agricultural higher education. He received an Engineering degree from the National Agronomic Institute of Tunisia (INAT) in 1988, a graduated engineering degree from the National Polytechnic Institute of Toulouse (INPT) and a Diploma of Advanced Studies from the National Polytechnic Institute of Lorraine (INPL). He received his PhD in mechanical and energetics from the National Polytechnic Institute of Lorraine (INPL) in 2001. From 1991 to 1993 he worked at the University Institute of Technology of the University of Lorraine as a temporary substitute teacher, and from 1993 to 1995 he worked at the European School of Engineering in Materials Engineering, National Polytechnic Institute of Lorraine as a temporary substitute teacher. His research interests are heat and mass transfer with applications for environmental physics, bioclimatology and remote sensing.

Capitalizing on Geospatial Technologies to Solve Urban Waste in Akure Nigeria

T. Oniosun, I. A. Balogun and P. Solis

Abstract

At the sustainable development summit in 2015, UN Member states adopted the 2030 agenda for sustainable development, which includes a set of 17 Sustainable Development Goals (SDGs). In response to this, YouthMappers emphasizes the creation and utilization of open data and open source software for geographic information directly related to development objectives in unmapped places of the world where the US Agency for International Development (USAID) works to end extreme poverty. We are living in an era full of trash, and it is clear the problem is massive, growing constantly and varies considerably by region in Africa. To get a handle on it, starting from Akure, Nigeria, geospatial technique is being used to locate and identify illegal dump sites and find suitable sites to locate dumping sites. Principal sub-criteria used for spatial analysis include slope, built-up area, road networks, geological maps, etc. This is important because it is not enough to identify illegal dumping sites, we need to give a complete solution guideline to the appropriate bodies. KoboToolBox was used to collect data from the field, data collected include: picture of the dump site, coordinate, site description, type of waste (special waste, liquid waste, hazardous waste, restricted solid waste, general solid waste [Putrescible], general solid waste [Non-Putrescible]), proximity to residential or water bodies (less than 10 m, 10–30 m, 30–80 m, 80–150 m, greater than 150 m), size of site and

T. Oniosun (✉)
Surulere Lagos, 24, Ekololu Street, Surulere Lagos, Nigeria
e-mail: temidayo.isaiah@africanews.space

I. A. Balogun
Federal Univerisity of Technology, Akure, Nigeria
e-mail: iabalogun@futa.edu.ng

P. Solis
Texas Tech University, Lubbock, TX, USA
e-mail: patricia.solis@asu.edu

A. Froehlich (ed.), *Space Fostering African Societies*, Southern Space Studies,
https://doi.org/10.1007/978-3-030-32930-3_3

accessibility (either motorable or not). Navigation through the city for data collection was done using Bicycles/Motorbike while the processed information is made available openly on UMap. In identifying suitable locations to site dumping sites, ARCGIS was used for the Satellite Imagery analysis and Land Use Land Cover of the area was done. From the analysis, the most common type of waste is the General Solid Waste (non-putrescible) which mostly contain household waste, over 80 percent of the sites can be accessed using trucks and over 50% of the dump sites are less than 10 meters from buildings/water bodies which pose a serious threat to the health of residents in the city.

1 Introduction

On 25 September 2015, the 194 countries of the UN General Assembly adopted the 2030 Development Agenda titled Transforming our world: the 2030 Agenda for Sustainable Development. Out of the 17 goals signed, this project idea will help achieve goal 3 (good Health), goal 6 (clean water and Sanitation), goal 11 (Sustainable Cities and Communities) and goal 14 (life below water) in cities where it is being implemented. According to Mercer Human Resource Consulting's 2007 Health and Sanitation Rankings, as part of their 2007 Quality of Life Report, they ranked 215 cities worldwide based on levels of air pollution, waste management, water potability, hospital services, medical supplies and the presence of infectious disease.[1] 16 out of the top 25 are African cities. This includes: Port Harcourt, Nigeria, Maputo, Mozambique, Luanda, Angola, Niamey, Niger, Nouakchott, Mauritania, Conakry, Guinea Republic, Lome, Togo, Pointe Noire, Congo, Bamako, Mali, Ouagadougou, Burkina Faso, Bangui, Central African Republic, Dar es Salaam, Tanzania, Ndjamena, Chad, Brazzaville, Congo, Addis Ababa, Ethiopia and Antananarivo, Madagascar.

Urban waste is one of the major threats to global environment in the world today and also, it is one of the causes of communicable diseases and airborne diseases. As global civilization keeps improving, there is increase in commercial, residential and infrastructure development due to the population growth which has a negative impact on the environment if not properly planned, monitored and maintained. An average city in Africa is dirty; people dump refuse everywhere. Because waste bins are not distributed across the places, people drop waste anywhere and anytime not considering the environmental effect. Because there are no laws against this, it makes it difficult to keep track of it and almost impossible for government to organize clean ups.

[1]https://www.forbes.com/2008/02/24/pollution−baku−oil−biz−logistics−cx_tl_0226dirtycities_slide.html#6ec2fcab1ef2.

We know there is trash lying around. But in order to clean it up, we first have to get a clear picture of the situation. Mapping (Using Remote Sensing and GIS) can give us the exact location, amount, and type of garbage. This is essential to organize the logistics and the handling for clean-up by the Government, other agencies and NGOs. After it is gone, this data is great for tracking our progress, and keeping tabs on these dump sites for the future. But that is not all. Perhaps the most important reason to map is to spread the message. If seeing is believing, then looking at a virtual map of overflowing trash sites will help open our eyes to the problem. And, hopefully, inspire people around to join us and do something about it.

Every year, 20 million tons of garbage is added to our oceans, 80% of it from mainland waste. That is like dumping over 710,000 Boeing 737 airplanes into the ocean…each year. Getting an exact measurement of how much waste has ended up in nature is difficult. But think about this. Out of the 1.3 billion tonnes of household waste generated per year, only about 258–368 million tons of trash end up in one of the 50 largest dumpsites. Where is the rest of it? (about 0.50 kg of waste is generated per day per person in Africa).

We are living in an era full of trash, and it is clear the problem is massive, growing constantly and varies considerably by region. To get a handle on it, we need to locate and identify illegal dump sites and then use Surveying and Geoinformatics techniques tools to find suitable sites to locate dumping sites in these cities. Principal sub-criteria that will be used for spatial analysis include slope, built-up-area, road networks, geological maps, etc. This is important because it is not enough to identify illegal dumping sites, we need to give a complete solution guideline to the appropriate bodies. With this project, better plans and more structure waste system can be made and enforced. Which is one of the major steps to take in promoting a better and improve the environment and atmosphere. This will assist the government and NGOs in clearing of dumping site and make our city a better place to live in. This project started at the YouthMappers chapter of the Federal University of Technology, Akure Ondo State Nigeria.

1.1 Project Site Description

Akure is a city in south-western Nigeria, and is the largest city and capital of Ondo State. The city has a population of over 550,000. The residents of the City are primarily of the Yoruba ethnic group and the state has one of the highest literacy rates in Nigeria. It lies in the southern part of the forested Yoruba Hills and at the intersection of roads from Ondo, Ilesha, Ado-Ekiti, Benin and Owo (Fig. 1).

Akure is an agricultural trade centre for the yams, cassava, corn (maize), bananas, rice, palm oil and kernels, okra, and pumpkins grown by the Ondo branch of the Yoruba people. Although cocoa is by far the most important local commercial crop, cotton, teak, and palm produce are also cultivated for export. The town's industries include electronics manufacturing, soft drink, bottling among

Fig. 1 Oja Oba, Akure Metropolis

others. Akure lies about 7°25' north of the equator and 5°19' east of the Meridian. It is about 700 km Southwest of Abuja and 311 km north of Lagos State. Residential districts are of varying density, some areas such as Arakale, Ayedun Quarters,

Fig. 2 Alagbaka Roundabout

Ijoka, and Oja-Oba consist of over 200 persons per hectare, while areas such as Ijapo Estate, Alagbaka Estate, Avenue and Idofin have between 60-100 people per hectare.

The town is situated in the tropical rainforest zone in Nigeria (Fig. 2).

2 Methodology

KoboToolBox was used to collect data about the illegal dumping sites from the field. Data collected include:

1. Picture of the dump site
2. Coordinate
3. Site description
4. Type of waste
5. Proximity to residentials or water bodies
6. Size of site and accessibility

Navigation through the city for data collection was done using Bicycles/Motorbike while the processed information is made available openly on UMap (Fig. 3).

Fig. 3 YouthMappers during mapping

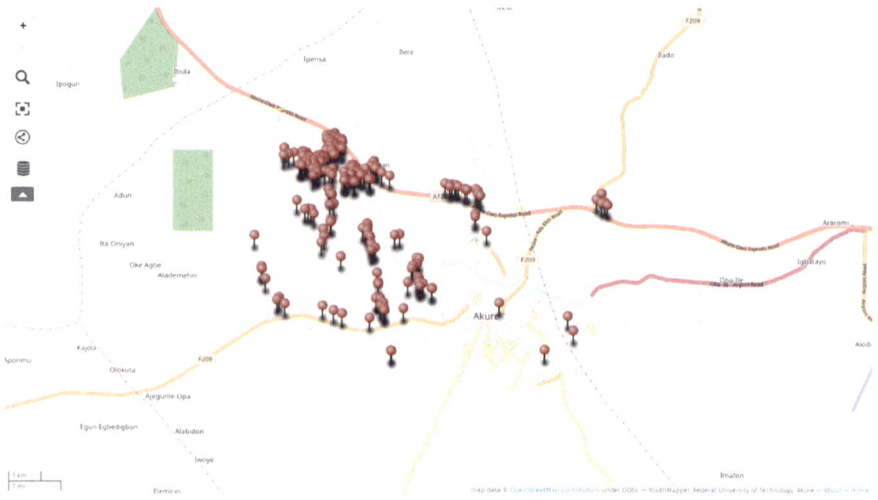

Fig. 4 Project Data on Umap

In identifying suitable locations to site dumping sites, ARCGIS was used for the Satellite Imagery analysis. Land Use Land Cover of the area was done, different guidelines for design and operation of waste landfills site[2] were also considered.

2.1 Open Source

The data collected is made available online through UMap[3] (Figs. 4 and 5).

2.2 Waste Classification

The following classes of waste are defined in clause 49 of Schedule 1 of the Protection of the Environment Operations Act 1997 (POEO Act)[4]:

1. special waste
2. liquid waste
3. hazardous waste
4. restricted solid waste
5. general solid waste (putrescible)
6. general solid waste (non-putrescible).

[2]https://www.iswa.org/index.php?eID=tx_iswaknowledgebase_download&documentUid=3159.
[3]http://u.osmfr.org/m/185030/.
[4]https://www.epa.nsw.gov.au/resources/wasteregulation/140796−classify−waste.pdf.

Fig. 5 Project Data on Umap showing one of the datasets

2.2.1 Special Waste

'Special waste' is a class of waste that has unique regulatory requirements. The potential environmental impacts of special waste need to be managed to minimize the risk of harm to the environment and human health. Special waste means any of the following:

1. clinical and related waste
2. asbestos waste
3. waste tyres
4. anything classified as special waste under an EPA gazettal notice.

The only exception to this is where special waste is mixed with restricted solid or hazardous waste. In these circumstances, the waste must be classified as special waste and restricted solid or hazardous waste (as applicable), and managed as both of those classifications.

2.2.2 Liquid Waste

1. Liquid waste means any waste (other than special waste) that:
2. has an angle of repose of less than 5 degrees above horizontal
3. becomes free-flowing at or below 60 degrees Celsius or when it is transported
4. is generally not capable of being picked up by a spade or shovel
5. is classified as liquid waste under an EPA gazettal notice.

If the waste meets the criteria outlined above, it is classified as liquid waste, and no further assessment for classification is required.

2.2.3 Restricted Solid Waste

Currently, no wastes have been pre-classified by the EPA as 'restricted solid waste'.

2.2.4 General Solid Waste (Putrescible)

The following wastes (other than special waste, liquid waste, hazardous waste or restricted solid waste) have been pre-classified by the EPA as 'general solid waste (putrescible)':

1. household waste that contains putrescible organics
2. waste from litter bins collected by or on behalf of local councils
3. manure and night soil
4. disposable nappies, incontinence pads or sanitary napkins
5. food waste
6. animal waste
7. grit or screenings from sewage treatment systems that have been dewatered so that the grit or screenings do not contain free liquids
8. any mixture of the wastes referred to above.

In assessing whether waste has been pre-classified as general solid waste (putrescible), the following definitions apply:

1. Animal waste includes dead animals and animal parts and any mixture of dead animals and animal parts.
2. Food waste means waste from the manufacture, preparation, sale or consumption of food but does not include grease-trap waste.
3. Manure includes any mixture of manure and biodegradable animal bedding, such as straw.

2.2.5 General Solid Waste (Non-putrescible)

The following wastes (other than special waste, liquid waste, hazardous waste, restricted solid waste or general solid waste (putrescible)) are pre-classified as 'general solid waste (non-putrescible)':

1. glass, plastic, rubber, plasterboard, ceramics, bricks, concrete or metal
2. paper or cardboard
3. household waste from municipal clean-up that does not contain food waste
4. waste collected by, or on behalf of, local councils from street sweepings
5. grit, sediment, litter and gross pollutants collected in, and removed from, stormwater treatment devices and/or storm water management systems, that has been dewatered so that they do not contain free liquids

6. grit and screenings from potable water and water reticulation plants that has been dewatered so that it does not contain free liquids
7. garden waste
8. wood waste
9. waste contaminated with lead (including lead paint waste) from residential premises or educational or child care institutions
10. containers, previously containing dangerous goods, from which residues have been removed by washing or vacuuming
11. drained oil filters (mechanically crushed), rags and oil-absorbent materials that only contain non-volatile petroleum hydrocarbons and do not contain free liquids
12. drained motor oil containers that do not contain free liquids
13. non-putrescible vegetative waste from agriculture, silviculture or horticulture
14. building cavity dust waste removed from residential premises or educational or child care institutions, being waste that is packaged securely to prevent dust emissions and direct contact

3 Data Analysis

3.1 Data Retrievals

The most common type of waste is the General Solid Waste (non-putrescible) which mostly contain household waste. Since majority of the sites are very close to residential areas (Check Fig. 7), this is expected (Fig. 6 and Table 1).

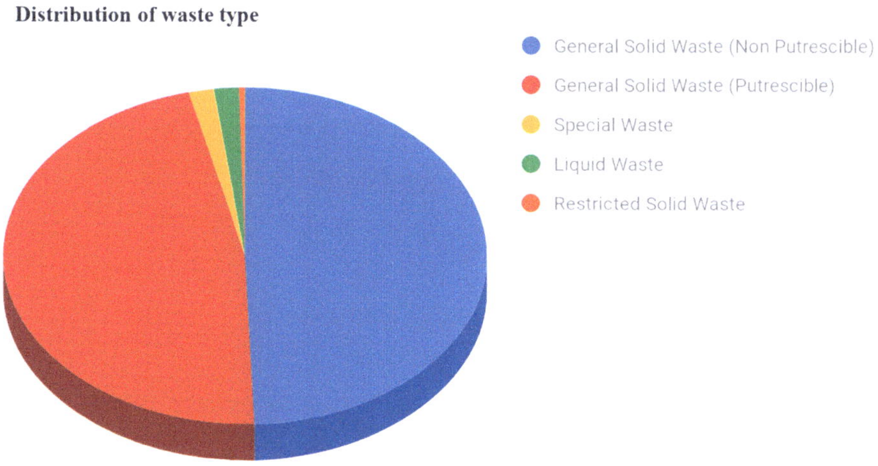

Fig. 6 Distribution of waste type

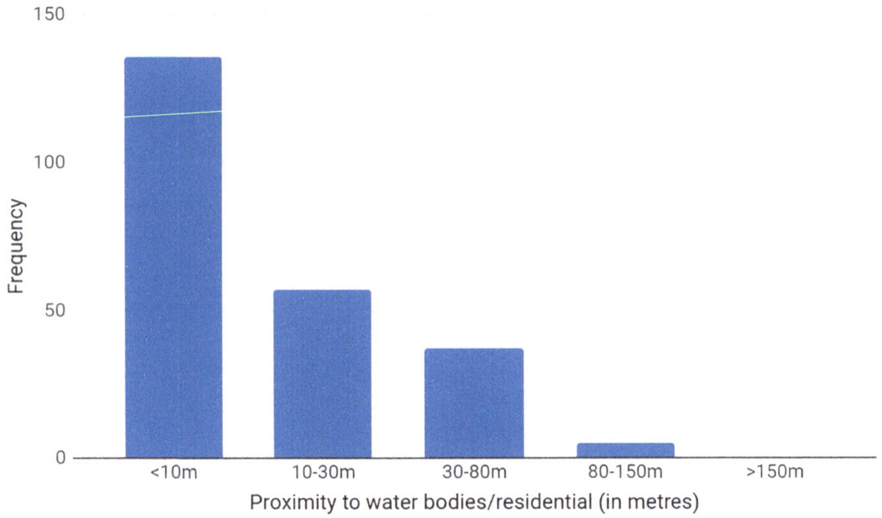

Fig. 7 Distribution of site proximity to water bodies/residential

Table 1 Table showing the frequency of the various waste types

Value	Frequency	Percentage
General solid waste (Non Putrescible)	116	49.36
General solid waste (Putrescible)	110	46.81
Special waste	4	1.7
Liquid waste	4	1.7
Restricted solid waste	1	0.43

3.2 Proximity

The Proximity of the sites to buildings or water bodies were classified into the following:

1. <10 m
2. 10 m – 30 m
3. 30 m – 80 m
4. 80 m – 150 m
5. >150 m

Features such as residential buildings, industrial infrastructures, hostels, rivers, water dams and other buildings were taken into consideration. These sites are very close to residential areas and are detrimental to the health of the residents. It also shows these sites mainly contain household wastes (Table 2).

Table 2 Table showing the proximity of the sites

Value	Frequency	Percentile
<10 m	136	58.87
10–30 m	57	24.26
30–80 m	37	15.74
80–150 m	5	2.13
>150m	0	0

3.3 Size of Site

The sizes of the sites were classified into the following:

1. Low
2. Normal
3. Large
4. Very Large

It is important to note the following:

1. Low: less than or equal to 15 square meters
2. Normal: between 15 square meters to 30 square meters
3. Large: Between 30 square meters to 50 square meters
4. Very Large: Over 50 square meters (Fig. 8 and Table 3)

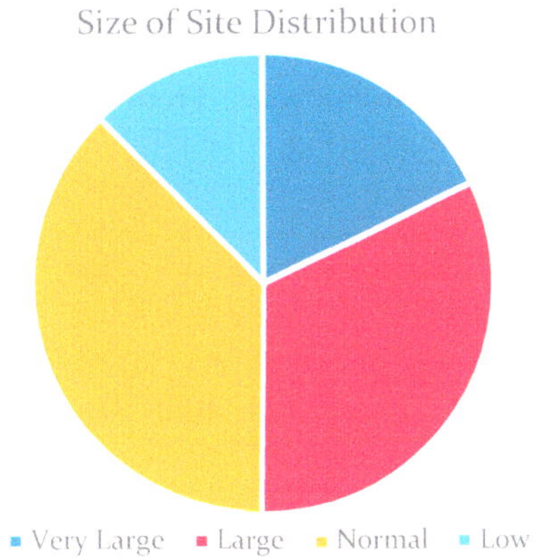

Fig. 8 Size of site distribution

Table 3 Table showing the size of the sites

Size	Frequency	Percentage
Very large	42	17.87
Large	76	32.34
Normal	88	37.45
Low	29	12.34

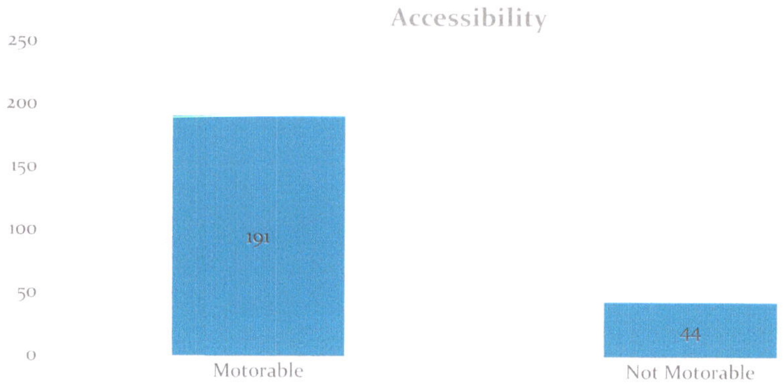

Fig. 9 Distribution of site accessibility

Table 4 Table showing the accessibility of the sites

Accessibility	Frequency	Percentage
Motorable	191	81.28
Not Motorable	44	18.72

3.4 Accessibility

It is important for the government or the NGOs who will organize clean-ups of these dump sites to be aware of how accessible they are. A motorable site will allow cars or trucks to access the place while the not-motorable sites can only allow motorcycles/bicycles. From the analysis, over 80% of the sites can be accessed using trucks (Fig. 9 and Table 4).

3.5 GIS Analysis

Land Use Land Cover analysis of the city was done using ARCMAP. Features include Bare Land, Built Up, Mixed Vegetation, Outcrop and Water Bodies (Fig. 10).

Landfill Site Selection in order to select an appropriate site for a landfill, several issues are to be considered:

1. Neighborhood (distances from residential areas, from waterways and water bodies, and from airports)
2. Geological and hydrogeological conditions in the area
3. Seismic conditions in the area

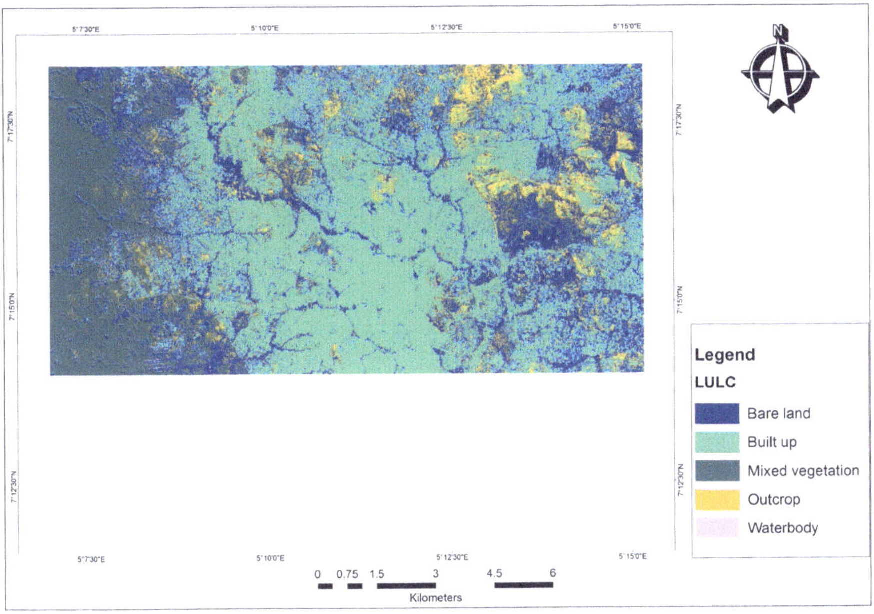

Fig. 10 Land use land cover of Akure

4. Existence of groundwater and its current (and future) utilization
5. Risk of flooding, subsidence and landslides
6. Transport distances and existing infrastructure (e.g., access roads)
7. Access to intermediate and final cover material
8. Topography of site

In this study, a maximum distance of two kilometers (2 km) was used in which five hundred meters (500 m) was considered the most suitable, based on the level of development in the city. This would create a protection zone around the buildings and other facilities to sitting waste disposal system. This was done in order to avoid pollution, ecological disturbance, and other health related issues and concerns. From analysis, over 50% of the dump sites are less than 10 meters from buildings/water bodies (Fig. 11).

The data from illegal dumping sites is overlaid on the LULC image (Fig. 12).

In the figure below, the black dots represent the illegal dumping sites mapped while the red dots represent the newly discovered suitable locations for siting dumping sites (Fig. 13).

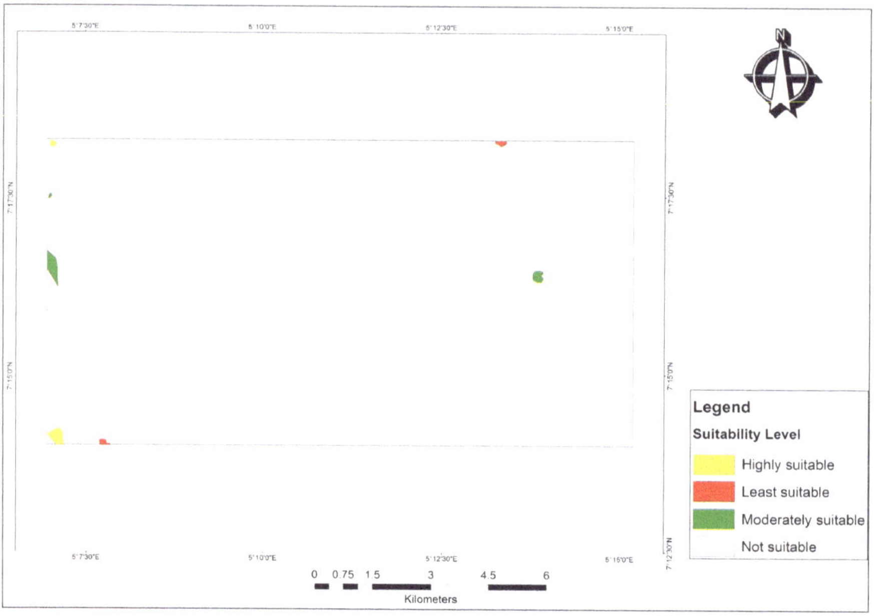

Fig. 11 Map showing suitability of areas for siting dump sites

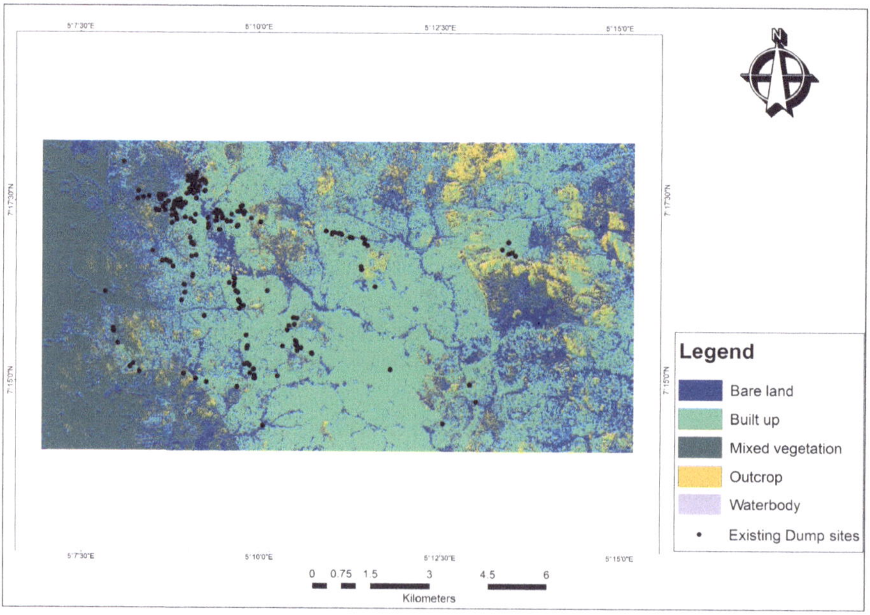

Fig. 12 Map overlaying illegal dump sites over the LULC of the city

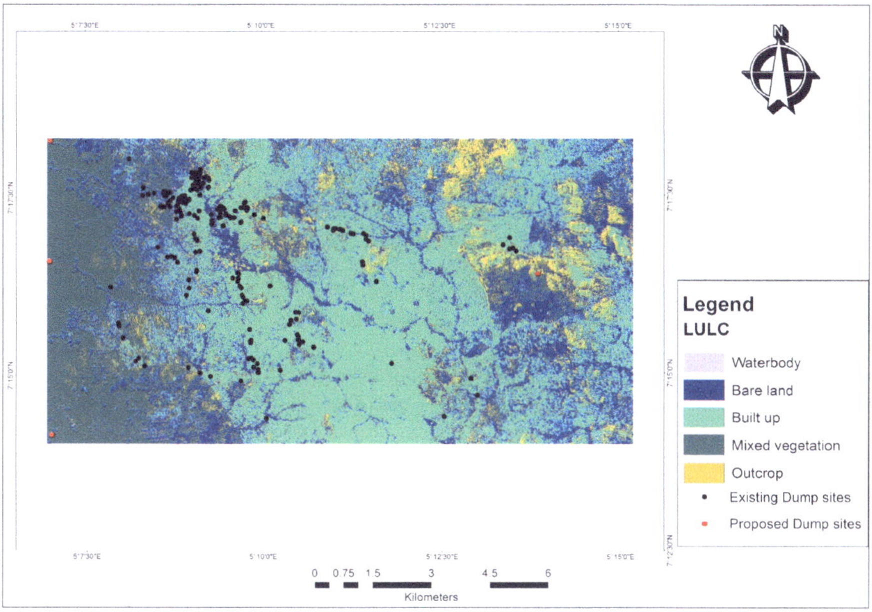

Fig. 13 Final map showing the locations of illegal dump sites (black dots) and proposed locations for dump sites (red dots)

4 Conclusion

In one of the developing cities of Nigeria, it is obvious that the number of illegal dumping sites are enormous, this calls for immediate action for the government.

5 Recommendation

The outcome of this project will be presented to the State Ministry of Environment in coming weeks for implementation. We recommend the state government should pass an anti-litter bill to stop dumping of refuse in different parts of the major city. Upon implementation, this project will be transposed to other cities in the country. We also encourage NGOs to make free use of the open source data from this project.

Acknowledgements All pictures used were taken by the project team for the sole purpose of this research. This project acknowledges the support/donations from:

- YouthMappers
- Digital Globe
- USAID

T. Oniosun is a multiple award winning space scientist. A YouthMappers fellow and former research scientist at the Centre for Space Research and Applications, Federal Univerisity of Technology, Akure. He is currently the Managing Director for Space in Africa, the authority on news, data and business analytics for the African Space Industry.

Dr. I. Balogun is a senior lecturer at the Department of Meteorology and Climate Science, Federal University of Technology, Akure. His current research focus is on understanding multifaceted roles of cities in modifying its landscape, weather, climate, air quality and quality of life of inhabitants (man, animals, plants and organisms).

Prof. P. Solis is currently the Executive Director, Knowledge Exchange for Resilience, at Arizona State University and the Director and Co-founder of YouthMappers.

Economic Growth Through Investment into Space Science and Technology: The African Colocation Programme

Carla Sharpe

Abstract

Space infrastructure, both ground-based and space-based, provides a cornerstone to economic development. If one considers sectors such as communications, banking, security, weather services, agricultural monitoring, scientific research or municipal services, it is self-evident that these sectors all rely on seamless information gathering and exchange, via space-based infrastructure. The information provided by space-based infrastructure enables governments to make real time, informed and strategic decisions towards growth, security and risk mitigation. The African Colocation Programme is a programme designed to colocate space science and technology infrastructure alongside the radio telescopes of the AVN programme, within the SKA Africa partner countries. The programme is designed to grow expertise, industry, innovation and academia in the partner countries, creating larger African networks for data, fibre, satellite ground stations and science instrumentation. In other words, the AVN colocation is a proposed central and southern African network of space based industry and science infrastructure. The network will provide augmented solutions to users through collocated sites designed to promote sustainability, industry development, training, and scientific research collaboration. In addition to this is the longer term view of providing African built satellite infrastructure and services.

C. Sharpe (✉)
South African Radio Astronomy Observatory (SARAO), University of Cape Town (UCT), Rondebosch, South Africa
e-mail: carla@ska.ac.za

A. Froehlich (ed.), *Space Fostering African Societies*, Southern Space Studies,
https://doi.org/10.1007/978-3-030-32930-3_4

1 Introduction

Africa is a continent of stark contrasts. It is well documented that the continent is faced with many harrowing challenges in the post-colonial 21st century. Most Africans are faced with poverty, hunger, poor education, ill health, corruption and violence, with more growing numbers living in urban slums each year. However, the continent possesses undoubted, and as yet unrealised potential for greatness. The continent's growth trajectory is uniquely positioned to reap the benefits of the Fourth Industrial Revolution—a phenomenon that has been relatively exclusionary to Africa thus far.

In Africa, the general trend of the post-war period in which independence and revolutionary movements led to the establishment of self-governing states across the continent, has been an economic growth rate that is far too low, compounded by insufficient overall industrial development. While it is true that enviable growth rates were experienced by several African countries in the early 2000s, this was substantially driven by rapidly growing investment into abundant, but unstable, resource sectors. Inevitably, the big challenges have re-emerged, as they have in many developing regions where they are yet to be addressed holistically: inequality; unemployment; aid dependence; the stresses of providing infrastructure for expanding populations and the externalities which accompany the process.

It is clear that change is needed. More importantly, however, it is time for Africans to change Africa through educating their people, creating opportunities, value chains and industries that are capable of sustaining the growth of a vibrant and integrated African economy. It is well understood that innovation and technology development are drivers of sustainable economic growth. This coupled with improved education, collaboration and good governance, African partners can work together to overcome the difficulties faced by our continent.

Furthermore, the fostering of technology infrastructure can directly impact the lives of 1.2 billion Africans by contributing to the improvement of education, provision of services, industry development and employment opportunities, as well as by supplementing situational awareness efforts and enhanced regional economic integration. The focus of this paper is on the utility and implementation of space infrastructure across Africa, as one of the main sectors of technology infrastructure, towards positively impacting economic growth and societal welfare.

Space infrastructure, both ground-based and space-based, provides a cornerstone to economic development. If one considers sectors such as communications, banking, security, weather services, agricultural monitoring, scientific research or municipal services, it is self-evident that these sectors all rely on seamless information gathering and exchange, via space-based infrastructure. The information provided by space-based infrastructure enables governments to make real time, informed and strategic decisions towards growth, security and risk mitigation.

A practical application of the knowledge gained from this infrastructure is in the provision of food security—a key global challenge which is particularly pertinent in those developing countries whose landscapes have become increasingly vulnerable to the effects of natural disasters. The vast majority of the world's food-insecure population lives in developing countries, where thirteen percent of their populations are undernourished. Sub-Saharan Africa is the region with the highest prevalence of hunger, and one in four people is undernourished across this region.

Daily satellite observations offer a wide variety of environmental monitoring applications including rainfall, temperature, vegetation health, ocean productivity and soil moisture conditions. Many individuals and organisations are able to use this data to better understand and diagnose the impact of a multitude of factors on agricultural productivity. For example, extreme weather events trigger food insecurity by reducing the supply of food and compromising the incomes of households working within the agriculture sector. Satellite data contributes to the prevention, detection and mitigation of these potentially disastrous impacts. Other ecosystem-based assessments can also be used to inform decisions regarding long-term food security and human health outcomes due to changes in access to basic resources.[1]

If it is underpinned by collaboration and complementary value chain development, Africa's economic growth can develop significantly, allowing the continent to become competitive within the global market. A handful of countries currently own independent African space infrastructure, but as a continent, Africa should be able to offer, own and develop end-to-end solutions. These include space-based assets such as satellites, as well as ground-based assets such as receiving stations and communications networks, which include fibre networks and data facilities.

2 Space Science and Technology Development in Africa

The sector of space science and technology provides a practical focus area for distinguishing between, on the one hand, public funding into science-based programmes with purely science-based outcomes as their primary goal; and on the other hand, science-based programmes with primary outcomes that are industrial and revenue-based.

Public investment into space science and technology can serve as both a provider of purely science-based programmes as well as revenue-based industry-driven programmes. This is a unique arena, where space science programmes and the space industry may overlap, although they are often structured differently and operated independently. This paper is presenting a cooperative solution of both space science and space industry towards sustainable development.

The space industry was launched, quite literally, in the 1950s and, for the next forty years, it was dominated by governments and more importantly their militaries. Only in the last few decades has there been a significant move of space industry research, manufacturing and even travel, to the commercial arena and more

[1]Brown, Molly. (2015). Satellite Remote Sensing in Agriculture and Food Security Assessment. Procedia Environmental Sciences. USA.

specifically, the private arena. This has created many opportunities for countries, markets and companies to interact in the space arena, where this had been impossible in the past. The most significant limiting factor has been the associated costs, which pose a major barrier to entry. Many space agencies and government initiatives now have it in their mandate to assist companies in overcoming these barriers.

Technological development is indeed a driver of economic growth, provided there is political suitability, sound institution, human capital and a positive investment environment. This is true, too, for space technology development. There are, however, some unique challenges when developing space capability and capacity as compared to other technology sectors. For a start, the space arena is subject to a legal framework termed 'space law', which is essentially based on five international treaties.

Space law, consisting of a number of legal instruments negotiated at international, regional and national levels, is aimed at regulating human activities in outer space. It has been successful in preventing military confrontation in outer space and in facilitating relatively harmonious international cooperation in space projects. Space law has developed in two stages, each of them influenced by international political and economic factors: the first phase occurred during the competitive, militarily-dominated era of the Cold War; the second within the more collaborative paradigm that has emerged since the early 1990s. This is clearly illustrated in manned spaceflight—particularly in the context of space stations.[2]

The space arena, moreover, provides a resource, not only for the aforementioned technological development and innovation, but also for critical services such as those provided by remote sensing and Earth observation activities. The output of space programmes yields private goods, public goods, social goods and strategic goods. For clarity, we define a private good as an item that yields positive benefits to people that are excludable, that is, its owners can exercise private property rights, preventing those who have not paid for it from using the good or consuming its benefits. In contrast, a commodity or service provided without profit to all members of a society, by a government, a private individual or an organisation, is a public good. For example, research in space sciences and most meteorological services are public goods. The Global Positioning System (GPS) is a space-based navigation system that provides location and time information across the globe and as such provides critical capabilities to military, civil, and commercial users around the world. The United States government created the system, maintains it, and it renders it freely accessible to anyone with a GPS receiver.

A good or service that benefits the largest number of people in the largest possible way is defined as a social good, an example of which is literacy. Strategic goods such as goods for military purposes are goods which one is not allowed to export, import or transit to certain countries or only under certain conditions, for reasons of security and international agreements.

[2]Sharpe, C., Tronchetti, F., Von der Dunk, F. (2015). Handbook of Space Law. Elgar.

Space science and space exploration are generally undertaken for the benefit of humanity at large and thus are normally regarded as a public good provided at a cost to national governments. The space industry, however, although utilising space science and providing for further scientific development, is revenue-driven and provides goods and services to clients, both private and public. For the purposes of this chapter, 'space-based activities' refer to all public and private space infrastructure, academic and industrial research, as well as commercial activities.

Successful space-related industrial and manufacturing development provides revenue streams and the means for economic expansion. These activities include, but are not limited to, the design, development, testing, manufacturing and launching of spacecrafts. In addition, applications such as those of satellite services, are provided to users commercially. This includes a number of private goods, such as Direct To Home (DTH) television broadcasting.

Other information from space-based activities includes familiar Google products, such as Google Earth or Google Maps, which are examples of freely available satellite imagery. Commercially available products and services, such as those provided by local companies in South Africa whose business model is centred on space-based assets capable of providing high resolution, day or night, all-weather surveillance worldwide for a variety of purposes. These uses can vary from defence and disaster management to agriculture.

Participation in this arena is very costly, and the barriers to entry can appear to be insurmountable, particularly for developing nations. This leads to a practical dilemma for these nations. The need for space services and infrastructure to ensure a nation's growth, its involvement in the global space industry, and its development of a knowledge economy for economic expansion can come at a large cost, which may detract from the provision and delivery of other more basic necessities to the population.

It is essential to resolve this paradox, to explore and understand the ways and the extent to which space-based activities can benefit developing nations. The potential benefits can be seen by looking at established national space programme activities. These activities are used to support national and global solutions to current and emerging problems, including food security, resource management, climate change impacts, and disaster mitigation and response.

Furthermore, to support sustainable economic development, space science and technology activities are often implemented as part of a long-term capacity-building plan. Naturally, this needs to be supported with the appropriate policy and legal frameworks, institutional development, community participation and human capital development if it is to stand a reasonable chance of achieving the desired long-term outcomes.

In terms of space activity in Africa, the African Union (AU) has been driving a collaborative effort to enhance space development across the continent. The AU is an intergovernmental organization formed to help nurture security, economic development and economic integration among its 54 member states The launch of the AU was a major event in the contemporary history of the continent. It began with the Extraordinary Session in Sirte in 1999, when the Assembly of Heads of

State and Government of the Organization of African Unity (OAU) decided to accelerate the process of economic and political integration of the continent. They formulated the fundamental vision of the AU as: "An integrated, prosperous and peaceful Africa, driven by its own citizens and representing a dynamic force in the global arena."

In March 2010, the Fourth African Ministerial Conference on Science and Technology (AMCOST IV) recommended setting up a Space Working Group to develop a draft African Union Space Policy and Strategy. The objectives of the African Space Policy are to address user needs; access space services; developing the regional markets; adopting good governance and management; coordinating the African space arena; and promoting international cooperation.

As outlined earlier, the rule of law and institutions play an important role in fostering a suitable economic environment in order to stimulate economic growth. This applies not only to the countries themselves, but also within industry sectors such as the space sector, which are subject to a global market under the jurisdiction of international laws. Of the more than fifty countries that account for the African continent, only a handful have a national space policy. The move towards a continental African Space policy, that would be applicable to all 54 member states of the AU, is a positive step towards pan-African collaboration and growth.

Developing a first generation Earth Observation (EO) satellite remains the point of entry for countries with aspirations to acquire space-related technology and industry capability. Emerging programmes commonly partner with established manufacturers in other regions to gain know-how through technology transfer and localisation, facilitated by lower cost COTS solutions.

There are several reasons why countries invest in EO, namely to attain autonomous manufacturing capabilities to contribute towards the development of a space science programmes; EO has a relatively low cost of entry compared to other satellite applications such as communications and exploration; and importantly the need for assets to support the information requirements of the government such as for defence, monitoring, agriculture, disaster management, marine domain awareness and other solutions.

Small satellites are an emerging class of spacecraft that extensively incorporates recent software and hardware improvements, as well as the benefits from the resulting high capability that is feasible in smaller packages. Compared with traditional satellites, small satellites typically have shorter development cycles, smaller development teams, and consequently a lower cost, both for the development and for the launch of these satellites. CubeSats have the additional benefit of a standardized form-factor, enabling mass production and easier launch vehicle integration, which can further lower costs. These lower-cost satellites boast augmented expendability, faster refresh, and simultaneous deployment in large numbers which enables greater risk-taking, experimentation and creation of new applications that were not feasible with larger satellites.

As a result, small satellites are making inroads in almost every area of space, namely communication; remote sensing; technology demonstration; and science and exploration. Through this prodigious industry expansion, they have come to be operated by an ever-growing base of users.

Due to this growing number of increasingly affordable satellite services, new uses for "space data" are opening up across many industries. In farming, satellite data can be used to monitor factors which influence crop yield. In real estate, areas prone to flooding or sinkholes can be more accurately identified, impacting property developments and prices. In retail, foot traffic around shopping centres can be monitored in real-time, invaluably adding multiple dimensions to the overview of consumer behaviour in such industries.

Few African countries operate their own EO satellites. Not only is it a challenge to manufacture and operate the satellites, but the required ground-based infrastructure is also inadequate at present. Countries including Algeria, Nigeria, and South Africa have developed satellite capability, while other countries such as Kenya, Ghana and Morocco have now established remote sensing agencies with more development, and germination of skills progressing at a university level. As this development continues with greater adoption of satellite applications and capacity building in African countries, so the revenues will grow, both in data acquisition and the resultant offering of value added services (Fig. 1).

Estimates provided by market analysts can be seen in Fig. 2, showing the significant annual growth of commercial EO data and value added services (VAS). VAS are expected to grow at a greater rate than for the data itself, as the raw data—

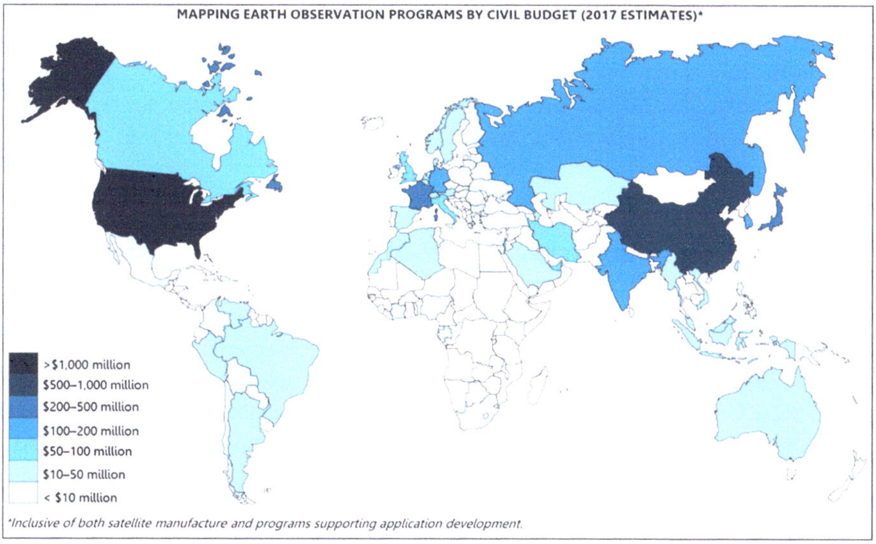

Fig. 1 Global government funded earth observation programs Credit: Euroconsult, 2018

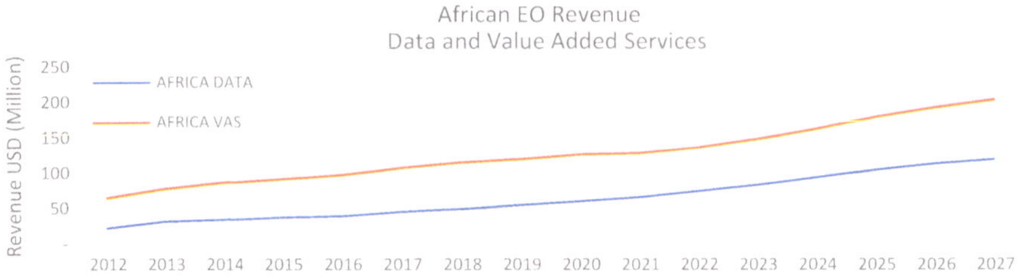

Fig. 2 African earth observation data revenues versus value added services revenues

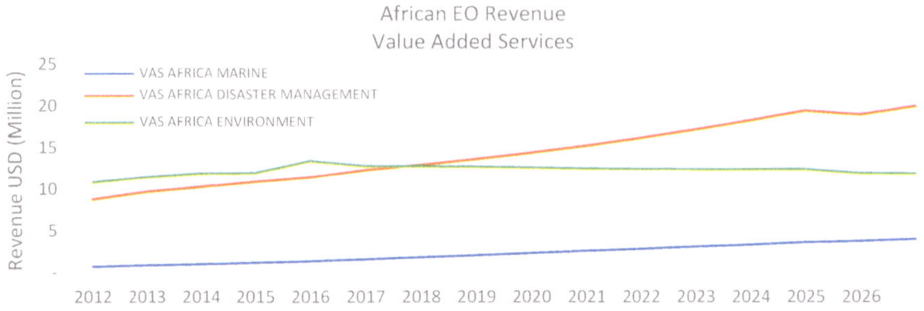

Fig. 3 African earth observation types of value added services revenues

without being packaged as user-friendly information—is not as desirable to end users. In addition, there is also a larger profit margin on VAS than the less price-elastic raw data. VAS may be divided into categories of uses and in the African context, the largest growth areas in VAS are expected to be disaster management and marine domain awareness solutions, as shown in Fig. 3.

3 Space Science in Africa: South Africa and the Square Kilometre Array

The Square Kilometre Array (SKA) is the largest space science project in Africa, and is set to become the largest radio telescope in the world.

The SKA Organisation (SKAO) consists of twelve member countries, of which South Africa, the UK and Australia, are both members and host nations. The SKAO is headquartered at Jodrell Bank in the United Kingdom. The member and associate member nations include both developed and lesser developed countries; Australia, Canada, China, France, India, Italy, the Netherlands, New Zealand, Portugal, South Africa, Spain, Sweden, and the United Kingdom. The SKA represents an unrivalled opportunity for the development of high-level skills and expertise in Africa and will allow the continent to become a significant contributor to the global knowledge economy.

The SKA radio telescope will operate over a wide range of frequencies. Its size will enable it to be fifty times more sensitive, and up to ten thousand times faster, than the most advanced radio telescopes that exist today (SKA, 2015). Moreover, it will produce resolution superior to that of the Hubble Space Telescope by a factor of fifty.

Ending in 2011, an international bidding process was carried to determine the optimal location for hosting the site of the telescope. This decision was contingent on a number of factors pertaining to the prospective locations, such as the availability of a radio quiet area with a reliable power supply, coupled with the host nation's relevant economic indicators and science capacity. The two short listed bidders were South Africa and Australia. As part of the bidding process, both countries undertook to build and contribute precursor telescopes. South Africa committed to the construction of the MeerKAT telescope, and Australia committed to construct the ASKAP telescope, respectively.

On the 25th May 2012, the Board of Directors of the SKAO announced that the SKA telescope would not be hosted in only one country, but rather that it would be split over the proposed sites in both South Africa and Australia. The portions of the telescope to be hosted in South Africa and at the other African sites are the mid frequency arrays (SKA MID), whereas Australia will host the low frequency array.

Over the last decade, South Africa has designed and constructed the MeerKAT Telescope, the precursor to the SKA Telescope, which was inaugurated in July 2018. South Africa also plays a significant role in the SKA Project as a host country, member of the SKA International Governmental Organisation (IGO) and importantly, in the design and engineering of the SKA Mid Telescope.

The SKA telescope will be constructed in several phases to manage risk and to allow for technological advancement over the lengthy construction and commissioning periods. SKA MID Phase 1 will include the construction of 133 dishes, due for completion in 2026. Thereafter, SKA MID Phase 2 construction will commence, contingent on meeting the funding requirements, with expected completion in the early 2030s. It is envisaged that the full SKA telescope will consist of approximately 2000 antennas. The word 'antennas' here refers to what are more commonly termed 'dishes' or 'radio astronomy dishes'. Each dish will have a diameter of 15.5 m and the array's baselines will span up to a length of 120 km. The SKA MID Phase 1 has a cost cap of EURO 650 million and Phase 2 is anticipated to cost up to EURO 2 billion.

The core site in South Africa is located in the Karoo area of the Northern Cape province, about 80 km northwest of Carnarvon. SKA Phase 2 is intended to include remote arrays in African Countries namely, Botswana, Ghana, Kenya, Madagascar, Mauritius, Mozambique, Namibia and Zambia. A vital part of this effort towards building SKA on the African continent over the next decade, is the development of the African Very Long Baseline Interferometry (VLBI) Network (AVN), to be outlined in the proceeding section.

4 Space Science in Africa: The African VLBI Network[3]

Over the last decade, South Africa has worked to achieve growth targets through technology development and innovation, as laid out by the Information and Communications Technology (ICT) roadmap, the Medium Term Strategic Framework and the ten year innovation plan of the Department of Science and Technology (DST), the purpose of which is to aid the procession of South Africa's continued transformation towards a knowledge-based economy, in which the production and dissemination of knowledge would lead to an advanced and ameliorated social economy.

The implementation of this innovation plan has included many programmes including the SKA South Africa (SKA SA) project. This mega-science project contributes towards infrastructure-led growth, direct economic impact, innovation and commercialisation, and induces positive externalities in science and education.

In 2017, the SKA SA project merged with the HartRAO facility in South Africa, and now operate as the South African Radio Astronomy Observatory (SARAO), a national facility of the National Research Foundation (NRF).

To date, SARAO has awarded approximately 1000 bursaries with almost R500 million invested into Human Capital Development programmes over the last decade. Through this project, R200 million has been invested into industry development, with approximately R200 million spent in communities surrounding the construction site.

With support from the Department of International Relations (DIRCO) through their African Renaissance Fund (ARF), the SKA SA project initiated the AVN programme with the eight African Partner countries (APC): Botswana; Ghana; Kenya; Madagascar; Mauritius; Mozambique; Namibia; and Zambia.

The programme aims to establish VLBI-capable radio telescopes in the SKA Africa partner countries through the conversion of redundant telecommunications antennas, new-build telescopes or through the establishment of training facilities with training telescopes. Developing a network of VLBI capable radio telescopes on the African continent will allow for the transfer of knowledge and technology, as well as the development of the necessary transferable skills within participating countries. In addition to this, collaboration will bring new science, innovation, and ICT-related opportunities to participating countries on a relatively short time scale.

Very-long-baseline interferometry (VLBI) is a type of astronomical interferometry used in radio astronomy. In VLBI, a signal from an astronomical radio source, such as a quasar, is collected at multiple radio telescopes on Earth. The time difference between the arrival of the radio signals at different telescope dishes, as well as the distance between the radio telescope dishes, is then used for calculations. This allows for simultaneous observations of an object by many radio telescopes to be combined, emulating a telescope with a size equal to the maximum separation between the telescopes.

[3]www.ska.ac.za.

The key objectives of the AVN project are to develop a network of VLBI-capable radio telescopes on the African continent; transfer knowledge and technology and develop the necessary skills in participating countries in Africa to operate these telescopes independently; bring new science opportunities to participating countries on in a short time scale; and to enable participation in SKA pathfinder technology development and science.

However, the AVN implementation remains a challenge on a number of levels. Whether participating in radio astronomy or other technology programmes, the same basic elements are required and these include stable electricity supply, uninterrupted access to fibre networks, data storage infrastructure, data transport infrastructure, data analytics capability, scientific capacity, governance structures and operational funding. The same requirements need to be met regardless of the programme if sustainable technology and science advancement is to be achieved.

The first AVN telescope was officially launched in Ghana at the Kuntunse site in August 2017. This was a great step forward for the AVN programme, however the timeframe and cost to completion were a challenge to the implementation of the overall programme. Lessons learned, as well as new and unique solutions, have had to be developed for the programme moving forward. This paper will not examine these lessons learned, as that is a paper in itself, but rather seeks to present the colocation programme and the importance of collaboration to development on the continent.

5 The African Colocation Programme

The African Colocation Programme is an evolving programme designed to colocate space science and technology infrastructure alongside the radio telescopes of the AVN programme within the partner countries. The programme is designed to grow expertise, industry, innovation and academia in the partner countries, in turn creating larger African networks for data, fibre, satellite ground stations and science instrumentation. In other words, the AVN colocation is a proposed central and southern African network of space-based industry and science infrastructure. The network will provide augmented solutions to users through collocated sites designed to promote sustainability, industry development, training, and scientific research collaboration.

In order to identify and manage expectations, it was essential to develop a proposal for buy-in by both local and African Partner stakeholders, with a roadmap that realises quick wins as well as long-term solutions and importantly, sustainable income-generation capabilities. The roadmap addresses the basic requirements for technology advancement in a phased approach, and this will be achieved through the implementation of the colocation of science instruments, satellite data receiving ground stations, passive tracking radar for southern and central African aircraft security and data processing infrastructure. The colocation of all these services and infrastructure on one site allows for site operations and infrastructure to be shared.

In addition, some of these implementations will be based on low cost, robust computing solutions developed for the MeerKAT telescope, as well other locally-developed technologies.

Each APC Site would be able to address Human Capital Development (HCD), African technology infrastructure and science goals, as well as industry development and revenue streams.

This programme has been designed in order to provide:

- Sustainable revenue for programmes and ongoing operations and maintenance of colocation sites
- Employment, education and training opportunities in areas such as radio astronomy, related technology development, data administration and management skills
- Value added data solutions from data applications to earth observation data with many applications including but not limited to agricultural and environmental monitoring, illegal fishing monitoring, poaching mitigation, fire tracking, disaster management, resource management and more
- New radar solutions that not only contribute to conventional aircraft tracking but also aid in security concerns such as addressing illegal trafficking activities

In essence, the colocation of science, technology and industry solutions on one site allows for cost and resource sharing for ongoing operations and maintenance of the sites. Through academia, government and industry activities sharing not only a geographic site, but skills, knowledge and innovation opportunities, one creates a fertile ground for growth and development.

No one partner owns the programme, and each collaboration within the programme, and specific to each site, will develop their own partnership and sharing agreements. Each colocation site will be provided with a basic implementation to build on over time, this would provide a physical satellite receiving dish[4] the relevant hardware and software required, basic satellite data contracts relevant to the country's initial needs, value add services to address the information requirements and a distribution solution. The full implementation would also be supported through a thorough training solution. It is thus a dynamic programme that allows for unique solutions to partnership and development and an overview of the programme is expressed below in Fig. 4. The immediate implementation requires that the partner country establishes the lead agency mandated to govern space related activities. This phase should lead to data and information solutions that would be provided to countries through centralised government procurement and distribution. In addition, this phase should see the parallel implementation of skills development. Once these activities have been successfully implemented, the satellite receiving station can be implemented and this will greatly reduce the cost of, and access time to data.

[4]The satellite receiving dish is in addition to the radio astronomy dish, the business case is not assuming a dual use dish but rather colocated instruments.

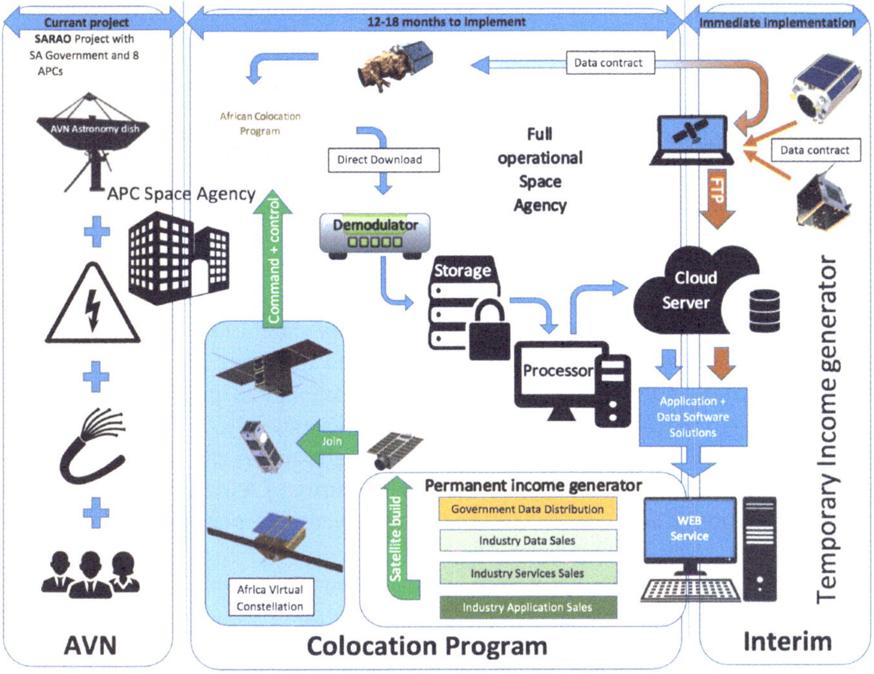

Fig. 4 Overview of the African colocation programme

The African Colocation Programme partners can also make use of alternative funding models with a combination of development, loan and grant funding. One such model, referred to as a Financial Assistance Programme, allows government to develop solutions or provide services, on a shared risk and shared cost basis in collaboration with compliant commercial partners. This model was developed by SKA SA in order to enable industry to participate in the self-funded design phase of the SKA Project, as well as to become competitive internationally. The programme has been centred on a South African collaboration interacting internationally, in which industry partners act with government and academia to provide the deliverables to the SKAO. The shared cost is funded by government and the returns are provided by industry growth and industry contract awards or similar benefits. In this case, it would be an investment towards industry stimulation and skills development.

6 Conclusion

The successful implementation of large scale science and technology programmes within the African context, must offer a well-defined programme, HCD and skills development benefits, long-term sustainability, industry development, innovation potential and sustainable revenue streams. The programme must address the national science goals, national education and skills development plans, innovation strategies and the industry development strategies of the participating countries.

Africa should focus on developing market places that are powered by African buyers and sellers, with development programmes based on collaboration, synergies and complementary services. African partners should look towards building value chains in a collaborative fashion, whereby each partner provides a link into the chain, avoiding the scenario where partners attempt to duplicate efforts and expenditure on replicating the full value chain independently.

Collaboration is also the solution to developing regional environmental awareness and management, as well as enhancing the security of African resources by, for example, providing an African marine domain awareness platform for protecting coastal waters.

The African Colocation Programme provides an African platform for collaborative growth in space science and technology, providing sustainable solutions and progressively extend benefits to multiple sectors and participants.

In the developing nation context, it is essential that technology and data solutions whether they be science instruments, infrastructure or information services, are not provided to these nations without also providing the skills, training, hardware and software towards independent ownership, further development and management of these solutions.

Carla Sharpe has an Executive MBA from the International Space University and is currently completing her Ph.D. in Space Studies at the University of Cape Town. She is a founder of the South African Space Association, the Foundation for Space Development and currently serves on the Management Committee of both, as well as Women in Aerospace Africa. Carla is responsible for African Programme Management at the South African Radio Astronomy Observatory (ska.ac.za).

Internet by Satellite for Connecting the African Continent: A Glance on the Partnership Between Rwanda and the Private Company OneWeb

Anne-Sophie Martin

Abstract
This contribution will address the importance of Internet by satellite for the benefit of African citizens, and in particular, for education which is part of the "aspirations" established in the African Union's Agenda 2063. In order to connect the African Continent, the cooperation between States and private companies is of utmost importance with the goal of developing further the space market. Moreover, this contribution will emphasize how space applications can be deployed to support the development of African societies. In a more specific way, how these applications can be increasingly used to bolster development and education in rural zone. It will take as reference the case of Rwanda which recently launched a satellite in partnership with the private entity OneWeb. Finally, the African Union should continue to support the development of space programs at national and regional levels but also by setting up a clear space policy strategy.

A.-S. Martin (✉)
Department of Political Sciences, Sapienza University of Rome,
Piazza Aldo Moro 5, 00185 Rome, Italy
e-mail: annesophie.martin@uniroma1.it

A. Froehlich (ed.), *Space Fostering African Societies*, Southern Space Studies,
https://doi.org/10.1007/978-3-030-32930-3_5

1 Introduction

The African Union's Agenda 2063[1] is a strategic framework for the socio-economic transformation of the African continent over the next 50 years. The Agenda is based on past and present initiatives, such as the New Partnership for Africa's Development (NEPAD), the Nigerian treaties of Lagos and Abuja of Economic Community of West African States (ECOWAS) and the African Economic Community (AEC). The foundations of the African Union provide general aspirations for "*an integrated, prosperous and peaceful Africa, led by its own citizens and representing a dynamic force in the international arena*".[2]

The African citizen is therefore the target of the Agenda that in fifty years has promised to totally change Africa, overcoming the current fragmentation.

It is composed of seven "aspirations" based on growth and sustainable development of Africa, as well as a continent politically stable, prosperous and united, with the respect for human rights. The Agenda highlights the importance of developing a peaceful and secure continent with a strong cultural identify which represents the African people in order to become a prominent global player.

It is a necessity to eradicate poverty and build prosperity through social and economic transformation of the continent as recalled on the first "aspiration". Moreover, it is of utmost importance to provide well educated and skilled citizens, supported by science, technology and innovation for a knowledge society where no child misses school due to poverty or any form of discrimination. It this sense, space applications, and in particular telecommunication by satellite providing Internet, have a role to play in order to achieve these objectives. Indeed, modernized infrastructures will provide the people to have access to Internet at work, at school, at home and in a more general way to use Information and Communications Technology (ICT).

A central point of this contribution is the term "education" which can be found in several "aspiration" of the Agenda 2063. Indeed, investments in science, technology, research and innovation are very important in order to provide better infrastructures for education at all levels.[3] The "sub-aspiration 14" precises that Africa's human capital will be fully developed as its most precious resource, through sustained investments based on universal early childhood development and basic education, and underpinned investments in higher education, science, technology, research and innovation, and the elimination of gender gaps at all levels of education. Access to post-graduate education will be expanded and strengthened to

[1]African Union's Agenda 2063: https://www.un.org/en/africa/osaa/pdf/au/agenda2063.pdf (last accessed April 24, 2019).

[2]*Lo Spiegone*, January 7, 2018: https://lospiegone.com/2018/01/07/agenda-2063-unafrica-senza-confini/ (last accessed April 24, 2019).

[3]See Osborne (J.), Ebert (M.), Georgakas (S.), Hernandez Villatoro (R.), Johnson (C.), Modi (H.), Topham (R.), *Assisting Development Across the African Continent Using Space Applications*, 63rd IAC, 2012, ISU, pp. 7–8; Space in Africa, *Why You Need to Bother Investing in the African Space Technology Sector*, October 24, 2018: https://africanews.space/why-you-need-to-bother-investing-in-the-african-space-technology-sector/ (last accessed April 24, 2019).

ensure world-class infrastructure for learning and research and support scientific reforms that boost the transformation of the continent.

Furthermore, the point 32 of the Agenda emphasizes that a culture of peace and tolerance shall be fostered in Africa's children and youth through peace education.

The access to education and information is also important and all harmful social practices will be ended and barriers to quality health and education for women and girls eliminated.[4]

The African Union's Agenda promotes actively science, technology and research in order to build knowledge, capabilities and skills to drive innovations. The aim is to build and to expand an African knowledge society through transformation and investments in universities, science, technology, research and innovation, and through the harmonization of education standards and mutual recognition of academic and professional qualifications. In this context, the access to technology will raise the standards of higher education.

Africa is growing in the space industry, bringing more opportunities to all the communities in the continent by creating capacity building and bringing long-term sustainability in all space activities.[5]

The African space sector is currently subject to a phase of rapid and dynamic expansion with new actors entering the field and with space applications increasingly used to support the continent's social, economic and political development.[6] In this context, it is important to develop further the African space policy and strategy in order to address user needs, the access to space services, to develop the regional market, to adopt good governance and management, and to promote international cooperation as it will be hereinafter emphasized with the case of Rwanda and the private company, OneWeb.[7]

It should be noted that the African Union Heads of State and Government during their Twenty-Sixth Ordinary Session on 31 January 2016 in Addis Ababa adopted the African Space Policy and Strategy as the first step to realize an African Outer Space Program, and as one of the major programs of the AU Agenda 2063.[8]

---

[4]Point 51 of the AU Agenda 2063.

[5]Akinyede (J.O.), Adepoju (K.A.), *Prospects and Challenges of Building Capacity for Space Science and Technology Development in Africa*, ISPRS Commission VI Mid-Term Symposium, 2010, pp. 82–91; UN Doc, *Report on the United Nations/South Africa Symposium on Basic Space Technology: Small Satellite Missions for Scientific and Technological Advancement* (A/AC.105/C.1/2018/CRP.9), February 8, 2018: http://www.unoosa.org/res/oosadoc/data/documents/2018/aac_105c_12018crp/aac_105c_12018crp_9_0_html/AC105_C1_2018_CRP09E.pdf (last accessed April 24, 2019); *CNN*, May 16, 2018: https://edition.cnn.com/2017/08/10/africa/africa-space-race/index.html (last accessed April 24, 2019).

[6]AfricaNews, March 12, 2019: https://www.africanews.com/2019/03/12/tracking-africa-s-race-to-space/ (last accessed April 24, 2019).

[7]12th EUMETSAT User Forum in Africa, *African Space Policy and Strategy*, September 13, 2016: https://ufa.eumetsat.int/userfiles/file/African%20Space%20Policy%20and%20Strategy.pdf (last accessed April 24, 2019).

[8]African Union, *African Union Heads of State and Government Adopts the African Space Policy and Strategy*, January 31, 2016: https://au.int/en/pressreleases/20160131-3 (last accessed April 24, 2019).

Space science and technology play a central role for the Africa's socio-economic development and for the achievements of the Agenda 2063.

The African Telecommunications Union (ATU)[9] and the regional organizations, such as the Regional African Satellite Communications Organization (RASCOM),[10] have also a role to play by promoting initiatives for investment in access and backbone cross-border information infrastructure through smart partnerships between public, private and not-for-profit sectors, in order to achieve universal access and full inter-country connectivity.[11]

The main goal of the ATU is to foster investments into information infrastructure in order to meet the 'aspiration' challenges' of global access and solid inter-connectivity. To that end, it is important to facilitate regional initiatives into ICT infrastructure, and to coordinate feasibility studies on integrated infrastructure.

In this context, the contribution will address in particular the case of Rwanda which recently launched a satellite of telecommunication (1) in partnership with the private company, OneWeb (2). Through these two points, we will observe the importance of the cooperation between States as it is mentioned in art.I of the Outer Space Treaty 1967 "*[…] States shall facilitate and encourage international cooperation […]*",[12] but also with the private sector so as to respond to the various "aspiration" of the African Union's Agenda in order to reduce the disparity of access to internet in Africa and to promote education.

2 Connecting Rural School to Internet: The Case of Rwanda

Last February, Rwanda launched a satellite named Icyerekezo, in partnership with OneWeb, a global communications company[13] and the Rwandan government. The satellite was launched to "bridge the digital divide"[14] in rural schools in Rwanda

[9]African Telecommunications Union, ICT Infrastructure: http://atu-uat.org/ict-infrastructure/ (last accessed April 24, 2019).

[10]http://www.rascom.org/ (last accessed April 24, 2019).

[11]African Telecommunications Union, *op.cit.*

[12]Treaty on Principles Governing the Activities of States in the Exploration and Use of Outer Space, including the Moon and other Celestial Objects (1967), 610 UNTS 2015.

[13]Republic of Rwanda, Ministry of Education: http://mineduc.gov.rw/media/news/details-news/?tx_ttnews%5Btt_news%5D=903&cHash=83bd471ed45a74c47383b19f03919ed3 (last accessed April 24, 2019); Space in Africa, Rwanda and OneWeb partner to launch Icyerekezo communication satellite, February 28, 2019: https://africanews.space/rwanda-and-oneweb-partner-launch-icyerekezo-communication-satellite/ (last accessed April 24, 2019).

[14]Ospina (S.), *The Digital Divide and Space Activities in the Southern Hemisphere: A General Overview of Africa and South Africa*, Proceeding of the 54th IISL Coll. on the Law of Outer Space, Cape Town, South Africa, Vol. 54, 2011, pp. 266–283.

where internet access is non-existent.[15] In particular, it will bring internet to students at Nkombo Island, in Lake Kivu.

OneWeb seeks to "deliver connectivity for everyone, everywhere, through a global satellite constellation".[16] Its founder Greg Wyler said "*Connecting remote schools to bridge the digital divide that still impacts half the population of the world is at the heart of OneWeb's vision. We are delighted to partner with the Rwandan Government and particularly the students of Nkombo. The connectivity we can provide them will allow them to realise their dreams and allow Rwanda to become a hub for technological innovation*".[17]

The Rwandan government sees the use of technology has a solution to solve challenges the country faces. Indeed, it was the first country in the world to use drones to transport blood to remote clinics.[18]

The launch of Icyerekezo has far more reaching impacts than just internet access. Rwanda's Minister of Education Dr. Eugene Mutimura emphasized that "*Connecting schools is a foundational aspect and driver of transformative learning. Rwanda's ICT in education Master Plan outlines ambitions of interventions to connect schools and empower Rwandan children with immense opportunity, notably research, support our competence-based learning, ease to access and share digital content, support systems to monitor and evaluate process among others. This partnership with One-Web to use satellite technology and connect all schools is a huge opportunity to support and allow us to leapfrog the current process in bid to connect all schools in the country in the next 3 years*".[19] The Minister of ICT and Innovation, Paula Ingabire highlighted that the Government of Rwanda has made outstanding efforts to invest in broadband connectivity and sees this as a great opportunity to continue connecting underserved communities. He stresses that "*Rwanda's choice to invest in space technologies is part of our broader mission to bridge the digital divide by providing equal digital opportunities to rural and remote communities. We are delighted to partner with OneWeb in this transformative initiative which presents us a huge opportunity to leverage satellite connectivity, using OneWeb's constellation, providing low-latency and high-speed internet to schools in remote communities of Rwanda. […] This partnership responds to our intention of becoming a regional Technology Innovation Hub,*

[15]AllAfrica, *Rwanda Launches Icyerekezo Satellite to Connect Rural Schools to the Internet*, March 1, 2019: https://allafrica.com/stories/201903010506.html (last accessed April 24, 2019).
[16]OneWeb: https://www.oneweb.world/ (last accessed April 24, 2019).
[17]IGL, Rwanda Launches Satellite to Connect Remote School to Internet, February 27, 2019: https://www.infosgrandslacs.info/productions/rwanda-launches-satellite-connect-remote-school-internet (last accessed April 24, 2019).
[18]The Guardian, 'Uber for Blood': How Rwandan Delivery Robots Are Saving Lives, January 2, 2018: https://www.theguardian.com/global-development/2018/jan/02/rwanda-scheme-saving-blood-drone (last accessed April 24, 2019).
[19]Hiiran, Rwanda Launches Satellite to Connect Remote Schools to Internet, March 2, 2019: https://www.hiiraan.com/news4/2019/Mar/162530/rwanda_launches_satellite_to_connect_remote_schools_to_internet.aspx (last accessed April 24, 2019).

opening new pathways for connectivity, providing better education and creating new opportunities for our innovators".[20]

Rwanda has placed space technology at the forefront as a major feature of national development and transformation[21] in order to wipe out poverty and build prosperity.

In that sense, OneWeb is partnering with other African countries like Sierra Leone to make schools have more access to the internet.[22]

This is an objective that should be realized at regional level with the aim to map school connectivity status globally, not just at national level. For instance, Rwanda joined the UNICEF Project Connect[23] which the data obtained are major input for designing sustainable programs in order to connect schools across Rwanda and across the continent. This is an important point also for the education.

The internet connectivity will transform the education sector in the African Continent. And it is necessary to find a solution that will fill the digital divide in an inclusive and sustainable way.[24]

Moreover, one can argue that the lack of connection means has harmful effect on economies, education, and development which stagnate and fall behind in a significant way compared to connected areas. For instance, Nkombo Island sits in the middle of Lake Kivu that makes extremely costly and inefficient to reach the area with standard fiber connections while a satellite is the perfect solution to provide internet connectivity.[25] The benefits are beyond internet access, enabling the whole community to access Government online services, providing access to global educational content to students and educators, and enhancing the research in space technologies. This partnership will enable orbiting satellites to connect more remote schools across Rwanda. However, this is vision has to be extended at the African Continent.

[20]AllAfrica, *Rwanda Launches Icyerekezo…, op.cit.*: https://allafrica.com/stories/201903010506.html.

[21]Republic of Rwanda, Ministry of Education: http://mineduc.gov.rw/media/news/details-news/?tx_ttnews%5Btt_news%5D=903&cHash=83bd471ed45a74c47383b19f03919ed3 (last accessed April 24, 2019).

[22]Liberty Writers Africa, Rwanda Launches Satellite to Connect Rural Schools to the Internet, March 1, 2019: https://libertywritersafrica.com/rwanda-launches-satellite-to-connect-rural-schools-to-the-internet/ (last accessed April 24, 2019).

[23]UNICEF, Project Connect: https://www.projectconnect.world/ (last accessed April 24, 2019). "*Map School Connectivity Globally and Eliminate the Digital Divide, Increasing Opportunity for Every Community*".

[24]ITU, *Bridging the Gender Digital Divide*, January 2018: https://www.itu.int/en/Lists/consultationOct2017/Attachments/59/1801%20Input%20EQUALS%20Action%20Map%20CWG%20Internet%20Consultation.%20v4.pdf (last accessed April 24, 2019); see also ITU, *The State of Broadband: Broadband Catalyzing Sustainable Development*, September 2017: https://www.itu.int/dms_pub/itu-s/opb/pol/S-POL-BROADBAND.18-2017-PDF-E.pdf (last accessed April 24, 2019).

[25]Rwandaful, The Hague, Issue 83, February 2019: http://www.netherlands.embassy.gov.rw/fileadmin/user_upload/RwandaFul-TheHague-83.pdf (last accessed April 24, 2019).

3 Towards More Cooperation with the Private Sector: The Case of OneWeb

Icyerekezo represents the first satellite of a large constellation that will allow global broadband coverage.[26]

Internet by satellite already exists but the connection is very slow due to their positions. Indeed, most of them are located in the geostationary orbit (GEO) that is 36,000 km from the earth's surface, and this means that the signal is not very reactive since it must achieve a consequential distance with significant delays in data transmission.

The idea of OneWeb is instead to place smaller satellites, in the Low Earth Orbit (LEO)—at about 1200 km in height—so as to avoid this problem. Obviously, the closer the devices are to the Earth, the smaller the portion of the earth's surface covered. For this reason the company's plan involves the construction of as many as 900 satellites, of which 650 will have to be in orbit by mid-2021.

OneWeb is not the only company that is working on this front. Similar projects are being considered by various aerospace companies in particular SpaceX. *Starlink*, this is the name of the ambitious satellite network, is based on the same idea: to create a huge satellites constellation to provide broadband internet all over the world.

Once active and secure, the OneWeb constellation could reduce the dependence on the Internet cable system that crosses the earth and the oceans, currently bringing 97% of intercontinental data.[27] The almost total dependence on these submarine carriers also creates fears related to security and defense. Considering, for instance, if an unfriendly country decides to cut off internet cables to the detriment of another country, this act would have very dramatic consequences at all levels.

Furthermore, one can observe that global Internet coverage would be an important and relevant socio-economic catalyst for those areas that are currently isolated and therefore for the people who live there: all schools in the world could be connected to the network and therefore having access, among other things, to distance learning programs. Accessing to internet would also mean greater gender equality since many studies tell us that the majority of the population without access to the network is female (about 58% of whom 60% live in rural areas).[28]

[26]AMIStaDes, *L'Idea di OneWeb: Internet dallo Spazio*, March 25, 2019: https://www.amistades.info/blog/l-idea-di-oneweb-internet-dallo-spazio?fbclid=IwAR0ncHJK1k7qJ6_XMkjsSJnL-GAFWSNHpACdMcX4RYUuBAxd2OuijIZ3dis (last accessed April 24, 2019).

[27]Hemispere Cable Company SL, *Intercontinental Submarine Cable Communication in High Demand*: https://www.hccwasace.com/2018/12/02/intercontinental-submarine-cable-communication-in-high-demand/ (last accessed April 24, 2019).

[28]Antonio (A.), Tuffley (D.), *The Gender Digital Divide in Developing Countries*, Future Internet, 6, 2014, pp.673-687; See also European Parliament, Policy Department for Citizens' Rights and Constitutional Affairs, *The Underlying Causes of the Digital Gender Gap and Possible Solutions for Enhanced Digital Inclusion of Women and Girls*, PE 604.940, March 2018: http://www.europarl.europa.eu/RegData/etudes/STUD/2018/604940/IPOL_STU(2018)604940_EN.pdf (last accessed April 24, 2019).

All of this will be possible thanks to technological developments and by the miniaturization of satellites that are increasingly smaller and easier to control.[29]

However, these advances bring with it many concerns related to congestion of the space environment. These huge constellations would in fact add to the already exorbitant number of space objects, increasing the risk of collisions. [30]

Thus, NASA, after having published a report which describes the effects of these mega-constellations on the safety of the space environment, invites all companies in the aerospace sector to devise return systems, so that satellites, once completed their activity, they return to the atmosphere.[31]

The successful first launch of OneWeb's satellites, enable the company to accelerate the development of the first truly global communications network by 2021. OneWeb is now moving from the planning and development stage to deployment of a full constellation.[32]

It is important to highlight the power of connectivity to change and improve lives by creating new opportunities for people everywhere. This represents a response to the "aspiration" of the African Union's Agenda 2063 by providing investments in science, technology, research and innovation in order to provide a better level of education at school and University. OneWeb's priority rights on a wide spectrum and a LEO constellation will provide an unequaled combination of high speed and global service by enabling connectivity for rural communities and schools as well as for business and industries. Internet by satellite will allow to develop and to bolster a wide range of emerging applications that require real-time communication and collaboration.

4 Conclusion and Future Perspectives

The African Union's Agenda 2063 and its "aspirations", in particular with regard to education, has the means to meet the new challenges of the African Continent, to support the development of space infrastructures and to foster a broader cooperation

[29]Paul B. Larsen, Small Satellite Legal Issues, Journal of Air Law and Commerce, Vol. 82, Issue 2, 2017, pp. 275–309.

[30]Francis Lyall, Paul B. Larsen, Space Law A Treatise, 2018, Routledge, p.239 ss.

[31]NASA archive, *Best Practices for Operations of Satellite Constellations*: https://ntrs.nasa.gov/archive/nasa/casi.ntrs.nasa.gov/20080039173.pdf (last accessed April 24, 2019); The Verge, *As Satellite Constellations Grow Larger, NASA Is Worried About Orbital Debris*, Sept 28, 2018: https://www.theverge.com/2018/9/28/17906158/nasa-spacex-oneweb-satellite-large-constellations-orbital-debris (last accessed April 24, 2019); NASA – Quarterly News, *Orbital Debris*, Vol.22, Issue 3, Sept. 2018: https://www.orbitaldebris.jsc.nasa.gov/quarterly-news/pdfs/odqnv22i3.pdf (last accessed April 24, 2019).

[32]*Space in Africa, OneWeb raises $1.25 billion from Rwandan Gvt and others; to mass-produce high-speed internet satellites*, March 18, 2019: https://africanews.space/oneweb-raises-1-25-billion-from-rwandan-govt-and-others-to-mass-produce-high-speed-internet-satellites/ (last accessed April 24, 2019).

between States and private entities in particular in the field of connectivity and Internet as highlighted previously.

Over the past decade, african investments in space applications and technology have increased significantly, driven by Earth observation programs such as in Algeria, Egypt, Nigeria, Gabon and South Africa, and satellite telecommunications in Angola, Rwanda and Congo.[33]

However, the lack of political and financial support may be an obstacle to the development of such program.

The Regional African Satellite Communications Organization might be a solution to respond to the new challenges, and to permit the access to Internet across the African Continent. Nevertheless, space projects lead by African leadership have been mixed results.[34] For example, the outcomes from the African Resource Management Satellite Constellation (ARMC) project between Nigeria, Algeria, Kenya, and South Africa are yet to be determined.

The access to ICT and in particular Internet all over the african continent is of fundamental importance to face the Continent's problems in the long term. There is a need of more visibility, leadership in regional space programs and capacity to adequately address issues. For instance, South Africa, Kenya, Egypt, Ghana, Angola, Algeria, Morocco, Rwanda and Nigeria have launched their own satellites.[35]

The 2012 Khartoum Declaration[36] recommends that the African Union Commission develop a space policy for the continent in collaboration with relevant stakeholders by taking into account the benefits that space applications, such as telecommunication by satellite, could bring to education which represents a core part of the AU's Agenda 2063 "aspirations".

The development of space programs in Africa should enable each country to enhance its scientific and technical knowledge and experience in space science by addressing African citizens' needs. The case of Rwanda and OneWeb is a good example.

Boosting the development of new space infrastructures at national and regional levels is a key element in order to concretely address the AU's Agenda 2063 objectives as well as to exploit all possible opportunities available in the short, medium and long term so as to ensure in particular a better access to education within the next 50 years.

---

[33]Aganaba-Jeanty (T.), *Towards Realising a Regional African Space Program*, Africa Portal, March 26, 2019: https://www.africaportal.org/features/towards-realising-regional-african-space-program/ (last accessed April 24, 2019).

[34]*Ibidem*.

[35]Space in Africa, *Ethiopia to Acquire a Communication Satellite Through Ethio Telecom*, January 7, 2019: https://africanews.space/ethiopia-to-acquire-a-communication-satellite-through-ethio-telecom/ (last accessed April 24, 2019).

[36]African Union Conference of Ministers in charge of Communication and Information Technologies (CITMC-4), 4th Ordinary Session, Khartoum, Sudan, 2–6 September 2012: https://au.int/sites/default/files/documents/30935-doc-declaration_khartoum_citmc4_eng_final_2.pdf (last accessed April 24, 2019).

Anne-Sophie Martin is a Doctor of Law specialized in Public International Law and Space Law. Her doctoral research focused on the legal aspects of dual-use satellites. She received her LL.M in Space Law and Telecommunications Law from the University of Paris-Sud XI (France) and her PhD from Sapienza University of Rome (Italy). On August 2017, she attended the Centre for Studies and Research of The Hague Academy of International Law. Since 2019, she is a visiting researcher within the Centre for a Spacefaring Civilization and she is part of the For All Moonkind's Legal Council. She is a member of the International Institute of Space Law, Space Generation Advisory Council, International Institute of Space Commerce, European Centre of Space Law, American Institute of Aeronautics and Astronautics, French Society of Air and Space Law, and Institute of Space Law and Telecommunications.

How to Set Up a Space Nation on the Example of Ghana

Anna Fogtman, Christine Müller and Moses Oketch

Abstract

Ghana is one of the emerging countries in sub-Saharan Africa. The country's economic power is mainly based on export goods such as gold or cocoa. All in all, Ghana is experiencing economic growth and is focusing now on strengthening the education system to position it to contribute more effectively towards the development of the country—also to become more independent from fluctuating commodity prices in the world market. Education in STEM subjects and the establishment of Technical Universities are among recent initiatives developed to strengthen skills in information and communication technologies, engineering and natural sciences. In recent years, Ghana also has taken first steps into becoming a space nation: In 2012, the Ghana Space Science and Technology Institute was launched as Ghana's Space Agency to promote space science and technology. Advantages in the field of agriculture, weather forecasting and environmental degradation seem to justify larger investments. To establish a research institution for the training of young scientists, the institute was the coordinating organization for the conversion of a telecommunication dish to the Ghana Radio Astronomy Observatory in 2017. The first Ghanaian satellite was sent into space in the same year. This article profiles Ghana's beginning journey into becoming a space nation.

A. Fogtman
European Astronaut Centre, ESA Cologne, Cologne, Germany
e-mail: anna@fogtman.eu

C. Müller (✉)
University of Bonn, Bonn, Germany
e-mail: christine.mueller@uni-bonn.de

M. Oketch
University College London (UCL), London, UK
e-mail: m.oketch@ucl.ac.uk

A. Froehlich (ed.), *Space Fostering African Societies*, Southern Space Studies,
https://doi.org/10.1007/978-3-030-32930-3_6

1 Introduction

This chapter does not seek to provide comprehensive analysis of space science in Ghana because the sources of material upon whose material it draws, does not have enough information and data. Nevertheless, it is possible to draw attention to some of the ongoing efforts and the importance of developing space science and technology in Ghana, in comparison with some of the more common trends of space science development across the globe which have been well analysed in the literature. It is also possible to note some of the challenges Ghana faces in establishing space science and technology, but also the opportunities that exist in pioneering this area of knowledge and innovation. There is no doubt that global demand for skills in space science and utilisation of space science and technology enabled information and data to address human development challenges will intensify in the coming decades. While the most developed regions of the world have experienced greater demand for skills related to space science, not least as a result of the realisation of the importance of such knowledge in boosting market competition in the so called knowledge economy, less developed countries, particularly those in sub-Saharan Africa such as Ghana, cannot afford to be left behind or to marginally engage in space science research and operations.

With the intention of giving these requirements a political framework, the African Union adopted the "African Space Strategy—Towards Social, Political and Economic Integration" during the 2nd ordinary session for specialized technical committee meeting on education, science and technology in Cairo in October 2017.[1] The strategy highlights four key areas: earth observation, navigation and positioning applications, satellite communication applications and space science and astronomy and is aiming to express significant political support for the growth and development of high-technology sectors, including the space sector, and for the establishment of national and regional space programmes.[2]

In the face of these new realities of knowledge necessary for modern human development, benchmarks are necessary to be set for how education systems and training is thought to be capable of supporting a functional space science technology. These considerations allow us to analyse, albeit in very general terms, some of the common trends in space science in Ghana's tertiary education system. An important aspect of this analysis is the provision of Technical and Vocational Education for "high skills".

[1]The African Union: African Space Strategy—Towards Social, Political and Economic Integration, 2017, https://au.int/sites/default/files/newsevents/workingdocuments/33178-wd-african_space_strategy_-_st20445_e_original.pdf (31.03.2019).
[2]Ibid.

2 Trends in Value, and Purpose of Space Science to Human Development

Since humanity started to explore the outer space more than fifty years ago, the quality of life on Earth significantly improved through the direct and indirect benefits of spaceflight. The space sector has played a crucial role in the development of modern societies and their economic advancement, by generation of scientific knowledge, promotion of innovation, creation of markets, inspiration of people around the globe and agreements forged between the countries engaged in space exploration.[3] It is well represented by the growth of space nations' economies.

According to OECD, space exploration initiated the economic development of space that today, year after year, delivers high returns for invested funds in space.[4] Resources spent on space allow creating new resources and this translates in success at survival. Although space research is designed to explore space, many of the technological innovations may be used in everyday life on Earth. Over the past years, the European Space Agency (ESA) has 'spun-off' over 150 separate technological innovations from the space industry into non-space applications. The programme has fostered the creation of around 20 new private-sector companies, creating or saving nearly 2,500 jobs across Europe in these or established companies.[5] The most recent assessment of ESA's participation in the International Space Station (ISS) Programme shows a very good return on investment of public funding, with significant direct and indirect impacts in all ESA's Member States. Investments in this programme contribute, as well as in other sectors of the space industry, to industrial competitiveness and to the development of a knowledge-based society.[6]

The estimated value added (direct + indirect + induced benefits) generated over the entire period of participation in the ISS Programme (1995–2016) is €14.6 billion, while the total ESA funding is some €8 billion. This represents a multiplier of 1.8 which is at the upper end of the range of equivalent multipliers for manufacturing industries in Europe. The incremental government revenues originating from the ISS programme amounted to some €7 billion in total over 1995–2016.[7] Also, ISS in an exceptional example how working together in space can help overcome political and cultural differences. This partnership has consolidated solid relationships between space agencies and this collaboration has generated benefits

[3]International Space Exploration Coordination Group: Benefits Stemming from Space Exploration, September 2013, page 7, https://www.nasa.gov/sites/default/files/files/Benefits-Stemming-from-Space-Exploration-2013-TAGGED.pdf (28.04.2019).

[4]See: OECD (Ed.): Handbook on Measuring the Space Economy, 2012, and OECD (Ed.): The Space Economy at a Glance, OECD, 2014.

[5]European Space Agency: http://www.esa.int/Our_Activities/Space_Science/Spin-off_technologies, 28.04.2019.

[6]Assessment of the socioeconomic impact of the ESA participation to the International Space Station (ISS) Programme, ESA-IPL-PLH-MT-LE2016-182, PwC, 2016.

[7]Ibid.

along the entire value chain, including exchanging know-how and development and adaptation of innovative technologies.

This shows that investing in space increases the opportunities for start-ups and emerging economies to get into the space sector, but it means governments should keep up their spending on space research and development, which can yield big returns in the form of new technologies, and invest in industry niches where they can be competitive in this new space race.

2.1 Ghana—Overview and General Figures

The Republic of Ghana is located along the Gulf of Guinea and the Atlantic Ocean in the subregion of West-Africa and bordered by the Ivory Coast, Burkina Faso and Togo. It is home of approximately 28 million inhabitants, with about 57% of the population under the age of 25.[8] A growth of up to 45 million inhabitants by 2050 is expected by the UN world population prospect.[9] In 2018, Ghana was ranked 140 of 189 countries in the Human Development Index (HDI).[10] The official language is English, while there is a linguistic diversity of 79 languages and idioms. Ghana is considered as an anchor of stability in the region. Since 1992 democratic elections have taken place, change of government is peaceful and proof of democratic strength of the country.

Economically, there has been a strong upswing since the early 2000s, then a period of weak growth until 2016. This was caused by low world market prices for raw materials, but also by the economic policy. Since 2010, Ghana is considered a low-middle income country. Economic strength results mainly from the export of products such as gold, oil and cocoa. This focus makes the country vulnerable to fluctuations in the world market, so economic planning is difficult. There are major developmental differences between the economically strong coastal region and the north of the country. Nevertheless, there is success in achieving the millennium development goals for 2015: Ghana was the first country in sub-Saharan Africa to halve poverty in its population compared to 1990 levels. The government under President Nana Addo Dankwa Akufo-Addo is now aiming to increase value at home, considering a key role for the private sector. In 2017, Ghana had a GNI per capita of 1,951 (current US$), the region sub-Saharan Africa had 1,522.[11]

[8]The World Factbook, https://www.cia.gov/library/publications/the-world-factbook/geos/gh.html (12.04.2019).

[9]United Nations Desa/Population Division: World Urbanization Prospects 2018, https://population.un.org/wup/Country-Profiles/ (30.04.2019).

[10]United Nations Development Programme: Human Development Reports, http://hdr.undp.org/en/2018-update (14.04.2019).

[11]The World Bank: Database, https://databank.worldbank.org/data/reports.aspx?source=1296&series=NY.GNP.MKTP.PC.CD# (26.04.2019).

2.2 Educational System

Education is one of the priorities of Ghanaian politics. Public education expenditure accounts for 4.4% of GDP or 20% of total government spending, but only a share of 0.38% of GDP for research expenditure. In 2008, Ghana was ranked 110 on the Knowledge Economic Index (KEI) by the World Bank Institute, that took the four pillars of the knowledge economy into account: "1. Economic and institutional regime: The country's economic and institutional regime must provide incentives for the efficient use of existing and new knowledge and flourishing of entrepreneurship, 2. Education and skills: The country's people need education and skills that enable them to create and share, and to use it well. 3. Information and communication infrastructure: A dynamic information infrastructure is needed to facilitate the effective communication, dissemination, and processing of information. 4. Innovation system: The country's innovation system—firms, research centers, universities, think tanks, consultants, and other organizations—must be capable of tapping the growing stock of global knowledge, assimilating and adapting it to local needs, and creating new technology."[12]

The literacy rate for adults over the age of 15 is 90.6%. The government of Ghana wants to increase the proportion of girls and women with higher education and provide rural communities with better access to education. Compulsory education is eleven years. It is divided into a two-year preschool (kindergarten), a six-year primary school and a three-year junior high school, which concludes with the Basic Education Certificate (BECE).[13]

Primary education may be followed by a three-year secondary school education (SHS) and lead to the West African Secondary School Certificate Examination (WASSCE). There is also the option of attending a vocational or technical institute with a four-year training period after BECE. Vocational schools do not yet exist. At the beginning of 2017/2018 school year, the government introduced free secondary school, one of its key election promises. Financing the program—which includes the abolition of all school fees and free board and lodging at boarding schools—is one of the major challenges facing the government. In the first phase, around 90,000 students will benefit from the program every year.

2.3 Higher Education

The university landscape of Ghana is in a phase of massive expansion. In 2018, the tertiary education sector has a total of 205 universities, colleges, polytechnics and other public and private institutes recognized by the National Accreditation

[12]"Measuring Knowledge in the World's Economies. Knowledge Assessment Methodology and Knowledge Economy Index", http://web.worldbank.org/archive/website01030/WEB/IMAGES/KAM_V4.PDF (26.04.2019).

[13]Auswärtiges Amt: Ghana–Kultur und Bildung, https://www.auswaertiges-amt.de/de/aussenpolitik/laender/ghana-node/kultut-bildung/203400, (16.03.2019).

Board.[14] The universities are higher education institutions that offer a wide range of subjects and research opportunities in addition to training. The two- to three-year polytechnics offer post-secondary education in the technical field. At colleges of education students can acquire national diplomas in various subjects: languages, agriculture, nursing professions, etc. These diplomas can either be used to enter the profession or can be used for further studies at a university.[15]

A similar upward trend applies to Ghana's student numbers, rising from just over 20,000 to nearly 420,000 between 1994 and 2016.[16] These developments are accompanied by an increasingly strong political focus on education. The Ghanaian government's higher education priorities—as stated in the "Education Strategic Plan 2010–2020", the strategy document by the Ministry of Education (MoE) and the National Council for Tertiary Education (NCTE)—include improved access to tertiary education, securing higher education funding, combating the gender gap, targeted promotion of engineering and economics study programs.[17]

This rapid growth in student numbers reinforced the weaknesses of the previous training system. "These are poor and inadequate machines and equipment to train students, lack of qualified lecturers, ill-defined academic progression and career advancement pathways, as well as unfair salary and remuneration packages for lecturers. Others are poor infrastructural development and facilities, poor funding, discrimination on the products of the system, and poor curriculum development."[18]

The Ghanaian government followed the suggestions of Amedome, Agbezudor and K Sakyiama: Ghanaian polytechnics are currently being converted into Technical Universities. The government hopes it will boost science and engineering.[19] "African universities are subject to the new information-driven global economy, the impact of new information and communication technologies and the growing expectation for higher education institutions to survive and thrive as market-like organisations while facing resource constraints," said Ghana's Vice-president Kwesi Amissah-Arthur while opening the second Times Higher Education Africa Universities Summit in 2016.[20]

In order to improve the also practical and vocational skills of the graduates, the Ministry of Education started not only to raise the Ghanaian polytechnics to university level, but also tries to switch to the model of a German University of Applied Sciences ("Fachhochschule"): The "German-African University Platform

[14]National Accreditation Board, http://www.nab.gov.gh/public-universities (26.04.2019).

[15]DAAD-Ländersachstand: Kurze Einführung in das Hochschulsystem und die DAAD-Aktivitäten 2018, https://www.daad.de/medien/der-daad/analysen-studien/laendersachstand/ghana_daad_sachstand.pdf (26.04.2019).

[16]Ibid.

[17]Ibid.

[18]Amedorme, Sherry, Agbezudor, Kowu, K Sakyiama, Frederick: Converting Polytechnics into Technical Universities in Ghana Issues to Address, page 524; In: International Journal of Education and Research, Vol. 2, No. 5, May 2014, 523–528.

[19]Kokutse, Felix: Ghana's vice-president calls for move from liberal arts, in: University World News, 28.04.2016, https://www.universityworldnews.com/post.php?story=20160428194405486 (26.04.2019).

[20]Ibid.

for the development of entrepreneurs and Small/Medium enterprises" funded by the German Ministry of Economic Cooperation and Development (BMZ) and the German Academic Exchange Service (DAAD) is a powerful example of North-South and South-South cooperation. It is carried out by the Bonn-Rhein-Sieg University of Applied Sciences in Germany (BRSU), the University of Cape Coast in Ghana (UCC) and the University of Nairobi in Kenya (UoN).[21] The goal is more practical orientation, improvement of university management and the promotion of didactic innovations.

3 STEM Education and Space Science and Technology in Ghana

STEM stands for science, technology, engineering and mathematics, but a wider range of academic disciplines fall under this description like aerospace engineering, astronomy, biochemistry, biology, chemical engineering, chemistry, civil engineering, computer science, electrical engineering, mechanical engineering, physics, psychology and many others. These fields have resulted in significant improvements in different areas like communication, transportation and health. Therefore, graduates of STEM training and degree programs are sought-after technology experts and find attractive career paths—not only in companies in metal or electronics industry, in chemical or in information technology industry, but also increasingly in service industry like finance and insurance companies. In order to significantly increase the number of qualified applicants for STEM training occupations and study courses and thus to ensure prosperity in the future, all sources of talent must be exhausted and educational barriers systematically dismantled. All over the world, there is a need to improve STEM teaching and training quantitatively and qualitatively at school and university.

Although the benefits of education in STEM subjects are obvious and the Ghanaian government took appropriate action and promised support, cultural change from Kindergarten to primary and secondary school to university level, needs a lot of time and persuasiveness. For 2016, "a report by the National Council for Tertiary Education showed that only 39% of the 127,502 students who were admitted into the eight public universities in the 2012/2013 academic year were admitted into Science-related programs. This percentage falls incredibly short of the 60% mandatory Science-related admissions policy of the Ministry of Education."[22]

To change this, there are several ongoing projects to raise awareness in STEM education at school, especially for girls. An example of a project to empower girls in STEM is the "UNESCO-HNA Partnership for Girl's and Women's Education":

[21]German-African University Platform for the Development of Entrepreneurs and Small/Medium Enterprises Hochschule Bonn-Rhein-Sieg: https://www.german-african-entrepreneurship.org.

[22]Hassan, Ferdinand: Rethinking STEM education in Ghana, 24.08.2016, https://ghscientific.com/gh/rethinking-teaching-stem-ghanaian-schools/.

The project started in 2015 and seeks to strengthen global and regional advocacy, networking and communication to ensure girls' right to education.[23] It shall develop national capacities for gender-responsive education in Ghana and Ethiopia, the other African partner countries in the network. It is coordinated by UNESCO Headquarters with technical support from UNESCO International Institute for Capacity Building in Africa.[24] In Ghana, the project supports the Girls' Education Unit (GEU) of the Ghana Education Service and aims to increase girls' participation in science-related subjects. This includes capacity-building training for GEU staff members and the organization of STEM-Clinics, one-day events that have inspired over 1,550 girls to follow STEM courses and careers through hands-on activities and interactions with female role models.[25] These measures receive a lot of media attention, thus supporting the government's goal of getting more school children and students excited about STEM.[26]

Despite all efforts, change is slow: lack of equipment and a growing shortage of STEM teachers, especially due to a general embargo of the government on public sector employment[27] are still problems in STEM education in school. It is obvious that capacity building in STEM training takes its time, but also higher education capacity is still insufficient. There are only 10 public universities in the country, 3 polytechnics and 8 technical universities. The University of Ghana is the oldest and largest university in Ghana. It is located northeast of Accra in the suburb Legon. Founded in 1948, the university was originally closely associated with the University of London. As University of Ghana it became independent in 1961. Initially, the research and teaching focus was on the humanities and social sciences, agriculture and medicine, but was extended as part of a national education reform on technology-oriented sciences in the recent years. Kwame Nkrumah University of Science and Technology (KNUST) is the country's second university after the University of Ghana. It was founded as Kumasi College of Technology in 1952 and named after the first president of Ghana, Kwame Nkrumah. In 1955, the Faculty of Engineering signed an agreement with the University of London, which enabled students to pass the Bachelor's degree in engineering with an external exam by the University of London. The college was granted full university status in 1961. Nowadays both, the University of Ghana and Kwameh Nkrumuah University of Science and Technology, started courses in astronomy and space science.

[23]UNESCO-HNA Partnership for Girl's and Women's Education, 2015–2020, https://unesdoc.unesco.org/ark:/48223/pf0000247748 (28.04.2019).

[24]UNESCO: Building teachers' capacity to promote gender equality in education, here: https://en.unesco.org/themes/education-and-gender-equality/hna-partnership (30.04.2019).

[25]Ibid.

[26]Ghanaian girls empowered to study STEM subjects, UNESCO 07.02.2019, YouTube: https://www.youtube.com/watch?v=Xuj0AK3QiVo. See also: Space in Africa: Space Science System Research Institute to inspire 1000 girls in Ghana with Space education, 15.02.2019, in https://africanews.space/space-science-system-research-institute-to-inspire-1000-girls-in-ghana-with-space-education/.

[27]Hassan, Ferdinand: Rethinking STEM education in Ghana, 24.08.2016.

Another institution became increasingly known for its efforts in space science and technology: the All Nations University College (ANUC), which was founded in Kofuridua in the Eastern Region in 2002. It became an accredited university in Ghana in October 2005. It is affiliated to Kwame Nkrumah University of Science and Technology and collaborates with SRM Institute of Science and Technology in India. Here, Ghana launched its first satellite into space. It was developed by students from ANUC in a two-year project, costing 500,000 US$, and was supported by the Japanese Aerospace Exploration Agency (JAXA) and a NASA engineer. GhanaSat-1 was send into orbit from the International Space Station (ISS) in July 2017. Richard Damoah, director of the first university Space Systems Technology Laboratory, said it marked a new beginning for the country: "It has opened the door for us to do a lot of activities from space," he told the BBC. He said it would "also help us train the upcoming generation on how to apply satellites in different activities around our region." For instance, [monitoring] illegal mining is one of the things we are looking to accomplish."[28]

4 The Creation of Ghana Space Science and Technology Institute (GGSTI) and the Ghana Radio Astronomy Observatory

In the preceding section we noted some of the progress that Ghana has made in political stability, economics and its ambitious education ideas which are introducing German University of Applied Sciences-model. It is also apparent that Ghana has been preparing for space science and technology.

The willingness to set up space research programs is growing in Ghana as in other African countries, even though there is still scepticism and criticism about spending in a country where poverty still affects about 20% of the population.[29] For this reason it is one of the main challenges to gain acceptance from the Ghanaian people for government spending on space research and technology.

A big step has been reached when Ghana's Space Science and Technology Institute (GGSTI) was opened officially on May 2, 2012 as Ghana's first space science, space exploration, astronomy and technology space agency. It is one of the institutes of the Ghana Atomic Energy Commission (GAEC) under the Ministry of Environment, Science, Technology and Innovation (MESTI). It has four centers: 1. The Satellite Communications and Development Centre, 2. the Instrumentation and Engineering Services Centre, 3. the Remote Sensing, GIS and Climate Centre, that coordinates all activities on Climate, Remote Sensing (RS) and Geographical Information Systems (GIS) in the country, and 4. the Radio Astronomy and Astrophysics Centre. The objectives are to develop and manage research and training infrastructure for Space Science and Technology as well as high level

[28] BBC News: Ghana launches ist first satellite into space, 07.07.2017, https://www.bbc.com/news/world-africa-40538471.

[29] The Worldbank, Database: https://data.worldbank.org/country/ghana.

human capacity in Space Science and Technology, to apply Space Science and Technology for socio-economic development, to create public awareness in Space Science and Technology through outreach programmes.[30]

In 2015, the government allocated GHC$ 38,5 million, about 10 million US$, to nuclear and space science technology to further space education and benefit from their own satellite data.[31] One of the first projects of GGSTI was the conversion of a telecommunication dish into a Radio Astronomy Observatory: "The Kutunse dish outside Accra is the first in a number of planned conversions, heralding a new era in African astronomy and skills development."[32] Three more are planned in Zambia, Madagascar and Kenya. The long-term plan is to hook these up with new telescopes in Namibia and Botswana, as well as existing facilities in South Africa, to form the African Very Long Baseline Interferometry (VLBI) Network. "Ghana's location just north of the equator means it can view the entire plane of the Milky Way galaxy, thus filling a gap in global VLBI observations. Kuntunse could also become one of the receiving stations of the Square Kilometre Array (SKA), and will also play an important role in training African astronomers to be ready to spring to work when the entirety of the SKA network comes online in a few years' time."[33]

In August 2017, the Ghana Radio Astronomy Observatory was officially launched by Ghana's president Akufo-Addo. "Recognising the role of science and technology in the socio-economic development of the country, President Akufo-Addo said he had charged the Ministry of Education, and Ministry of Environment, Science, Technology and Innovation to step up efforts in developing a potent science, technology, engineering and mathematics (STEM) education model for Ghana. That, he said, would stimulate the interest of pupils and students in engineering sciences and technology."[34]

Although Ghana's Radio Astronomy Observatory will not be part of the SKA, it will transmit radio signals, increase signal connection of waves and help to study celestial objects. But the real goal seems to be the creation of a research site, that will lead to ‚brain gain', to bring talent back home. Since 2005, South Africa has trained about 800 students in radio astronomy, several of them coming from Ghana. Until now, they had no telescope to go home to. Now the situation has changed, and the government tries to push the universities to offer programs in radio astronomy: "If you train one astronomer, you train a mathematician, an astronomer, a computer scientist and an engineer [in one]."[35]

[30]Ghana Space Science and Technology Institute: Objectives, https://gssti.gaecgh.org/objectives/ (26.04.2019).
[31]Matthews, Chris: Why Ghana started a space program, 05.02.2016, https://motherboard.vice.com/en_us/article/nz7bnq/why-ghana-started-a-space-program.
[32]Wild, Sarah: Old dish gives Ghana new taste of the African sky, in: Business Day, 07.09.2017.
[33]Nordling, Linda: The accidental astronomer, Institute of Physics, http://www.iop.org/careers/working-life/articles/page_70020.html (26.04.2019).
[34]GNA: President launches Ghana's Astronomy Observatory, http://www.ghana.gov.gh/index.php/news/3933-president-launches-ghana-s-astronomy-observatory (27.04.2019).
[35]See above: Wild, Sarah: Old dish gives Ghana new taste of the African sky, in: Business Day, 07.09.2017.

In April 2018 Ghana's Minister of Environment, Science, Technology and Innovation issued a press release, that Ghana is ready to host the African Spence Agency, an initiative of the African Union that will allow the continent to launch and explore the space for improved technological advancement. In the same press release, the minister announced that Ghana is set to launch a supercomputer facility at the Institute of Scientific and Technological Information (INSTI) of the Council for Scientific and Industrial Research (CSIR).[36] However, Ghana's space ambitions have experienced a setback at the turn of the year 2018/2019, after Ghana's application to host the Headquarters of the African Space Agency was rejected with the notice from the missed application deadline in autumn 2018.[37] In February 2019, the Executive Council of the African Union decided, that Egypt will host the African Space Agency Headquarters.

But there is still hope for Ghana, when the African Union will follow the example of Europe. The European Space Agency (ESA) has its headquarters in Paris, with the Director General and cabinet. But the largest ESA establishment, the European Space Research and Technology Centre (ESTEC) is located in the Netherlands. Two centres are located in Germany: the European Space Operations Centre (ESOC) and the European Astronauts Centre (EAC). Others are in Italy, Spain, in French Guiana, Belgium and the UK. If the centres of the African Space Agency will also be spread over the continent, Ghana is having a good chance to host one of them.

5 Conclusion

All regions of the world are today aware of and reaping some of the benefits associated with space science, including opportunities for trade and provision of more advanced goods and services supported by advances in technology supported by space science. Therefore, creating space science cannot be policy ambition of only the most developed countries. Nowadays, in many parts of the world's societies, there is evidence of space science utilisation everywhere, particularly the mobile telephone supported technology and satellite mapping technology for climate and weather, to mention only two. It is increasingly evident to countries such as Ghana, for instance, that to be space science illiterate is to be marginalised from the world of global development.

In this article, we observe the first steps of a country to become a space nation. Like most other countries in sub-Sahara Africa, Ghana faces challenges like food security, urbanisation, sustainable use of the environment and the need to educate a growing population. Therefore, Ghana has taken steps to tackle the problems:

[36]Space in Africa: Ghana shows readiness to host African Space Agency, 15.04.2018, https://africanews.space/ghana-shows-readiness-to-host-african-space-agency/.

[37]Adolga-Bessa, Delali: Ghana missed deadline for bid to host African Space Agency, 07.02.2019, https://citinewsroom.com/2019/02/07/ghana-missed-deadline-for-bid-to-host-african-space-agency/.

fostering STEM education in schools, the conversion of polytechnics into technical universities according to the German Universities of Applied Sciences-model and the launch of Ghana's Space Science and Technology Institute and its Ghana Radio Astronomy Observatory as a research site to educate the next generation of space scientists. Sustainable success can be measured by projects to inspire the Ghanaian people about STEM and space science and by young researchers who take part in international satellite programs. It seems Ghana is championing a positive role for space science.

Anna Fogtman has obtained her Ph.D. degree in Biology and currently is working at the European Astronaut Centre (EAC), European Space Agency (ESA). Her work is focused on health risk assessment of astronauts during exploration-class missions beyond low Earth Orbit. She is studying the biological effects of space radiation environment on the human body, especially in the context of individual sensitivity to ionising radiation. Before she joined EAC, she was working at the Faculty of Biology, University of Warsaw and the Institute of Biochemistry and Biophysics, Polish Academy of Sciences. Her research was focused on the influence of changing environmental conditions on cellular genetic networks. Earlier she worked at the Interdisciplinary Centre for Mathematical and Computational Modelling, where she participated in the project: "Complex Processes: Modelling, Simulation and Optimisation". She graduated both: University of Warsaw and University of Wroclaw.

Christine Müller studied East European History, Political Science and German Studies in Tübingen, Heidelberg, Warsaw and Krakow. She is working in science management since 2006. From 2008 until 2011 she was regional expert for Central and Eastern Europe, Asia and Africa at the University of Heidelberg. From 2011 until 2014 she was lecturer with specialized functions at the German Academic Exchange Service (DAAD) branch office in Warsaw, where she was responsible for scientific networking events within the framework of Poland's accession to the European Space Agency. From 2016 until 2018, she was in charge for the University of Siegen's partnership with Dedan Kimathi University of Science and Technology (DeKUT) in Nyeri, Keniya and she was one of the speakers at the symposium about the "Current Situation and Development of Further Education and Training in Sub-Saharan Africa" at the Namibia's University of Science and Technology (NUST) in collaboration with University of Rostock, held in Windhoek in August 2016. Since April 2018 Christine Müller is Head of the Department "European and International Networks" at the University of Bonn. She is working together with the Vice-Rector's office and participating to the position statement on cooperation with countries in the "Global South".

Moses Oketch is Professor of International Education Policy and Development at the Institute of Education at University College London (UCL). His research focuses on the connection between the theory of human capital and implementation of policies. He has argued that however compelling the theory of human capital is, it remains tacit if actual implementation of policies don't work. His areas of research interest include economics of education, education policy analysis, impact evaluation, mainly in sub-Saharan Africa. Previously he worked at Vanderbilt University, University of Illinois at Urbana-Champaign, and in 2012 he was a Visiting Professor at University of Pennsylvania. He has also contributed to and supported research capacity strengthening in Africa through his involvement with African Population and Health Research Center (APHRC) as a Senior Research Scientist and Director of Research (while on leave from university). During this period, he mentored several researchers on research proposal writing, implementation and management of large-scale projects, and co-authored research papers with them. He received his Ph.D. from the University of Illinois at Urbana Champaign focusing on economics of education.

Education System and Space Activities for Malawi

Patricia Helen Khwambala

Abstract
Space activities are all efforts that involve space with the aim of benefiting the nation, these activities can be military or civil. This is a broad term which encompasses space techniques, space technology application, satellite development, telecommunication system operations, just to mention a few. All these activities contribute to the nation's economic, social and cultural development. There is a number of factors that influence the advancement of space activities in a nation. Emergency that needs the use or application of space technology like a natural disaster. Financial capability of a nation to fund a space activity venture, like satellite development or space mission, leadership zeal and interest in space activities and scientific knowledge and understanding of the population are some of the factors. In all these factors education is at the core of them all. There are not much space activities taking place in Malawi. Like other developing nations, Malawi is realizing the need to develop its own space sector. In the early 2000 Malawi became a registered user of international charter for Space and major disasters, which allows the nation to access satellite information from the charter members in times of a major disaster event. In March 2018 discussions began between the Kyushu Institute of Technology (KYUTECH), Japan and Malawi's National Commission for Science and Technology—a government arm tasked with developing and supporting scientific innovations that address the nation's needs—to build Malawi's first satellite. While Malawi will not join KYUTECH's upcoming BIRDS project, this very discussion was a significant step in getting the government involved and aware of the benefits of space. BIRDS is a KYUTECH-based multinational project in which students from

P. H. Khwambala (✉)
Cape Peninsula University of Technology, Cape Town, South Africa
e-mail: khwambalap@cput.ac.za

A. Froehlich (ed.), *Space Fostering African Societies*, Southern Space Studies,
https://doi.org/10.1007/978-3-030-32930-3_7

various non space faring nations build and operate a nanosatellite in a period of two years. It thus trains these students, who can then return to their own nations to help in building the space industry there. It has been a very successful project with various developing countries like, Ghana, Nigeria, Bangladesh, Bhutan, Mongolia among the countries that have participated in and benefited from the BIRDS. This article is focusing on the correlation of basic education system, tertiary education and literacy on Space activities in Malawi. This information in this chapter will help unearth the hidden strengths and weaknesses of the education system in Malawi which is making it to derail in the space activities. It will help shed some light for the authorities to see the need to boost the education standards if they are serious in getting involved in space activities that play a very important role in the nation's economic and cultural development.

1 Introduction

Space studies is a scientific research, carried out in outer space by studying the outer space. This is a wide field that uses technology to observe the universe and what goes on in it. The different areas of study that are involved in space studies are Engineering, Medicine, Physics, Law, Economics, Material Science etc. All these fields can form part of the space research environment.

Basic education has a direct impact on how learners can conceive the idea of space studies and activities at an early age, this influences their perception towards the same when they are matured and responsible in different government positions. Apart from including the obvious, cause of the seasonal changes, studying the solar system and characteristics of the moon and the sun. The science general content and the way it is taught has significance in the way learners perceive the impact of space studies and activities in their lives.[1]

Strong basic science education background on space activities pave way to university qualification one would pursue to enhance the understanding and attain a skill to easily indulge in space studies, hence get involved in national space activities.

There are a number of benefits a nation can get from an advanced space activity on a national level. Space activities enhances communication. Sophisticated communications satellites are used in remote areas with the use of satellite phones and provide internet to areas that are unable to get internet in any other way. This is a huge advantage to developing countries to which a bigger percentage of its population is in remote areas.[2]

[1] R. Trumper, 2002, 'Teaching Future Teachers Basic Astronomy Concepts—Seasonal Changes at a Time of Reform in Science Education, Journal of Research in Science Teaching, 43(9): 879–906.
[2] Whitman, Ryan 2016, 'Connecting Remote areas to the Internet by Satellite', PC Magazine.

Space activities play a big role in improving global partnerships. In these relationships, developing countries successfully use the space technologies from developed countries to solve their national challenges, like managing disasters, climate monitoring, green systems and crop production monitoring. This helps to tackle workforce development issues and launch new space activities.[3]

National space technology and activities brings different global cultures together, bringing with it new lessons and understanding which brings advancement to the human global culture.[4]

1.1 Education Layout in Malawi

Malawi as a former British protectorate, has an education system background that was influenced by the British. The education official language is English, which starts way early in the elementary learning to tertiary education. English being the first official language for all official communication.

According to UNESCO Institute for Statistics March 2016, 65.75% of the adult population, aged 15 years and above, in Malawi, are able to read and write. For adult men, the literacy rate is 73% and for women it is 59%, The youth literacy rate, literacy amongst people age 15–24 in Malawi is 75%.[5]

1.1.1 Primary Education

70% of eligible children do not have access to any form of early childhood education, the average primary student to classroom ratio increased from 105:1 in 2011/2012 to 124:1 in 2012/13, the pupil qualified teacher ratio worsened from 92:1 in 2011/12 to 95:1 in 2012/13. This makes it not an easy task for a teacher to pass on knowledge to a learner as the student to teacher ration are very high. The pass rates for the Primary School Leaving Certificate of Education (PSLCE) have been declining each year between 2006/07 and 2011/12 from 74.4% in 2006 to 68.9% in 2011.[6]

Primary school education in Malawi is made up of eight years, referred to as Standard 1 to Standard 8. Although the official primary school age group in Malawi is categorised as 6–13 years, it is very common for students of varying ages to attend primary school, as many students have to repeat some primary school years. Primary school students in Malawi learn a variety of subjects, and take examinations in English, Chichewa, Maths, Science, and Social Studies.

Although more students now have access to education in Malawi, with free primary education, most primary schools in Malawi are under-resourced,

[3]Macleish et al., 2012, 'Global Partnerships: Expanding the Frontiers of Space Exploration Education', Acta Astronautica 80: 190–196.
[4]Harris, Phillip, R, 1986, 'The Influence of Culture on Space Developments', Behavioral Science, 31(1): 12–28.
[5]UNESCO Institute for Statistics March 2016.
[6]Global Partnership for education.

under-staffed, and under-funded, creating extremely challenging teaching and learning conditions for teachers and students alike.

1.1.2 Secondary Education

The percentage of the population aged 25+ years that have some secondary education is 15.93%. Secondary schools in Malawi are run in four years, referred to as Form 1 to Form 4. Students study English, Maths, Agriculture, Physics, Biology, Geography, History, Bible Knowledge, Social Studies, and Chichewa.

Students can choose to be tested on any combination of these subjects, and can drop their lower scores and keep their best six. However, they are required to pass English and Math in order to graduate. The MSCE is often considered an adequate credential for most jobs, as very few students in Malawi will proceed from secondary school on to university. The official secondary school age group is defined as 14–17.[7]

1.1.3 Higher Education

All higher education in Malawi is ultimately controlled by the University of Malawi, which was founded in 1964. Access to higher education is based on passing the Malawi Secondary Certificate of Education (MSCE). A student must earn at least five credits, including English. This exam may be taken after completing eight years of primary and four years of secondary education. Students wishing to be accepted by the university must achieve excellent scores in these exams, as selection to go to university is on merit and its limited number of students that are selected.

The first, or Bachelor's, degree is normally earned after four years of concentrated study in residence. It takes four years to complete courses in law, education, agriculture, and commerce, and five years to finish the full engineering programme. Honours degrees are awarded in some subjects. A professional qualification is awarded as a diploma after three years of study.

A second stage, or Master's degree, requires two years of full time study to complete. A third stage, or Doctorate degree, is awarded after finishing three to five years of study beyond the Master's degree, a successful defence of a thesis or dissertation, and at least six months in residence at the university.

As of July 2018, less than 1% of Malawi's population was enrolled in universities. Approximately 72% of all college students were pursuing degrees in education, 10.9% were taking degrees in the social sciences, 12.2% were pursuing science degrees, 3.9% were taking degrees in medicine, and 0.4% were pursuing degrees in the humanities. Given Malawi's growing need for high-powered labour, Malawi will be dependent on expatriate skilled labour far into the foreseeable future, unless the university system expands.[8]

---

[7]Ripple Africa, General Information About Education in Malawi, https://www.rippleafrica.org/education-in-malawi-africa/general-education-in-malawi-africa accessed 31/05/2019.

[8]Ripple Africa, General Information About Education in Malawi, https://www.rippleafrica.org/education-in-malawi-africa/general-education-in-malawi-africa.

In the curriculum, Agriculture is a compulsory subject for all students. Wood working, metal work, and technical drawing are encouraged for boys, and home economics is encouraged for girls. One of the biggest criticisms of secondary schools in Malawi is that they are too university-oriented and needs more technical skills taught. Most students immediately enter the workforce and need a different orientation. Therefore, Secondary schools do not produce as many graduates as the labour market demands. In fact, only one-fourth of Malawi's youth end up attending secondary school.[9]

2 Literature Review

The education background in Malawi as established with all its challenges, there are a number of variables which influences the learner to be successful, in this case, and become a participant in space activities in Malawi as a developing country.

2.1 Motivation

Motivation can be defined as a desire to achieve and represents an inner drive, such as wishes or goals to act or behave in a certain manner. There are the psychological mechanisms that enable some learners to thrive under challenge, while others of equal ability do not.[10] Blackwell, Trzesniewski, and Dweck, developed a motivational model which suggests that core beliefs can set up different patterns of response to challenges and setbacks.[11] In their research found that, having a belief that intelligence is malleable predicted an upward trajectory in grades over the two years of junior high school, while a belief that intelligence is fixed predicted a flat trajectory. This means it is of paramount importance to always assure learners of how incremental their intelligence is. This can be achieved by exposing them to messages and information that constantly reminds them that they are capable of achieving their goals regardless of the challenges.

2.2 Self-potency

Self-potency has an impact when it comes to understanding an individual's response in a learning situation. This can be defined as belief in one's own ability to

[9]Furlong, Andy (2013). *Youth Studies*. Abingdon, Oxon: Routledge. p. 233. ISBN 978-0-415-56479-3.

[10]Blackwell, L.S., Trzesniewski, K.H., & Dweck, C.S. (2007). Implicit theories of intelligence predict achievement across an adolescent transition: A longitudinal study and an intervention. Child Development, 78(1), 246–263.

[11]Dweck, C. S. (1999). Self-theories: Their role in motivation, personality, and development. Philadelphia: Psychology Press.

complete a specific task and reach one's goals.[12] Self-potency or self-efficacy shows how confident one is when performing specific tasks. It is predicted that individuals are more likely to engage in activities for which they have high self-efficacy and less likely to engage in those for which they do not.[13] Fencl and Scheel[14] in their studies indicated that self-potency is one of the most useful predictors of success and persistence. In fact, Fencl and Scheel reported that self-efficacy was influenced by a number of personal experience factors, such as teacher and parent encouragement, successful grades, etc. This helps the learners to clear their doubts on their education challenges and clear all the misinformation they may have associated with STEM careers as well as provide opportunities for STEM students to understand the day-to-day activities that are part of STEM-related careers.[15] Understanding more about STEM-related careers has been shown to shape the development of not only self-efficacy but also interest in STEM and long-term life goals.

2.3 Conceptual Framework

Space studies and space activities participation by students in STEM subjects can be extrapolated by the help of the deficit model, which discusses formal and informal barriers affecting career choice, and the difference model, which states that there are fundamental differences in peoples goals and perspectives. Barriers can be classified as personal and environmental factors. The personal factors attempt to explain the differences in beliefs about ability and self-potency of learners, while environmental factors refer to the interactions in the classroom. The deficit model prescribes intervention programmes which aim at the removal of the barriers, whilst the difference model requires deconstruction of STEM experiences. The introduction of the bridging courses reported in a research where both models were used (Fig. 1).[16]

It was found that introduction of the bridging courses that have lasting impression on the students. The course emphasised on the use of group work, life skills, which stress goal setting and developing self-esteem, study skills, career

[12]Ormrod, J. E. (2006). Educational psychology: Developing learners (5th ed.). Upper Saddle River, NJ: Pearson/Merrill Prentice Hall.

[13]van der Bijl, J.J., & Shortridge-Baggett, L.M. (2001). The theory and measurement of the self-efficacy construct. Scholarly Inquiry for Nursing Practice, 15(3), 189–207.2001.

[14]Fencl, H., & Scheel, K. (2006). Making sense of retention: An examination of undergraduate women's participation in physics courses. In J. Bystydzienski & S. Bird (Eds.), Removing barriers: Women in academic science, technology, engineering, and mathematics (pp. 287–302). Bloomington, IN: Indiana University Press.

[15]Levine, M., Serio, N., Radaram, B., Chaudhuri, S., & Talbert, W. (2015). Addressing the STEM gender gap by designing and implementing an educational outreach chemistry camp for middle school girls. Journal of Chemical Education, 92(10), 1639–1644.

[16]N. Mbano, K. Nolan, 2017, 'Increasing Access of Female Students in Science, Technology, Engineering and Mathematics (STEM) in the University of Malawi (UNIMA)', Science Education International, Vol 28(1), 53–77.

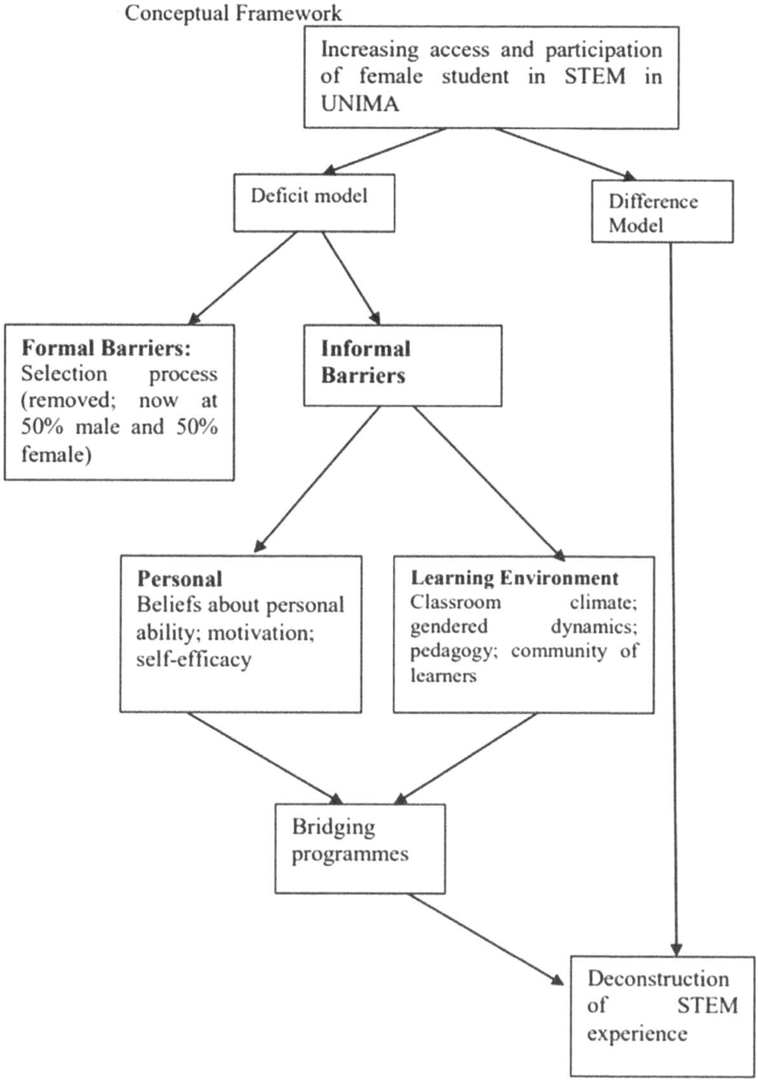

Fig. 1 Conceptual framework (N. Mbano, K. Nolan, 2017, 'Increasing Access of Female Students in Science, Technology, Engineering and Mathematics (STEM) in the University of Malawi (UNIMA)', Science Education International, Vol 28(1), 53–77)

guidance and the use of role models. when such approach is recommended and implemented in schools and colleges could assist students in developing confidence, reasoning and study skills through regular bridging courses in order to increase their access, success and retention in STEM.[17]

3 Conclusion

There is a great correlation between national education standards and national space activities. Not only should the knowledge and understanding of science be correctly imparted in people through education, but also education should be motivating and build confidence as they are rightly guided in their career paths. The basic education and the STEM curriculum should be extensive enough to include and take care of all other variables that builds a confident, knowledgeable learner who can fearlessly work towards attaining his or her goal. If such a learner follows a space studies career path, space activities will surely be in every government department. A developing nation like Malawi, needs a well-defined STEM careers for an increase in Space activities. This will eventually bring about the much needed economic development.

Patricia Helen Khwambala is a Malawian national pursuing a PhD in Space studies at the University of Cape Town and a lecturer at Cape Peninsula University of Technology. She has worked with various engineering companies in both Malawi and South Africa for over 15 years. Her research area of interest is in Space applications, particularly in how to use satellite aided tools to solve everyday challenges in developing countries. Patricia is currently working on what factors affect the non-adoption of satellite aided tools for disaster management in developing countries.

[17]N. Mbano, K. Nolan, 2017, 'Increasing Access of Female Students in Science, Technology, Engineering and Mathematics (STEM) in the University of Malawi (UNIMA)', Science Education International, Vol 28(1), 53–77.

Building Indigenous Space Capabilities as a Launchpad for Technological Advancement in Africa

Samuel Anih

Abstract

Most African countries are beginning to embrace the role space science and technology play in our everyday lives considering the potential it has in proffering solutions to some of the challenges being faced on the continent and the ability it has to foster development, growth and innovation in so many ways. However, space access and assets for African countries with satellites in space were attained mostly with the help of other countries outside the African continent while incorporating little indigenous input in the process. This article examines the role indigenous space capabilities can play in promoting development in Africa at a period the global community is heading towards Industrial Revolution 4.0 and Space 4.0. It also explores the experience of advanced space faring nations and the contributions space science and technology programmes have made on their development. Lessons learned could be adapted for development of local and regional space capabilities within the African continent to aid advancement of space technology and contribute to economic growth of the continent in general. Also discussed are the role of capacity building and partnerships through relevant institutions that could facilitate this process.

S. Anih (✉)
Spacelab, University of Cape Town, Rondebosch, South Africa
e-mail: anihsammy@gmail.com

S. Anih
ARCSSTE-E, OAU Campus, Ile-Ife, Nigeria

A. Froehlich (ed.), *Space Fostering African Societies*, Southern Space Studies,
https://doi.org/10.1007/978-3-030-32930-3_8

1 Introduction

The cutting-edge nature of space science and technology has made it to be at the fore-front of innovation and thereby contributing immensely to the advancement of humanity. It has played tremendous role in development of so many sectors not even related to space science technology at national and international levels through various spin-offs from space programmes and related projects.

Though African countries have benefited from these technologies, most countries on the continent have little or no contribution to the global quest for cutting edge innovations—most especially in space science and technology. One of the factors responsible for the back-seat role the continent has played over the years in this area is lack of adequate expertise in space science and technology, a result associated with lack of infrastructure and necessary proficiency in design, management and maintenance. Dr. Adigun Ade Abiodun listed several challenges faced by many developing countries in various aspects of space technology related areas[1]:

- limited funding available from governments
- shortage of qualified human resources at the local level
- shortage of education/training facilities
- lack of information about latest technological developments in space-related applications, products, services and equipment
- limited access to basic facilities (such as those needed for communication, information exchange, computing)
- lack of adequate information for formulating scientific and technological policy and for effecting an efficient implementation of space-related projects
- formulation, updating and efficient implementation of space-related projects.

These challenges have relegated most if not all African countries to the consumer-receiver status as they are not making significant contributions towards global efforts in the area of cutting-edge technological advancement in the various fields of science and technology especially those in the space sector.

1.1 The Nature of Space Technology

Most space related technologies are products of advanced science and technology innovations that place the possessor of such at an advantage either economically, militarily or both. A country like the United States of America leap frogged on the world stage from technological developments made after the Second World War in space science and rocket technology. The Space Race also provided an avenue for America to push further with the successful landing of astronauts on the moon

[1]Adigun Ade Abiodun, 'Space Education' (1997) 20 Advances in Space Research 1341.

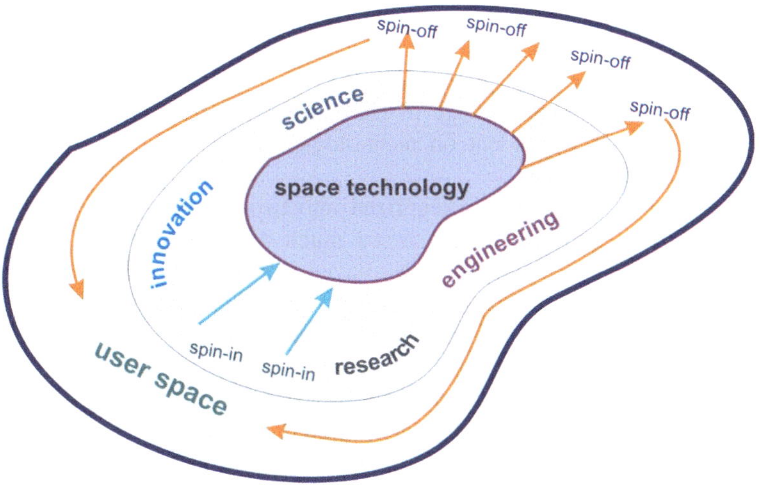

Fig. 1 Space technology and the spin-off spin-in cycle

between 1969 and 1972. Even as far back as late 1960s the technological benefit of the crewed lunar landing program was identified,[2] the tangible and intangible societal benefit of space exploration as well as the huge impact it has on human way of life were enumerated,[3] not to mention hundreds of spin-offs that came as a result of space exploration (Fig. 1).

This overarching effect means that space science and technology affects every part of our life is in one way or the other—from medicine, communication, transportation and so on.

African needs to innovate and make technology a strategic priority in order to advance and participate actively in the cutting-edge arena with focus on various areas of science and technology—a sector where space technology can make tremendous contribution. Several African countries are beginning to turn skywards to reap various benefits from space and also contribute to the space sector.

2 A Background to Space in Africa

A brief exploration into the history of African space experience and the effort made by several African countries to access space.

[2] Franklin A Long, 'The Industrial Impact of Apollo' (1969) 25 Bulletin of the Atomic Scientists 70.
[3] Stephen E Doyle, 'Benefits to Society from Space Exploration and Use' (1989) 19 Acta Astronautica 749.

2.1 African Space Experience and Dependence on Foreign Space Technology

Since the inception of the space age in 1957 Africa has prominently been on the side lines and mostly dependent on technologically advanced countries for any progress in the space sector.

African space history can be categorized into two epochs—passive and active periods.[4] The passive period encompassed much of the early part of the space program when several African countries played significant roles for space exploration—from hosting ground/tracking stations to providing other support services. These passive roles meant that Africa was receiving cutting-edge technologies then as "black boxes" without actively making direct contribution (in terms of technology export) to the technologies used in those missions!

In recent years the process has gradually changed from passive to somewhat active with attempts to play some roles in the space science and technology arena through participation in programmes in the space sector and acquisition of the space assets by few African countries.

However, the bulk of the technologies employed during the manufacture of the various African spacecraft have foreign origin except for South Africa which successfully built most of her satellites locally.

The implication is that in the game of space, African continent mostly plays the second fiddle and always have to try and catch up with advanced countries. This has overarching political and economic implications on the dynamics of the continent.

Space technologies due to their dual use—civil and military—are of tremendous national or regional interest and are considered objects of strategic value. This has also led to secrecy and lukewarmness on the side of partners that own these technologies to share vital/advanced space technology related information with their buying partners. Some of these technologies are regulated through policies such as International Traffic in Arms Regulations (ITAR) in the United States and similar regulations in other advanced countries that prevent export and acquisition by partners from developing countries. Space technologies that eventually filter to buyer countries in developing countries are those that are not of military or strategic consequence.

3 The role of space in African development

This section of the article looks at the role space technology can play in fostering and supporting development in the African continent through active participation and development of local/regional capabilities that ensure original and home-grown

[4]Samuel Anih, 'Passive to Active Space and the Role Space Assets in Sustainable Development' in Annette Froehlich (ed), *Embedding Space in African Society. Southern Space Studies* (Springer 2019).

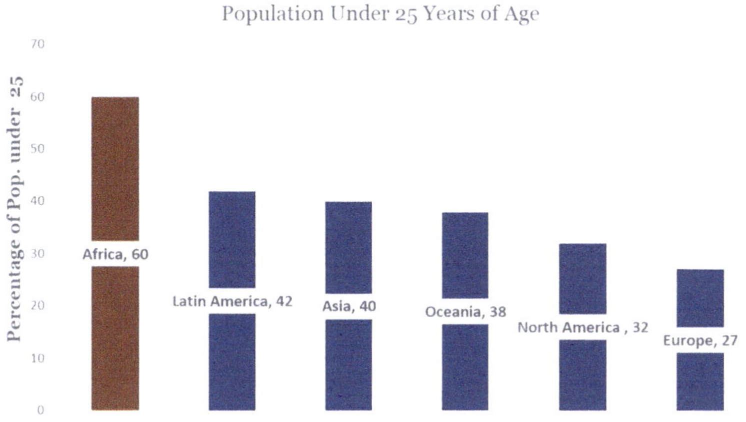

Fig. 2 Population of Africa under 25 years as at 2017

approach that focuses on solving peculiar challenges to the continent while promoting growth.

3.1 The Need for Space

A UN forecast shows that Africa's population is on the rise with the projection that half of the 2.2 billion people that would be added to the global population by 2050 will occur in Africa. Between 2017 and 2050 twenty-six African nations would expand to at least double their current size. As at 2017 about 60% of the African population is under 25 (Fig. 2), ahead of other continents.[5] This is an opportunity and a challenge—opportunity in terms of manpower, workforce and talent but challenge as this young population have to be catered for.

Apart from population explosion, Africa is also vulnerable to the effect of climate change that has led to devastating natural disasters and environmental degradation which further compound the economic and social challenges already faced by millions of Africans.

Space has played a huge role in so many areas in African growth, from economic development to disaster mitigation—and it is expected that this role would increase in coming years as many countries continue to embrace and enjoy the benefit of space applications including security, communications, environmental protection and natural resource management, education, disaster mitigation/management and agriculture.

[5]United Nations Department of Economic and Social Affairs Population Division, 'World Population Prospects: The 2017 Revision, Key Findings and Advance Tables. Working Paper No. ESA/P/WP/248.' (2017) <https://esa.un.org/unpd/wpp/Publications/Files/WPP2017_KeyFindings.pdf>.

3.2 Status Quo with Space Access in Africa

The motivations for launching satellites by African countries vary. According to Bertrand de Montluc, "*States like Nigeria, Algeria, Argentina, Malaysia or Indonesia have established space programmes, as symbols of independence, national pride and the desire to inspire their youth.*"[6] While de Montluc's submission is valid some other impetus for satellite acquisition such as the dynamics of partnership coupled with entrance of commercial space have opened opportunities not only for the states but also for industries and schools.

Few African countries have satellites in space and one of the reasons for this is that most African satellites are procured from manufacturers or institutions in advanced countries (Table 1). The launches of these satellites are also facilitated using foreign launchers as African countries currently have no the means of locally launching their satellites. However, there is a paradigm shift from that of total dependence on the advanced countries to that of collaboration with partners or institutions—mostly in the nanosatellite and microsatellite range. This kind of partnership allows for know-how transfer and training (KHTT) with scientists and engineers from affiliating countries. It has created platform for African engineers and scientists to participate in building some of the satellites African countries have launched so far. Table 1 shows African satellites, their manufacturers and launchers. From the list, only South Africa manufactured its satellites locally while the rest were manufactured outside the African continent. All the satellites were launched using foreign launchers.

The table also shows that the rate at which African countries acquire satellites is increasing. In the last 3 years 4 new African countries have launched satellites to space. Few have also started their own space agencies. Most of the new entrants however did not spend as much as budgeted by African space pioneers like South Africa and Nigeria at the beginning of their space programmes but opted for cheaper satellites offered by collaborating partners through bilateral programmes.

4 The Future of Technological Advancement and Space

This is a brief discussion on the new direction of technology with regards to 4th Industrial revolution and as Space 4.0 including how these are interwoven with the future of space.

[6]Bertrand de Montluc, 'The New International Political and Strategic Context for Space Policies' (2009) 25 Space Policy 20.

Table 1 Satellites launched by African countries with their manufacturers and launchers

Country	Satellite	Date launched	Manufacturer	Manufacturing country	Launcher	Launching country
Algeria	Alsat-1	28 November 2002	SSTL	UK	Kosmos-3 M	Russia
	Alsat-1B	26 September 2016	SSTL	UK	PSLV	India
	Alsat-2A	12 July 2010	EADS Astrium	Europe	PSLV	India
	Alsat-2B	26 September 2016	EADS Astrium	Europe	PSLV	India
	AlSat-Nano	26 September 2016	SSTL/UK Space Agency/ASAL	UK	PSLV	India
Angola	AngoSat 1	26 December 2017	RSC Energia	Russia	Zenit-3F/Fregat-SB	Russia
Egypt	Nilesat 101	28 April 1998	Matra Marconi	UK/France	Ariane	Europe
	Nilesat 101	17 August 2000	Matra Marconi	UK/France	Ariane	Europe
	Nilesat 201	4 August 2010	Thales Alenia Space	France	Ariane	Europe
	EgyptSat1	17 April 2007	Yuzhnoye Design Bureau	Ukraine	Dnper	Russia, Ukraine, and Kazakhstan
	EgyptSat2	16 April 2014	RSC Energia	Russia	Soyuz-U	Russia
	EgyptSat-A	21 Feb. 2019	RSC Energia	Russia	Soyuz-2	Russia
Ghana	GhanaSat-1	3 June 2017	Kyushu Institute of Technology's Birds	Japan	Falcon 9/Nanoracks	USA
Kenya	1KUNS-PF	2 April 2018	JAXA	Japan	Falcon 9/Kibo Module	USA
Morocco	Mohammed VI-A	November 8 2017	Thales Alenia Space and Airbus	Europe	Vega	Europe
			Thales Alenia Space and Airbus	Europe	Vega	Europe

(continued)

Table 1 (continued)

Country	Satellite	Date launched	Manufacturer	Manufacturing country	Launcher	Launching country
	Mohammed VI-B	November 21 2018				
Nigeria	NigeriaSat-1	27 September 2003	SSTL	UK	Kosmos-3 M	Russia
	NigcomSat-1	13 May 2007	CGWIC	China	Long March	China
	NigeriaSat-2	17 August, 2011	SSTL	UK	Dnper	Russia, Ukraine, and Kazakhstan
	NigeriaSat-X	17 August, 2011	SSTL	UK	Dnper	Russia, Ukraine, and Kazakhstan
	Nigcomsat-1R	19 December 2011	CGWIC	China	Long March	China
	Nigeria EduSat	June 3 2017	Kyushu Institute of Technology's Birds	Japan	Falcon 9	USA
South Africa	Sunsat-1	23 February 1999	Stellenbosch University	South Africa	Delta II	USA
	SumbandilaSat	17 September 2009	SunSpace	South Africa	Soyuz-2	Russia
	nSight-1	18 April 2017	SCS Space/QB50	South Africa	Atlas 5	USA
	ZACube-1	21 November 2011	F'SATI/CPUT	South Africa	Dnper	Russia, Ukraine, and Kazakhstan
	ZACube-2	27 December 2018	F'SATI/CPUT	South Africa	Soyuz-2-1a Fregat-M	Russia

SSTL—Surrey Satellite Technology Ltd, ASAL—Algerian Space Agency, CGWIC—China Great Wall Industry Corporation, F'SATI—French South African Institute of Technology, CPUT—Cape Peninsula University of Technology

4.1 The Fourth Industrial Revolution (4IR)

Whereas the first three industrial revolutions were characterized by and depended on advances in technology, the Fourth Industrial revolution is driven by advances in communication and connectivity and new technologies in machine learning and artificial intelligence, biotechnology and the emerging internet of things. Klaus Schwab describes 4IR as "*characterized by a fusion of technologies that is blurring the lines between the physical, digital, and biological spheres*" with "*The possibilities of billions of people connected by mobile devices, with unprecedented processing power, storage capacity, and access to knowledge, are unlimited. And these possibilities will be multiplied by emerging technology breakthroughs in fields such as artificial intelligence, robotics, the Internet of Things, autonomous vehicles, 3-D printing, nanotechnology, biotechnology, materials science, energy storage, and quantum computing*".[7] Space activities are beginning to see increase in the use of new technologies and methods that are part of 4th Industrial Revolution including additive manufacturing of rocket Engine components, 3D printing and so many other technology demonstrations some of which are already being used for human space exploration on board the International Space Station (ISS). These technologies would enable sustainable human spaceflight for much longer periods in space, a different scenario from current crewed space missions where astronaut have to depend on resupplies for survival. The opportunities that would be provided by 4IR should enable active participation with partners and also create avenues for countries and relevant institutions within the African continent to go beyond just space applications to full-fledged space exploration with partners across the globe.

4.2 Space 4.0

Space 4.0 was proposed by the European Space Agency (ESA) and "*represents the evolution of the space sector into a new era, characterised by a new playing field. This era is unfolding through interaction between governments, private sector, society and politics. Space 4.0 is analogous to, and is intertwined with, Industry 4.0, which is considered as the unfolding fourth industrial revolution of manufacturing and services.*"[8] Just like the 4th Industrial Revolution the Space 4.0 facilitates the interplay between invention, missions and innovation through basic research, applied research, technology development, project missions and market public application for the benefit of industry, society and the consumer[9] (Fig. 3).

[7]Klaus Schwab, 'The Fourth Industrial Revolution: What It Means, How to Respond' (*World Economic Forum*, 2016) <https://www.weforum.org/agenda/2016/01/the-fourth-industrial-revolution-what-it-means-and-how-to-respond/> accessed 12 February 2019.

[8]ESA Ministerial Council, 'What Is Space 4.0' (2016) <https://www.esa.int/About_Us/Ministerial_Council_2016/What_is_space_4.0> accessed 15 April 2019.

[9]Jan Wörner, '#SPACE4.0' (2016) <http://www.europarl.europa.eu/cmsdata/111744/ITRE2016-ESA-point13.pdf> accessed 7 April 2019.

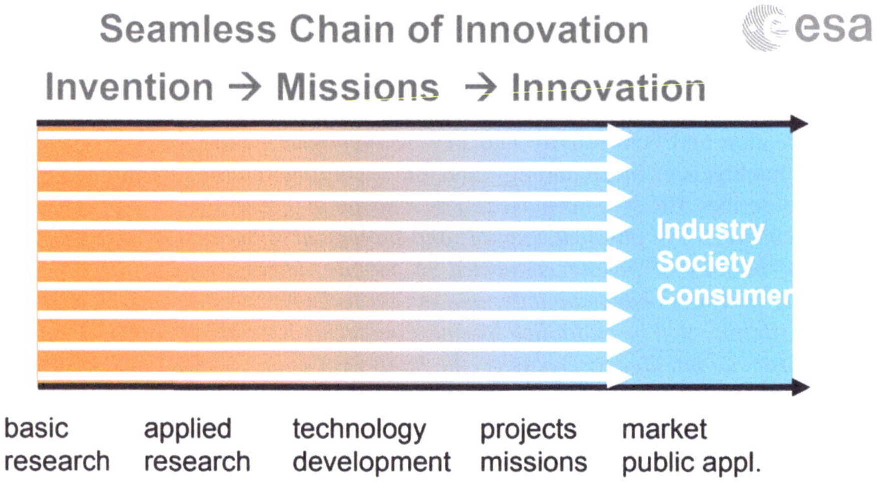

Fig. 3 Interplay between elements of Space 4.0

5 Towards Indigenous Space Capabilities

Discussion of possible methods that could be employed to achieve indigenous space capabilities with self sufficiency.

5.1 Taking the Right Step

Building indigenous space capabilities require concerted effort from all parties concerned—from the governments to higher institutions. Various areas have to be strengthened for progress to be achieved. More importantly, several steps have to be taken locally and internationally towards the goal of becoming self-sufficient within the African continent. There should also be an effort on local development of products. It is important to understand that it takes several steps to get to the final space related product. Different expertise and levels of collaborations are required for developing a product. This involves a plethora of processes that are well integrated to actualize the end goal of set objectives. Therefore, developing skillsets that will enable success—starting from development to the end of the chosen programme—is necessary if any country or institution is to eventually possess the final product. A remote sensing application development for example (Fig. 4), requires several well-articulated steps during the lifecycle.

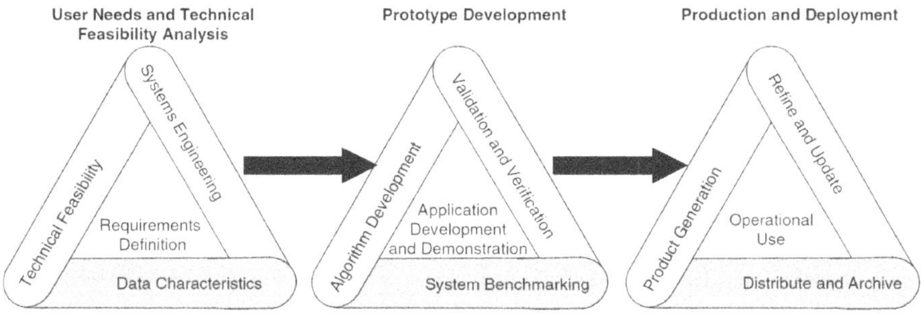

Fig. 4 From inception to final product—a remote sensing application development lifecycle (see footnote 10)

With the current push towards the commercialization of space, a great opportunity awaits Africa to play a important role. As expected the cheaper cost of production would encourage patronage from even advanced countries while booming the economies with cutting edge technologies.[10]

5.2 Multilateral Cooperation Within African Continent

Consultations among willing African countries could lead to multilateral agreements resulting in regional or/and continental space cooperation that covers technology transfer, joint satellite development and launching, educational, professional training and KHTT that involves exchanges of experts from participating countries. The proposed collaborative program would be executed in conjunction with related local/regional government institutions, research groups as well as the private sector.

The outcome of such cooperation would enhance security and help mitigate challenges particular to the concerned regions and the African continent at large. A follow up to this would be the establishment of necessary framework and institutions designed solely to achieve the cooperation objective of self-sufficiency with minimal technology import. This will considerably reduce dependence on foreign organizations and their expertise for the development of required technologies on the continent while boosting local economies and required manpower.

5.3 Local and Regional Drive

An example of a locally manufactured satellite was the Sumbandilasat satellite which was sponsored by the Department of Science and Technology with SunSpace as the prime contractor for the space and ground segment of the system while

[10]Peter T Gilruth and others, 'Measuring Performance: Moving NASA Earth Science Products into the Mainstream User Community' (2006) 22 Space Policy 165.

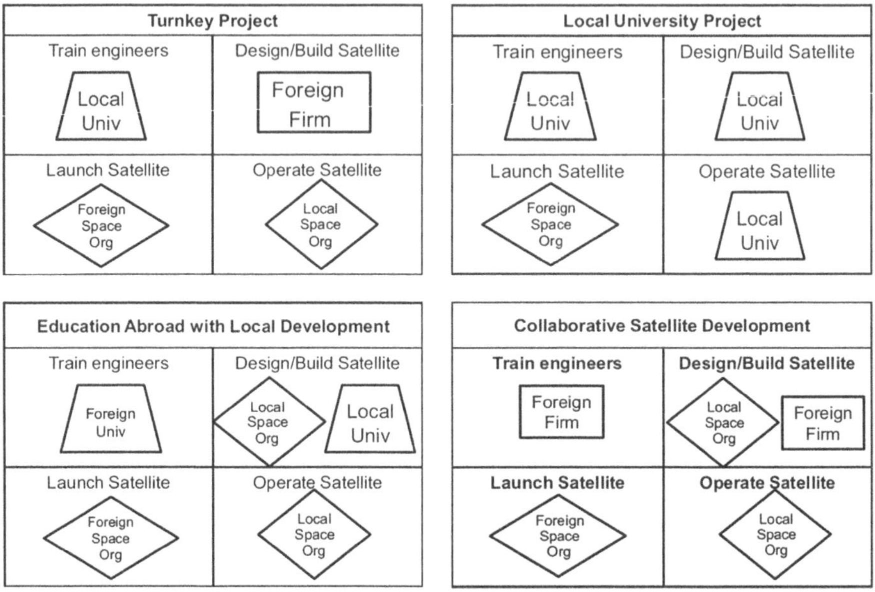

Fig. 5 The collaborative Satellite development project models from Wood and Weigel offer different benefits in terms of training and technology access with foreign participation in the loop

Stellenbosch University acted as the high-level programme manager.[11] The Sumbandilasat satellite project saw the participation of more than 40 local companies and the subsequent launch of the spacecraft in 2009. Other African countries can initiate similar programs designed to drive the local industries and partner with higher institutions. The skillset acquired from such venture would further strengthen the industry and provide infrastructure for future missions. Also, countries could develop specific capabilities which could be utilized by the continent instead of creating parallel competing institutions—for example, some regions could focus on launch capabilities while others concentrate on satellite design and so on.

The collaborative satellite development project models as described by Wood and Weigel[12] identified the roles played by foreign firms, local universities, local space organizations and foreign space organizations with the assumption that the intending customer country is interested in achieving four goals—training local engineers, designing and building a satellite, launching the satellite and operating

[11]Sias Mostert and others, 'Sumbandilasat—An Operational Technology Demonstrator.Pdf' (2008) 63 Acta Astronautica 1273.

[12]Danielle Wood and Annalisa Weigel, 'Architectures of Small Satellite Programs in Developing Countries' (2014) 97 Acta Astronautica 109.

the satellite with each of the models providing benefits in the areas of technology and management and training (Fig. 5).

The models—Turnkey, Local University, Education Abroad with Local Development, Collaborative Satellite Development projects—still tie satellite projects to foreign firms, especially with the launch services. While this is definitely the case in a globalized economy, creating an indigenous space capabilities would drastically reduce foreign dependence in the spacecraft development, launch and operations lifecycle.

6 Capacity Building

Capacity building is strategic to the development of the continent of Africa. The nature of space technology and current exponential growth of knowledge coupled with continuing developments in technology require concerted efforts from countries within the continent. Proper orientation and promotion of space science and technology at all levels including outreach programmes will ensure sustainability of local space capabilities within the African continent.

6.1 Building Indigenous Space Capability—The Way Forward

There has been a tremendous change in the space sector in Africa with the "rush" in recent times to have space agencies and also acquire space assets. According to Ibeh,[13] there has been a surge in the number of countries in Africa working towards becoming space-faring nations. In fact, this accelerated move by some African countries has seen the launch of 15 out of the total of 35 satellites happening in the last four years, a 43% increase after the first 1998 launch (Fig. 6).

Stephen Doyle described the required steps needed to acquire a functional national skills-building cycle,[14] while this applies to the space sector today, series of additional developments have taken place including the advent of off the shelf commercial space programs and private launch providers.

Space as an advanced and cutting-edge technology with strategic importance places the countries with related technologies at an advantage. Moghalu[15] submitted that there is no such thing as "transfer of technology" as no nation would willingly transfer its technological know-how to others since it forms the basis for competitive advantage—a point buttressed by the Porter Diamond Theory of

---

[13]Joseph Ibeh, 'A Breakdown of the 30 Satellites Launched by African Nations' (*Space in Africa*, 2019) <https://africanews.space/an-outlook-of-the-30-satellites-launched-in-africa/>.
[14]Doyle (n 4).
[15]Kingsley Chiedu Moghalu, *Emerging Africa: How the Global Economy's 'Last Frontier' Can Prosper and Matter* (Penguin 2014).

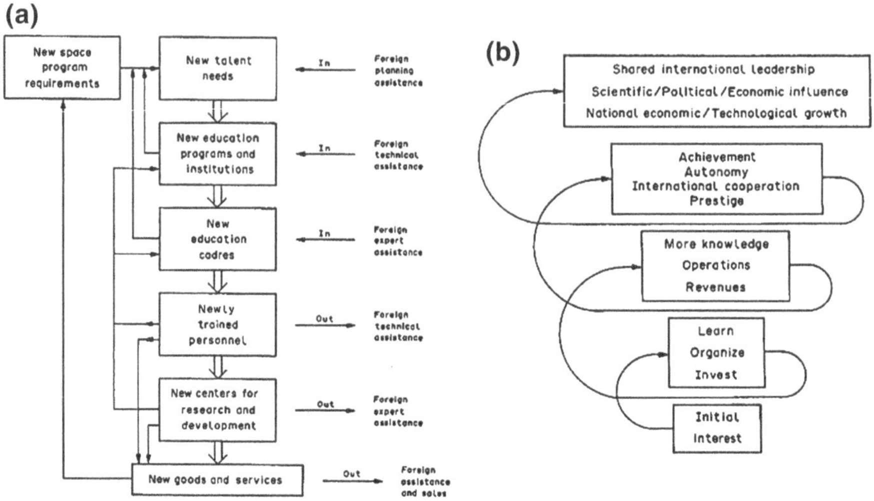

Fig. 6 Doyle's national skills building cycle (**a**) and the success spiral of space activity (**b**)

National Advantage a model proposed to understand the competitive advantage that nations or entities possess due to certain factors [technology inclusive] available to them.[16]

6.2 The Need for Education to Strengthen Space Science and Future Space Missions

So many proposals have been made concerning the sustaining of future space science missions in the African continent. J.O. Aseno proposed the involvement of scientists and engineers from Africa in order to strengthen the capacity building process while making sure the missions strive to meet the space science and technology needs in the African countries.[17] Scientists and engineers from Africa have to be involved in the planning of future space science missions in order to strengthen local capacity building as well as to ensure that they meet the expanding space science needs in the African countries. Steps to take towards achieving such goals should include—prior education and training, provision of funds from donor and space science agencies to facilitate this involvement, the introduction of space science education in relevant tertiary institutions. Also established African

[16]AJ Smit, 'The Competitive Advantage of Nations: Is Porter's Diamond Framework a New Theory That Explains the International Competitiveness of Countries?' (2010) 14 Southern African Business Review 105.
[17]JO Aseno, 'Space Science Education in the African Continent' (1997) 20 Advances in Space Research 1411.

Universities should offer postgraduate programmes in space science while other specialized training not available should be carried out abroad. It is expected that the assimilation capacity of the recipient of such training, which is dependent on the human capital of the recipient organization and the country, available infrastructure and the efficiency of the organization would promote and also contribute to the technology transfer process.[18]

6.3 Space Outreach Programmes to Schools, Policy Makers and the Public

Space technology has advanced technological breakthroughs, stretched the frontiers of science and engineering, expanded the understanding of the universe and tremendously propelled development of our world. This serves as inspiration for science, technology, engineering and math (STEM) related programmes for the public, upcoming professionals and schools.

Space Education outreach programmes are designed to inform and educate the public as well as students on the role of space while inspiring the young ones to acquire skills and take up careers in related fields.

In Africa, the inspirational power of space technology can be harnessed to encourage more people, most especially the younger ones, to focus on careers in STEM areas and thereby in the future make positive contribution to development in various fields of career within their home countries and the continent at large (Fig. 7).

Policies drive development of countries and very good policies are designed by leaders who have good understanding of problems/challenges and how to solve them. However, political leaders are not usually in touch with space science and technology and need to be updated. Therefore, building a strong STEM oriented programmes designed to also engage the leaders would help strengthen public understanding of the subject and enable a proactive participation of concerned government officials.

If properly carried out, space outreach programmes would provide unique opportunities for students, the general public and policy makers to contribute to the future of space science and technology within Africa in the areas of space applications, exploration and discovery. An approach would be to focus on space education outreach programmes to primary, secondary and tertiary institutions, encourage and provide participation in space/STEM related projects, support contribution to various forms of informal education and regularly evaluate the results of those programmes. Also making science and technology outreach participatory through citizen science projects would promote more public awareness and involvement.

[18]UM Leloglu and E Kocaoglan, 'Establishing Space Industry in Developing Countries: Opportunities and Difficulties' (2008) 42 Advances in Space Research 1879.

Fig. 7 Space education outreach programmes to secondary schools by the African Regional Centre for Space Science and Technology Education in English (ARCSSTE-E), a UN regional space centre in Nigeria *Photos* courtesy of ARCSSTE-E

6.4 A Call to Action

African countries should adopt a bottom-up evolutionary process where they build on existing structure to establish space infrastructures and institutions. This will allow space programmes to gradually mature while progressively building on experience and innovation towards the final space goals. It is expected that the number of space agencies in Africa will grow with time, but clear goals and roles should be set. However, the roles future African space agencies will play depend on the model they decide to adopt, for example the European Space Agency model focuses on societal benefits while the American model promotes commercialization. Whatever model chosen should allow for development of space programs suited for strategic and economic aspirations of the African people.

7 Conclusion

Space technology is cutting edge due to constant development of new products and expertise to tackle the challenges of spaceflight. However, not all products created during the spaceflight production process end up in space. Several advanced technologies spun-off from space process have terrestrial applications and have been designed to make our lives better. Indigenous space capabilities have the ability to transform how space programmes are carried out, provide opportunities for growth within various nations in the Africa and develop space technologies specifically tailored to the needs of each country or region. Also, regional and continental partnerships in space science and technology with proper frameworks and mechanisms will promote better use of space technology for development and mitigation of several challenges faced within the African continent while ensuring local space businesses and institutions are adequately protected.

Capacity building through proper education and adequate space outreach programmes to schools, general public and policy makers will provide a good launching pad to get the population enlightened, inspired and focused towards achieving the goal of building indigenous space capabilities within the African continent in the nearest future.

Samuel Anih developed a profound interest in space exploration during his high school days and later founded SpaceRovers which acted as a melting pot for fellow space enthusiasts as an undergrad at Obafemi Awolowo University, Nigeria. He received an M.Sc. from the International Space University (ISU), Illkirch-Graffenstaden (Strasbourg), France and later a graduate fellowship at NASA Ames, Moffet Field, California. He has more than 10 years working experience as a scientific officer at the African Regional Centre for Space Science and Technology Education in English (ARCSSTE-E), a United Nations affiliated centre. He is currently a Ph.D. candidate at SpaceLab in the University of Cape Town. Sammy is passionate about space science & technology, deep space exploration, most especially the future of crewed space missions beyond low Earth orbit, sustainability, planetary protection and space safety. He hopes to help build a new generation of space explorers and leaders.

Digital Africa: An Analysis of Digital Trends in Africa and Their Driving Factors

Christopher Yoon

Abstract

Today's space domain has been increasingly intertwined with the information domain, which is why digitization is an important issue within the context of space programs. New satellite systems, which are becoming more important for digital data communication, provide an opportunity to improve the access to the Internet. The following chapter analyzes the main trends of digitization in Africa and their driving factors. In doing so, the relationships between Internet usage and economic and technological developments in Africa will be explored by using quantitative data from the Worldbank and the Economist. The analysis will conclude that there are positive trends in the area of digitization, which are strongly affected by the access to electricity and economic well-being. Other variables examined in this study will be urbanization, literacy and "good" political management, which turned out to be important prerequisites for internet usage.

1 Introduction

With the emergence of modern information and communication technologies the information domain and the space domain have become more and more intertwined. In their early stages Internet providing satellite systems have played a rather marginal role for residentials with a speed of about few hundred kb/s and mainly provided a back-up technology for special needs such as banking communications or for users, who could not be reached by conventional access networks such as 3G,

C. Yoon (✉)
National Defense Academy Vienna, Institute for Higher Military Leadership, Vienna, Austria
e-mail: christopher.yoon@bmlvs.gv.at

A. Froehlich (ed.), *Space Fostering African Societies*, Southern Space Studies,
https://doi.org/10.1007/978-3-030-32930-3_9

ADSL or cable. Satellites and satellite systems are increasingly relying on Internet networks to guarantee their functionality. Moreover, satellites are of growing importance in Internet communications and other communication networks especially in the military and commercial sphere.[1] For African governments with space programs such as South Africa, Nigeria, Ethiopia, Kenya or Ghana, these programs are not only considered as a promising factor contributing to digitization, but are also expected to have a positive effect on technological advancement and economic growth.[2]

The following study provides an overview of general economic and digital trends in African countries and analyzes the driving factors behind digitization. The main research questions are, first, what are the main trends of internet usage in Africa and, second, what are the main reasons for internet usage at the structural level? In order to answer this research question, the study includes a data analysis based on quantitative data. As the analysis will conclude, digitization in Africa depends, at least to a considerable extent, on economic factors and structural prerequisites at the macro-level such as economic strength, access to electricity, urban areas, and "good governance" such as education or low corruption.

The first part of the study begins with a discussion of the main perspectives and theories in the academic literature. Here, the focus of attention lies on the digital divide and on theories that explain the digital divide such as the stratification and diversification hypothesis. The theoretical section presents the foundations of the theoretical argument and the empirical analysis and outlines the main hypothesis tested in the empirical part. The section on the theoretical foundations will be followed by a detailed description of the data used in the empirical part. The empirical investigation of digitization in Africa basically consists of two parts. The first part includes time series trends of digital, economic and social indicators. Using a correlation and regression analysis, the second part will explore the relationship between digitization and the main independent variables in more detail. Last not least, the final section summarizes and discusses the main findings of the theoretical and empirical analysis.

2 The Current State of the Art

One of main starting points in analyzing digitization in Africa is the theory of digital divide. Jan Van Dijk, one the most authoritative network theorists, defines the digital divide as "the gap between those who do and do not have access to

[1]Botta, Alessio & Pescapé, Antonio (2013) *New generation satellite broadband Internet services: should ADSL and 3G worry?*, in: The 5th IEEE International Traffic Monitoring and Analysis Workshop, pp. 399–404.

[2]See interview with Carla Sharpe cited in Giles, Chris (2018) *Africa leaps forward into space technology*, in: CNN, 16 May, https://edition.cnn.com/2017/08/10/africa/africa-space-race/index.html [accessed on 1.5.2019].

computers and the Internet".[3] The digital divide based on this definition is, in fact, closing at the global level mainly because most emerging countries in the periphery increasingly have access to broadband Internet and mobile phones.[4] In addition to this rather general definition, Van Dijk subdivides the digital divide into (1) inequalities in *mental access*, which is defined by a lack of elementary digital experience, (2) inequalities in *material access*, which means a lack of computers and network accesses, (3) inequalities in *skill access*, which describes a lack of knowledge, skills, and competencies, and (4) inequalities in *usage access*, which implies a lack of meaningful and profitable opportunities to use the digital space.[5]

According to the academic literature, digital inequalities reflect more traditional forms of inequality such as inequalities based on income, ethnicity, geography, gender, age or other social groups, which do not only persist at a global level, but within more advanced societies as well. In addition to the different definitions of digital divide, there are a couple of views on how and why these inequalities evolve. One of the main theories trying to explain the connection between the digital divide and other forms of inequalities is the so-called stratification hypothesis, arguing that traditional social inequalities cause or reinforce the digital divide because the access to digital networks and computers depends on the availability of financial, social and human capital.[6] Similar to the stratification hypothesis, which assumes that traditional social inequalities represent barriers for entering the digital sphere, the so-called complementary hypothesis claims that those, who are already in an advantageous position, can accumulate even more financial, social and human capital, for instance, by using online networks because privileged groups, as the argument goes, use online social media for complementary purposes and for enhancing their existing social network.[7]

At the global level too, the digital divide is often viewed as the result of economic disparities.[8] The socio-economic factors at the individual level, which are considered as the main reasons behind the digital divide, include variables such as income, employment status or education.[9] Additionally, studies found digital inequalities based on gender. For instance, previous research suggested that women

[3]Van, Dijk J. (2006) *The Network Society: Social Aspects of New Media*, 2nd edition, SAGE, London, pp. 178.

[4]Desta, Tedla (2018) *Comments on the Digitalization and Digital Divide in the Horn of Africa (HoA), Kenya and Ethiopia: The Media Perspective*, in: Global Media Journal, 28 February http://www.globalmediajournal.com/open-access/comments-on-the-digitalization-and-digital-divide-in-the-horn-of-africa-hoa-kenya-and-ethiopia-the-media-perspective.php?aid=86823 [accessed on 1.5.2019].

[5]Van Dijk, Jan. & Hacker, Ken (2003) *The digital divide as a complex, dynamic phenomenon*, in: The Information Society, Vol. 19, Nr. 4, pp. 315–326.

[6]DiMaggio, Paul & Garip, Filiz (2012) *Network effects and social inequality*, in: Annual Review of Sociology, Vol. 38, pp. 93–118.

[7]Litwin, Howard & Shiovitz-Ezra, Sharon (2011) *Social network type and subjective well-being in a national sample of older Americans*, in: The Gerontologist, Vol. 51, No. 3, pp. 379–388.

[8]James, Jeffrey (2003) *Bridging the Global Digital Divide. Edward Elgar*, Cheltenham, see Chap. 1.

[9]Witte, James C. & Mannon, Susan E. (2010) *The Internet and Social Inequalities*, Routledge, New York & London.

tend to have a lower frequency of Internet usage, a lower intensity of Internet usage and a narrower range of online skills and activities.[10] Other authors see a connection between the digital divide and ethnicity/race,[11] arguing that marginalized ethnic or racial groups coincide with a weak position in the digital space. While most theoretical studies view these factors as the causal variables explaining the digital divide, digital inequalities can also "reinforce existing social inequalities and even exacerbate them because they carry over preexisting differences in human capital into online setting".[12]

In contrast to the stratification hypothesis, the diversification hypothesis poses a more optimistic view and argues that digital networks can help in compensating weak positions in the digital space by transforming weak ties to other individuals into strong social capital. Furthermore, digital networks offer new opportunities in gaining information, for instance, when it comes to accessing job opportunities, by which marginalized individuals can empower themselves and overcome their underprivileged position.[13] As can be seen in the academic literature, there is a lot of evidence for these opposing theories, including quantitative studies as well as qualitative case studies. The problem though is that it is unclear under which circumstances these contradictory theories turn out to be true. While some studies suggest a prevalent linkage between gender and the digital divide, for instance, other researchers argue that the digital divide between men and women is closing or that age or a certain life-course are better predictors for digital adoptions than gender.[14]

In regard to the digital divide in Africa, large parts of the scholarly literature have portrayed a rather pessimistic narrative on digitization in Africa. The reason why the theory of digital divide is the most prominent perspective on digitization in Africa is that "Africa is the least-developed region in the world when income, school enrolment and life expectancy are taken into account [...] Thus, Africa is the continent most affected by poverty and underdevelopment. Africa also lags behind

[10]Robinson, Laura; Cotten, Shelia R.; Ono, Hiroshi; Quan-Haase, Anabel; Mesch, Gustavo; Chen, Wenhong; Schulzg, Jeremy; Haleh, Timothy M. & Stern, Michael J. (2015) *Digital inequalities and why they matter*, in: Information, Communication & Society, Vol. 18, No. 5, pp. 569–582.

[11]For an analysis of the connection between race/ethnicity and the digital divide see Nakamura, Lisa (2002) *Cybertypes: Race, Ethnicity, and Identity on the Internet*, Routledge, New York & London. Or see Campos-Castillo, Celeste (2015) *Revisiting the first-level digital divide in the United States: gender and race/ethnicity patterns, 2007–2012*, in: Social Science Computer Review, Vol. 33, No. 4, pp. 423–439.

[12]Robinson, Laura et al., *Digital inequalities and why they matter*, pp. 569–582.

[13]Mesch, Gustavo; Mano, Rita & Tsamir, Judith (2012) *Minority status and health information search: A test of the diversification hypothesis*, in: Social Science and Medicine, Vol. 75, No. 5, pp. 854–858.

[14]Yu, Rebecca P.; Ellison, Nicole B.; McCammon, Ryan J. & Langa, Kenneth M. (2016) *Mapping the two levels of digital divide: Internet access and social network site adoption among older adults in the USA*, in: Information, Communication & Society, Vol. 19, Nr. 10, pp. 1445–1464.

the rest of the world in terms of key indicators of the information society, such as subscriptions to the Internet."[15] Fuchs and Horak, for instance, argue that "the digital divide is a very pressing problem for Africa" and conclude that "most African countries are excluded from the information society".[16] Their study identified 57 African countries, in which less than 1% of the surveyed population had access to computers and Internet technology. The problem with the study by Fuchs and Horak is, however, that it was published in 2008 and the data examined in their analysis were collected in 2006, which means that it did not cover the rapid digital developments in Africa in the last ten years.

Whereas the traditional academic scholars tend to take a rather pessimistic view on digitization in Africa, arguing that the digital divide has persisted over the long-term, this account is not uncontested. In fact, there are a growing number of researchers favoring a more balanced view on digitization in Africa in recent years, especially proponents of the so-called empirical convergence theories or the catch-up theory. What most convergence theories have in common is that they emphasize the importance of economic development, technology and knowledge sharing in catching up with more advanced societies. Although the majority of African countries can be still classified as "digital desserts", Africa is a heterogeneous continent with an increasing number of emerging countries that become "digital oases".[17] There is strong evidence for a positive relationship between economic strength and digital access[18] and since a number of African countries have achieved promising growth rates, a closing digital divide is, at least to some extent, not surprising.

3 Theoretical Background

Whereas the traditional view on the digital divide in Africa is quite pessimistic, the data analysis will show that there has been a positive trend in terms of Internet users in Africa in recent years. But not only Internet usage, but also economic and social indicators have improved in recent years. The trend analysis will be composed of a descriptive examination of aggregated time series data of Sub Saharan African countries. Thus, the first hypothesis can be formulated as follows:

[15]Bornman, Elirea (2016) Information society and digital divide in South Africa: results of longitudinal surveys, in: Information, Communication & Society, Vol. 19, No. 2, pp. 264–278.
[16]Fuchs, Christian & Horak, Eva (2008) Africa and the digital divide, in: Telematics and Informatics, Vol. 25, pp. 99–116.
[17]Wentrup, Robert (2016) Digital oases and digital deserts in Sub-Saharan Africa, in: Journal of Science & Technology Policy Management, Vol. 7, No. 1, pp. 77–100.
[18]Evans, Olaniyi (2019) *Repositioning for increased digital dividends: Internet usage and economic well-being in Sub-Saharan Africa*, in: Journal of Global Information Technology Management, Vol. 22, No. 1, pp. 47–70.

- H_1: Countries in Sub-Saharan Africa show a positive trend in terms of Internet users, economic performance and social indicators in the last few years. The more time passed by, the higher the number of Internet users and the better the economic performance.

Since time is not a convincing causal variable in this regard, the second part of the empirical analysis consists of a correlation and regression analysis, in which the main factors behind digitization will be explored in more detail. The correlation and regression analysis will confirm the conventional view, arguing that the access to Internet depends on the availability of financial, social and human capital. The first independent variable will be the financial capital measured in GNI per capita. As indicated by the academic literature, the digital divide coincides, to a considerable extent, with economic strength because digital infrastructure and devices pose relatively expensive technologies especially for developing countries. Moreover, building up Internet infrastructure requires huge private and public investments. To put this theoretical relationship into a function, the hypothesis can be read as follows:

- H_2: Digitization heavily depends on financial capital. The greater the economic strength and the GNI per capita of a country, the higher the share of Internet users.

The question here is, however, which factors define economic strength or, to be more precise, which variables contribute to economic growth and, therefore, have a positive effect on digitization. While factors, which *define* economic strength, are relatively straightforward and include a combination of indicators such as GDP per capita, GDP growth, inflation rate or government debt, factors that *explain* economic strength are more complex and can include a variety of different considerations. One factor, which might explain both economic long-term growth and digitization, is the advancement of technology especially electricity. As historical analysis of the American economy shows, electricity and mechanization were main factors increasing the total factor productivity in the early 20th century.[19] As soon as energy production grew in the first few decades of the 20th century, electricity became cheaper. As a result, electrical machines increasingly substituted hand labor and steam based technologies in agriculture and industry and led to much higher productivity rates per input unit. Unskilled and expensive workers were replaced by mechanized processes on the factory floors and the production processes became more cheaper, accurate, efficient and faster not only in agricultural production but also in mining, manufacturing or transportation. In other words, access to electricity was crucial for technological adoptions and the development of the American economy.

---

[19]Oshima, Harry T. (1984) *The growth of U.S. factor productivity: The significance of new technologies in the early decadces of the twentieth century*, in: The Journal of Economic History, Vol. 44, No. 1, March, pp. 161–170.

The so-called energy-growth nexus could be observed in the vast majority of emerging countries and markets such as China, Turkey, Taiwan, South Korea and many other countries.[20] The Chinese government, for instance, opened up the energy sector for foreign capital in the late 1970s, which accelerated economic growth. There is strong evidence, however, that electricity consumption and economic growth decoupled in the 1990s mainly because investment into energy efficiency increased rapidly in the following years.[21] Similarly, several studies analyzed the relationship between access to energy and the economic development in Africa and found causal relationships between economic growth and energy consumption.[22] Electricity is not only a factor contributing to economic strength, but also an important predictor for the level of digitization because digital infrastructure and terminal devices depend on the availability of energy and electrical infrastructure. The second hypothesis, therefore, can be formulated as follows:

- H_3: Digitization heavily depends on electricity. The better the access to electricity, the higher the share of Internet users.

Other factors, which are often cited as main prerequisites for digital development, are the literacy rate and the educational level. As the theoretical literature on the digital divide argues, access to the digital world requires not only traditional skills such as reading and writing, but also digital skills such as knowledge about the usage of digital technologies.[23] Moreover, Internet content in local African languages is limited, which is why most Internet users require substantial English skills. The importance of literacy and education becomes even more obvious when it comes to the digital economy, which demands a deep understanding of the digital space such as technical skills or knowledge about doing business on the Internet. Based on these considerations, the fourth hypothesis can be formulated as:

- H_4: Digitization depends on literacy and education. The higher the literacy rate of a population, the higher the share of Internet users.

[20]For a general discussion see Ozturk, Ilhan (2010) *A literature survey on energy-growth nexus*, in: Energy Policy, Vol. 38, pp. 340–349. For specific case studies see Seung-Hoon & Jung, Kun Oh Jung (2005) *Nuclear energy consumption and economic growth in Korea*, in: Progress in Nuclear Energy, Vol. 46, No. 2, pp. 101–109. See also Oha, Wankeun & Lee, Kihoon (2004) *Energy consumption and economic growth in Korea: testing the causality relation*, in: Journal of Policy Modeling, Vol. 26, No. 8–9, pp. 973–981. For China see Cui, Huanying (2016) *China's Economic Growth and Energy Consumption*, in: International Journal of Energy Economics and Policy, Vol. 6, No. 2, pp. 349–355.

[21]Shiu, Alice & Lam, Pun-Lee (2004) *Electricity consumption and economic growth in China*, in: Energy Policy, Vol. 32, pp. 47–54.

[22]Odhiambo, Nicholas M. (2009) *Electricity consumption and economic growth in South Africa: A trivariate causality test*, in: Energy Economics, Vol. 31, pp. 635–640.

[23]Van Dijk, Jan. & Hacker, Ken, *The digital divide as a complex, dynamic phenomenon*, pp. 315–326.

In addition to literacy and education, another aspect, which is often linked to economic development, is urbanization—a hypothesis that dates back to the late 19th century.[24] The reason why researchers have emphasized the role of urbanization within the context of economic growth is that urban areas are concentrated areas, in which physical infrastructures such as transport, communication and managerial resources are easier to access than in rural areas. Furthermore, concentrated areas provide a stronger labor force for industries and imply more intense information and knowledge spill-overs.[25] It is important to note, however, that urbanization does not automatically lead to economic growth since cities can be overpopulated or grow too rapidly. Urbanization might be one of the prerequisites for industrialization, but it needs to be combined with other factors such as free markets, infrastructure, social stability and "good" political management. In this study, the connection between urbanization and economic growth will be applied to digitization, assuming that urbanization contributes to digitization mainly because of the very same reasons why urbanization can contribute to economic growth. Hence, the following hypothesis will be tested:

- H_5: Digitization requires concentrated areas with knowledge spill-overs, infrastructure and other economic factors. The higher the urbanization rate of a country, the higher the share of Internet usage of the total population.

The hypotheses, which were developed above, are considerably linked to the performance of governments and institutions. Research studies suggest that the economic development, infrastructure advancements or educational development depend on "good governance" and political management.[26] In a study on the relationship between electricity consumption and economic growth in Africa, it is argued that "Africa's energy problem is partly a result of the continent's macroeconomic mismanagement. Without improving the management of the economy and reducing the role of the state that has been blamed for Africa's economic ills, it is difficult to envisage how the energy challenges facing African countries can be addressed."[27] Since the Economist dataset does not provide indicators that measure the performance of governments directly, the empirical analysis will examine three different indicators which might be linked to governance, political management and institutions and which can be obtained from the dataset used in this study in particular the level of political stability, democracy and corruption. Therefore, the sixth and last hypothesis goes like this:

[24]Marshall, Alfred (1890) *Principles of Economics: An introductionary volume*, 8th Edition, MacMillan and Co., London.
[25]Henderson, Vernon (2003) The Urbanization Process and Economic Growth: The So-What Question, in: Journal of Economic Growth, Vol. 8, pp. 47–71.
[26]Evans, Olaniyi (2018) Digital government: ICT & public sector management in Africa, in: *New Trends in Management: Regional, Cross-Border and Global Perspectives*, Włodzimierz, S. (Ed.), London Scientific Publishing, London.
[27]Wolde-Rufael, Yemane (2006) *Electricity consumption and economic growth: a time series experience for 17 African countries*, in: Energy Policy, Vol. 34, pp. 1106–1114.

- H_6: Progress in digitization requires "good" political management. The better the political management of a country, the higher the share of Internet users.

In order to test these hypotheses, the empirical analysis consists of two parts: a trend analysis of the share of Internet users and economic indicators and a correlation and regression analysis estimating the impact of the different independent variables on the dependent variable. The regression analysis will comprehend ordinary least squares regression models (OLS), which will be based on the formula below. The purpose of the mathematical formula is to specify the argument, to insulate the main dependent and independent variables and to delineate the relationships between the variables. The following function will be tested in the regression models:

$$f(x) : Y = \beta_0 + X_1\beta_1 + \cdots + X_i\beta_i + \varepsilon$$

in which Y stands for the dependent variable share of Internet users, β_0 reflects the intercept of the regression line with the y-axis, β stands for the parameters, X_i mirrors the value of the independent variables, and ε reflects the error term of the regression. As the term X_i suggests, the impact of more than one independent variable on the dependent variable will be tested. In doing so, the regression analysis will use a step-by-step approach, by which single independent variables and covariates will be included one by one into a model depending on the strength of their correlation and their contribution to the determination coefficient. In a next step, specific indicators such as determination coefficients, standard deviations, the robustness of the model or a goodness-of-fit analysis will be investigated in order to evaluate the quality of the models.

As it will turn out in the empirical analysis, some relationships between variables are heteroscedastic, which occurs when the distances of the data points from the regression line are not constant. In order to deal with this problem, a two-step approach will be pursued. First, the models will be inspected by a Q-Q plot—a graphical tool that compares two probability distributions of the residuals of the regression models. The Q-Q plot also allows us to identify statistical outliers. Generally speaking, we can analyze whether the model fits the data using Q-Q plots. Second, the scatterplots of each of the relationships can give us an overview of the distribution of the data points.

If specific signs of distortions can be identified, the OLS regressions will be combined with robust regressions using MM-estimation. MM-estimation is a robust statistical method, in which the parameters of the regression formula are based on weights calculated by the distances between the observations and the regression line.[28]

[28]Susanti, Yuliana; Pratiwi, Hasih; Sulistijowati H., Sri; Liana (2014) *M Estimation, S Estimation, and MM Estimation in Robust Regression*, in: International Journal of Pure and Applied Mathematics, Vol. 91, No. 3, pp. 349–360.

In addition to heteroscedasticity, the issue of multicollinearity could be another problem in the empirical analysis. Regression models are usually interpreted as the value of change in the dependent variable as the independent variable increases by one unit. In multivariate regression models this logic involves that additional independent variables, which were included as control variables, remain constant. The problem of multicollinearity arises, if the independent variables in the regression models correlate with each other. As a consequence of multicollinearity, the precision of the estimated coefficients reduces and "inflates" the standard errors. Moreover, correlating independent variables imply that the predictors provide redundant information, which is why the coefficients cannot be interpreted as a change in the dependent variable if the independent variable increases by one unit.[29] Here too, a two-step approach will be applied to detect multicollinearity. First of all, the ordinary correlation coefficients of the relationships between the independent variables will be investigated. The higher the correlation coefficients between the independent variables, the more severe the problem of multicollinearity. Second, different indicators of the regression models such as the standard errors or the determination coefficients shed some light on the problem of multicollinearity.

4 Data

The empirical analysis in this study uses two different datasets, of which the first one was retrieved from the Worldbank and includes aggregated time series and cross-sectional data on economic, social and political indicators and of which the second dataset is the Inclusive Internet Index 2019 published by The Economist Intelligence Unit, which provides cross-sectional data on digital and economic aspects.[30] Whereas the general time series trends of digital and economic developments in the first part of the analysis treat Sub-Saharan Africa as a single entity, the analysis of the relationship between digitization, economic development and technology is based on country-level data of 28 African countries (including North Africa). Whereas the data in the trend analysis will be weighted based on the population size of each country, the data in the second part of the empirical analysis will not be weighted. The reason for this is the different nature of the research interests. While the trend analysis tries to assess trends of the population in Sub Saharan Africa, the correlation and regression analysis will investigate factors at the country-level.

[29]Yoo, Wonsuk; Bae, Sejong; Singh, Karan; Qinghua; Lillard Jr., James W. (2014) *A Study of Effects of MultiCollinearity in the Multivariable Analysis*, in: International Journal of Applied Science and Technology, Vol. 4, No. 5, pp. 9–19.

[30]The Worldbank dataset can be downloaded from https://data.worldbank.org/region/sub-saharan-africa?view=chart. The Inclusive Internet Index 2019 published by the Economist Intelligence Unit can be downloaded from https://theinclusiveInternet.eiu.com/.

The Worldbank dataset provides more than 1,500 cross-sectional variables on economic, social and political indicators combined with time series data for all African countries which are located south of the Sahara between 1960 and today, including Angola, Benin, Botswana, Burkina Faso, Burundi, Cabo Verde, Cameroon, Central African Republic, Chad, Comoros, Democratic Republic of Congo, Republic of Congo, Cote d'Ivoire, Equatorial Guinea, Eritrea, Eswatini, Ethiopia, Gabon, Gambia, Ghana, Guinea, Guinea-Bissau, Kenya, Lesotho, Liberia, Madagascar, Malawi, Mali, Mauritania, Mauritius, Mozambique, Namibia, Niger, Nigeria, Rwanda, Sao Tome and Principe, Senegal, Seychelles, Sierra Leone, Somalia, South Africa, South Sudan, Sudan, Tanzania, Togo, Uganda, Zambia and Zimbabwe.

The problem with the Worldbank dataset is, however, that time series data are not available for all indicators. Moreover, data are not available for all countries, which is why absolute numbers of some indicators do not reflect their actual magnitude. In addition to this problem, absolute numbers often vary from source to source because different datasets use different data collection or estimation methodologies. For instance, the Worldbank is more conservative in collecting and estimating data so that the figures tend to be lower than those provided by other sources. The share of Internet users is a perfect example. Whereas the Worldbank defines "Internet users" as individuals, who had access to the Internet in the last three months prior to the survey, other data sources have extended the time prior to the survey to one year. It is for these reasons why the analysis is not so much about absolute figures than about extracting general trends which can be expressed in percentage points.

The Inclusive Internet Index 2019, which was commissioned by Facebook and conducted by the Economist Intelligence Unit, covers 100 countries from all continents. For the purpose of this study non-African countries were removed from the dataset, making a total number of 28 African countries including Algeria, Angola, Benin, Botswana, Burkina Faso, Cameroon, Congo (DRC), Cote d'Ivoire, Egypt, Ethiopia, Ghana, Guinea, Kenya, Liberia, Madagascar, Malawi, Morocco, Mozambique, Namibia, Niger, Nigeria, Rwanda, Senegal, Sierra Leone, Sudan, Tanzania, Tunisia and Uganda. The dataset comprehends data on 86 different variables focusing on digital and economic indicators. As compared to the Worldbank dataset, the disadvantage of the Economist dataset is that it is less comprehensive in terms of the number of countries, variables and time. The advantage over the Worldbank data is, however, that the dataset focuses on digitization and comprises almost no missing data.

Table 1 depicts the main variables of the Economist dataset used in this study. Whereas the share of Internet users of a country's population will be the main dependent variable, the main independent variables include the factors GNI per capita, electricity access, literacy rate, urbanization rate and political management, of which the last one is measured by the peace index, democracy index and corruption index.

Table 1 Main variables of the Economist's III 2019 dataset examined in this study

Variable	Type	Description
Country	String	The country variable displays and identifies the country's name
GNI per capita	Numeric	Gross national income is measured in current US$ per person and is calculated by GDP plus net capital flows divided by population size. The data was obtained from the Worldbank dataset. In the correlation analysis, the GNI per capita will be logarithmized.
Electricity access	Numeric	Electricity access is measured in % of population and was provided by national household surveys
Literacy rate	Numeric	The level of literacy is measured in % of population and was obtained from the UNESCO
Urbanization rate	Numeric	The urbanization rate measures the share of the population living in urban areas and was provided by the Worldbank
Peace index	Interval	The data on the stability indicator was obtained from the Institute for Economics and Peace. The indicator has a scale of 1–5, where 5 means highest stability
Democracy index	Interval	Democracy index provides information on how democratic a country is. The data is provided by EIU Business Environment Rankings and consists of a scale between 1 and 10, in which 10 means highly democratic
Corruption index	Interval	The source of the corruption index is Transparency International and is measured by a scale of 1–100, in which 1 means the highest corruption

5 Empirical Results of the Trend Analysis

It is certainly true that Africa still lags behind most advanced countries in terms of economic well-being and digitization. As can be seen in Fig. 1, however, the pessimistic view of the traditional scholars on the digital divide is misleading. The share of individuals in Sub-Saharan Africa, who had access to Internet in the last three months, is slowly but steadily growing since the mid-2000s. The share of individuals using the Internet grew from 0.069% of the total Sub-Saharan African population in 1996 to 19.85% in 2015. This trend is even more astonishing considering the fact that the total African population south of the Sahara grew from half a billion in 1990 to more than one billion people in 2019.

The growth in mobile subscriptions has been even more rapid than the growth of Internet users and increased from 2.5% of the total population in 2000 to 74.6% in 2016. What is interesting though is the vast gap between mobile users and Internet users, indicating that the access to mobile phones does not necessarily imply an access to the Internet. One of the main reasons for this is that the majority of Africans still use flip phones or feature phones in contrast to more advanced countries, where smartphones are the most commonly used mobile device today. The challenge here is that most Africans can simply not afford a smartphone.

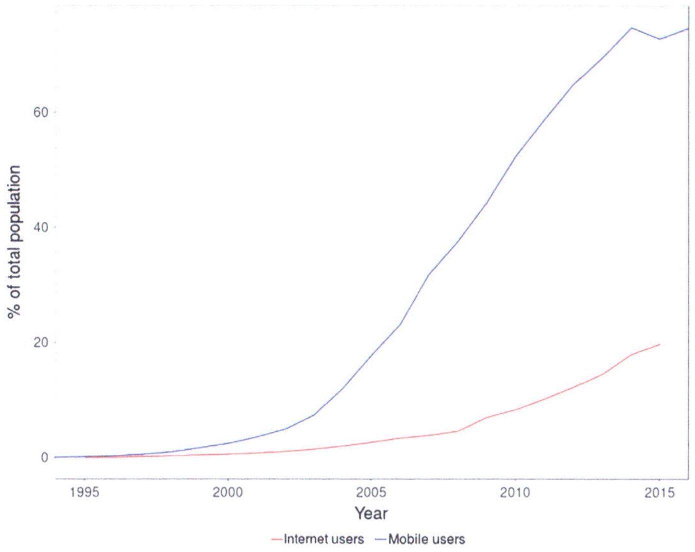

Fig. 1 This figure shows the long-term trends of individuals in Sub-Saharan Africa using mobile phones and the Internet measured in % of the total population of Sub-Saharan Africa. The numbers are weighted averages based on the population size of the countries (*Data* Worldbank)

Moreover, the Internet applications for flip and feature phones tend to be rudimentary or they have no Internet access at all. According to a technology report conducted by the Pew Research Center, South Africa is the only country in Sub Saharan Africa, in which the majority of the population (51%) uses smartphones.[31]

The growing number of Internet and mobile users went hand in hand with positive economic and social developments. If this was not a coincidence, it means that there is either a causal relationship between digitization and economic development or both developments have the same cause. As can be seen in Fig. 2, two key performance indicators of macro-economic development in particular the annual GDP growth rates and annual inflation rates have significantly improved over the last three decades. Between 2000 and 2016 the GDP of Sub-Saharan Africa grew on average 4.6% per year—much higher than in the 1990 and 1980s. One might argue that the higher growth rates can be explained by population growth. However, the real GDP per capita also increased by 1,000 US$ since the mid-1990s. Although the inflation rates between 2000 and 2016 were still quite high with an average annual inflation rate of 5.7%, inflation rates decreased significantly as compared to an average inflation rate of 9.69% in the 1990 and 8.07% in the 1980s.

[31]Silver, Laura & Johnson, Courtney (2018) *Internet Connectivity Seen as Having Positive Impact on Life in Sub-Saharan Africa*, Pew Research Center, 9 October, S. 12.

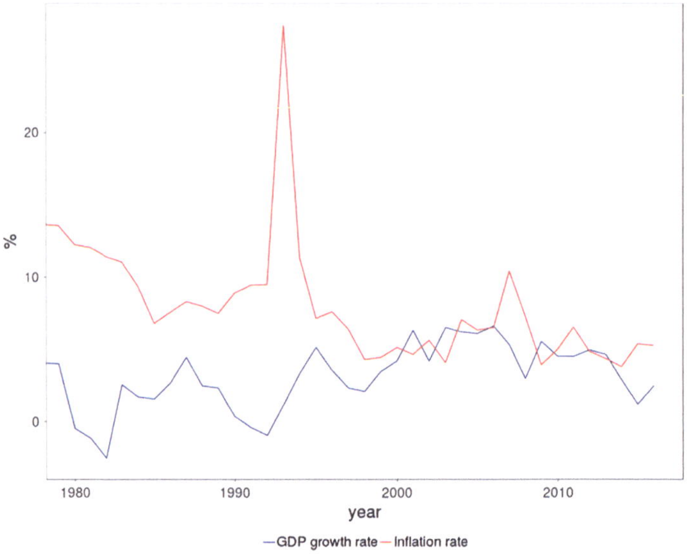

Fig. 2 This figure shows two main economic performance indicators of African countries south of the Sahara in particular the annual GDP growth rate and the annual inflation rate. The rates are weighted averages based on the population size of the countries (*Data* Worldbank)

Figure 3 shows the positive trends of two more important development indicators in particular the access to electricity and the literacy rate, both of which are viewed as essential preconditions to close the digital divide. Between 1990 and 2015 the amount of people having access to electricity grew from 17.48 to 42.79% of the total population of Sub-Saharan Africa. Also, literacy rates for individuals older than 15 years grew in the same time period from 52.23% of the total population in Sub-Saharan Africa to 64.35%. Here too, these figures are impressive given the growing population in Sub-Saharan Africa. What these figures do not show, of course, is the quality of electricity access or literacy. Although electricity access rates grew significantly, shortages, blackouts, brownouts and other deficiencies are still common in Sub-Saharan Africa even in countries such as South Africa with a relatively high electricity access rate.[32] Nevertheless, the trend is definitely moving into a positive direction.

As Figs. 1, 2 and 3 show, there is arguably a lot of evidence indicating that Sub-Saharan Africa is on the right trajectory—a fact that is usually overshadowed by bad news on Africa and pessimistic perspectives in academia. Whether it is digital access or economic strength, the data reveal a narrative quite different from the conventional views in the academic literature and the public discourse on the digital divide. The number of Africans with access to the Internet and mobile

[32]Brew Hammond, Abeeku (2010) Energy access in Africa: Challenges ahead, in: Energy Policy, Vol. 38, pp. 2291–2301.

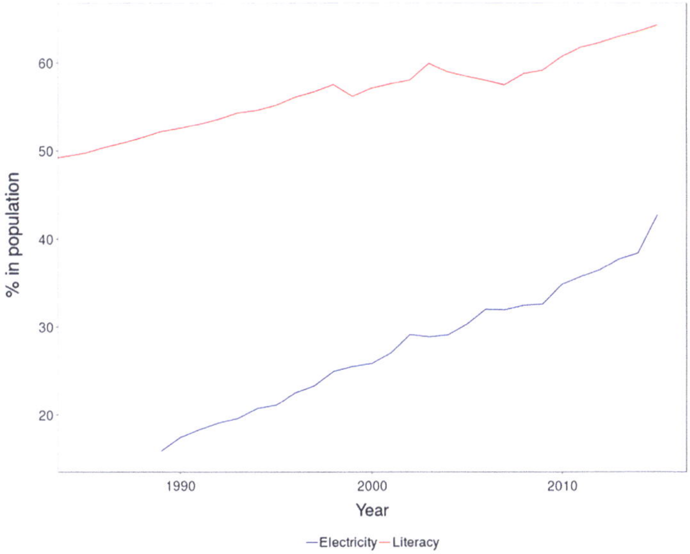

Fig. 3 This figure illustrates the trends of the access to electricity and the literacy rate. Both indicators are measured in % of the total population (*Data* Worldbank)

phones has increased dramatically in last three decades and is expected to continue to rise. Given the rapid population growth in African countries, these trends are remarkable. Furthermore, Sub Saharan Africa has followed the global trend of emerging countries and markets, where the situation of more and more people has significantly improved. Today, Sub-Saharan Africa has higher growth rates, significantly lower inflation rates, higher electricity access rates and higher literacy rates than in the 1990s and 1980s.

6 Correlation and Regression Analysis

To explore the variables described above in more depth, the following section includes a correlation and regression analysis estimating the effects of the main independent variables on the Internet usage. Before the correlation analysis is conducted, however, we will have a look at the graphs below. Figure 4a shows a highly linear relationship between the share of Internet users and access to electricity. Countries with higher electricity access rates also tend to have higher shares of Internet users. The leading countries in terms of electricity access and Internet users, at least among those countries included in the dataset, are the North African countries Morocco, Tunisia, Egypt and Algeria. Among the Sub-Saharan countries Botswana, Cote d'Ivoire and Ghana have relatively high electricity access rates and Internet users as compared to countries such as Liberia, Congo, Niger, Malawi,

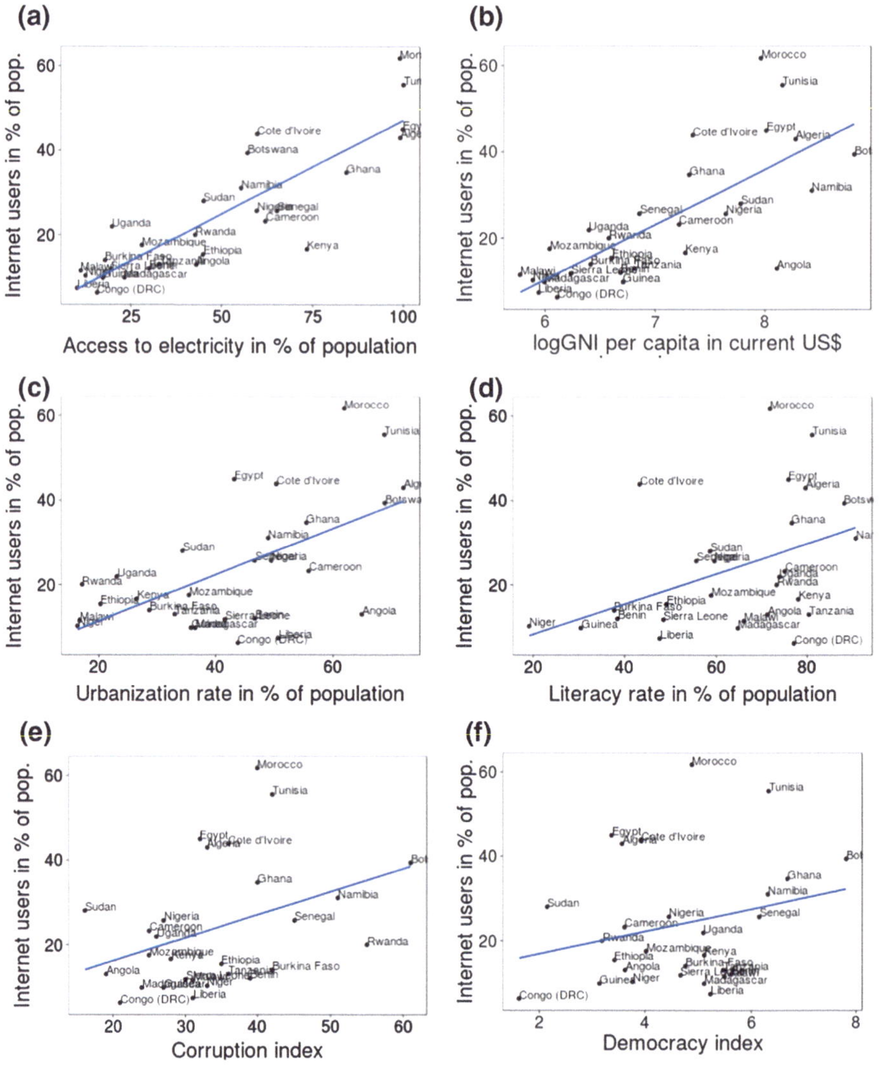

Fig. 4 a–e The following scatterplots illustrate the relationships between the dependent variable share of Internet users and the independent variables log GNI per capita, access to electricity, urbanization, literacy, corruption and level of democracy

Sierra Leone, Burkina Faso or Madagascar. The scatterplot in Fig. 4b represents the relationship between GNI per capita and Internet usage. The graph shows that the share of Internet users strongly depends on the availability of financial capital measured in GNI per capita. There seems to be a relationship between the two variables although the errors are heteroscedastic, which means that the deviation from the mean increases at the end of the spectrum.

As the data points and the regression line in Fig. 4c and d indicate more and higher deviations from the regression line, which means that the goodness-of-fit of the linear model is not that perfect and that the predictive power of the independent variable is worse than in 4b or to a lesser extent in 4a. The literacy rate is often cited as one of the main prerequisites for digitization. As the graphs indicate, the relationship between literacy and digitization is complex rather than linear. Several countries such as Tanzania, the Democratic Republic of Congo (DRC), Madagascar, Malawi, Angola or Kenya have relatively high literacy rates residing between 60 and 80%, but the share of Internet users is below 20% of the population. At the same time, there are a variety of countries such as Sudan, Senegal, Nigeria or Cote d'Ivoire, which have literacy rates lower than 60% but at the same time higher shares of Internet users than Tanzania, the DRC, Madagascar, Malawi, Angola or Kenya. As the graph in Fig. 4e indicates, there seems to be a relationship between the share of internet users and corruption, which was used as an indicator for the quality of political management. However, the distribution of the data points shows that the distances from the regression line are not constant at the end of the spectrum, which means that the data points are heteroscedastic.

In the next step the relationships in the scatterplots above will be analyzed by pairwise Pearson correlations. Table 2 contains the correlation coefficients and significance levels between the main variables examined in this analysis. Although correlation does not necessarily mean causation, correlation coefficients measure the strength of a relationship between two variables and their relative changes. It is calculated by the covariance of the variables divided by the product of their standard deviations. A positive coefficient means that if one variable increases, the other variable increases too. A negative coefficient would signal that if one variable increases the other variable decreases. High significance levels imply that the null hypothesis can be rejected, which means a very low probability that the correlation was a result of chance.

The correlation coefficients in Table 2 between Internet usage and access to electricity show the highest correlation with a coefficient of 0.87 and a very high significance levels with a p-value lower than 0.001. Another correlation, although to a lesser extent, can be seen between the share of Internet users and logGNI per capita with coefficient of 0.75 and a significance level lower than 0.001. We can also see a medium correlation between Internet usage and the variables urbanization rate (0.59) and literacy rate (0.43). Whereas the correlation between Internet usage and urbanization is significant at a 0.001 level, the p-value of the variable literacy rate is a bit higher and, therefore, the significance a bit lower. Moreover, there is a small but significant positive correlation between Internet usage and the corruption index, which implies that lower levels of corruption go hand in hand with higher Internet usage.

Apart from the main dependent variable Internet usage and its correlations with such variables as electricity, income, urbanization, literacy or corruption, the correlation matrix also reveals a connection between democracy, corruption and stability. With a highly significant coefficient of 0.77 the data shows that a better score in the democracy score is linked to higher stability. Moreover, the democracy index

Table 2 Pearson correlation matrix with ordinary correlation coefficients and significance levels

	Internet users	Literacy rate	Urbanization rate	log GNI per capita	Electricity rate	Peace index	Democracy index	Corruption index
Internet users	1.00	0.43**	0.59***	0.75***	0.87***	−0.05	0.24	0.38*
Literacy rate	0.43*	1.00	0.40*	0.56**	0.52**	−0.06	0.27	0.21
Urbanization rate	0.59***	0.40*	1.00	0.70***	0.60***	−0.15	0.28	0.15
logGNI per capita	0.75***	0.56**	0.70***	1.00	0.78***	0.02	0.22	0.32
Electricity rate	0.87***	0.52**	0.60***	0.78***	1.00	0.03	0.18	0.28
Peace index	−0.05	−0.06	−0.15	0.02	0.03	1.00	−0.77***	−0.53**
Democracy index	0.24	0.27	0.28	0.22	0.18	−0.77***	1.00	0.61***
Corruption index	0.38*	0.21	0.15	0.32	0.28	−0.53**	0.61***	1.00

*** < 0.001; ** < 0.01; * < 0.1

positively correlates with the corruption index, which means that a higher democracy score of a country goes hand in hand with less corrupted political elite. The correlation coefficient between democracy and corruption is 0.61 and is highly significant with a p value lower than 0.001.

In a next step we could use the variables associated with Internet usage to include them into a multivariate regression model. As can be seen in the correlation matrix, however, most of the independent variables such as electricity access, literacy rate, urbanization or GNI per capita correlate with each other, which is why they cannot be used as control variables due to the problem of multicollinearity. For instance, electricity access, which shows the highest correlation with Internet usage, also correlates with GNI per capita or urbanization. Since the purpose of this analysis is not to *predict* the dependent variable but rather to measure the impact of the independent variables, the problem of multicollinearity needs to be taken into account in the analysis of the following regression models.

Table 3 includes nine regression models estimating the impact of different independent variables on Internet usage. As can be seen in the regression coefficient in Model 1, the share of Internet users increases by 0.43 percentage points, if the share of those, who have access to electricity, increases by 1 percentage point. As can be seen in the last column, the coefficient is highly significant. In order to examine whether the model fits the data, a closer look at the standard deviation and R^2 is needed. Since no statistical function is perfect, the standard deviation gives us the average distance of the data points from the regression line. The statistical output of the regression model gives us the standard error (SE), by which we can calculate the standard deviation (SD). Using the formula $SD = \sqrt{n} \cdot SE$, in which n is the size of the sample size (28), we get an average deviation from the regression line of 0.249 percentage points. In addition to a very low standard deviation, the model's output include a relatively high R^2 of 0.7475, which means that 74.75% of the variance can be "explained" by the independent variable. In Fig. 5a we can check the distribution of the residuals of model 1, which had few small deviations, which is why robust models and indicators, in addition to the OLS models, were calculated and tested. However, changing the model from an ordinary OLS regression to MM estimation did not affect the numbers significantly, which means that the small deviations of the residuals did not have a large impact on the model.

Model 2 estimates the effect of the gross national income per capita on the share of Internet users of the total population. As the coefficient of model 2 shows, the share of Internet users increases by 0.006% if the GNI per capita increases by 1 US $. This would be a 0.6 percentage points higher share of Internet users with the GNI per capita increasing by 100 US$ per year. The null hypothesis can be rejected and the coefficient is highly significant with a p value below than 0.001. As compared to the variable access to electricity, however, the R^2 is with 0.413 lower than the R^2 in model 1, which means that the variable GNI per capita is less suitable for explaining the variance than the variable access to electricity. The relationship between GNI per capita and Internet usage is illustrated in Fig. 4b which reveals a heteroscedastic deviation of the data points. In this case, the distances between the regression line and the observations become longer at the end of the spectrum. But

Table 3 Regression models estimating the impact of different independent variables on the dependent variable internet usage

	Coefficient	Std. Error	t value	p value
Model 1				
Intercept	2.8757	2.6214	1.097	0.331
Electricity access	0.4316	0.0471	9.1642	0.0000000018***
Adjusted R^2: 0.747				
Model 2				
Intercept	12.2914	2.8153	4.336	0.0003
GNI per capita	0.006	0.0012	4.8941	0.000134***
Adjusted R^2: 0.413				
Model 3				
Intercept	0.8112	7.5048	0.1081	0.9478
Urbanization rate	0.5347	0.1647	3.2471	0.000842***
Adjusted R^2: 0.329				
Model 4				
Intercept	0.2521	9.0333	0.0279	0.9016
Literacy rate	0.3476	0.1375	2.5278	0.0225*
Adjusted R^2: 0.153				
Model 5				
Intercept	6.097	9.4168	0.6475	0.9016
Corruption index	0.4856	0.2653	1.8301	0.0458*
Adjusted R^2: 0.111				
Model 6				
Intercept	2.4426	2.6492	0.922	0.632
Electricity access	0.3801	0.0598	6.349	0.0000012***
GNI per capita	0.0019	0.0011	1.766	0.0896*
Adjusted R^2: 0.766				
Model 7				
Intercept	2.0693	4.2690	0.485	0.632
Electricity access	0.3778	0.0643	5.870	0.0000046***
GNI per capita	0.0018	0.0013	1.437	0.164
Urbanization rate	0.1429	0.1265	0.113	0.911
Adjusted R^2: 0.767				
Model 8				
Intercept	7.2465	6.5742	1.102	0.2818
Electricity access	0.3972	0.0669	5.933	0.0000047***
GNI per capita	0.0024	0.0014	1.717	0.0993*
Urbanization rate	0.0036	0.1267	0.028	0.9776
Literacy rate	−0.1035	0.1001	−1.034	0.3117
Adjusted R^2: 0.757				

(continued)

Table 3 (continued)

	Coefficient	Std. Error	t value	p value
Model 9				
Intercept	1.6052	8.6436	0.186	0.854
Electricity access	0.3895	0.0673	5.781	0.0000081***
GNI per capita	0.0018	0.0015	1.175	0.253
Urbanization rate	0.0342	0.1303	0.263	0.795
Literacy rate	−0.0958	0.1003	−0.955	0.350
Corruption index	0.1549	0.1541	1.005	0.326
Adjusted R^2: 0.757				

here too, the robust model only slightly reduces the impact of the variable GNI per capita, which indicates that the few deviations do not critically affect the model.

The third highest correlation following access to electricity and GNI per capita is the urbanization rate, whose impact on the share of Internet users is specified in regression model 3. The coefficient in model 3 shows, that the share of Internet users is expected to rise by 0.53 percentage points if the urbanization rate increases by one percentage point. The variable is highly significant with a p value below than 0.001 and has a determination coefficient of 0.329, which implies that the independent variable accounts for 32.9% of the variability. In the Q-Q plot in Fig. 5c few deviations can be identified. Comparing the ordinary OLS regression model with the MM estimation model, the effects of these small deviations on the model are negligible.

According to regression models 4 and 5, the share of Internet users increases by 0.35 percentage points, if the literacy rate increases by one percentage point and by 0.54 percentage points if the corruption index increases by one unit. However, both variables the literacy rate as well as the corruption level have much lower significance levels, higher standard deviations and lower determination coefficients, which means that the goodness-of-fit of the two models is worse than that of the previous models.

Summarizing the graphs and the results of the correlation and regression analysis, there is strong evidence that access to electricity is the most powerful predictor of the share of Internet users of the total population. The relationship is linear, highly significant, its standard deviation is low and the model is relatively robust. The second best predictor among the variables examined in this analysis is the independent variable GNI per capita. The GNI per capita has the second highest correlation with the dependent variable, a high significance level, a low standard deviation and the second highest determination coefficient. Although the distances between the data points and the regression line increase at the higher end of the spectrum, the results of the M-M estimation indicate that the model is relatively robust.

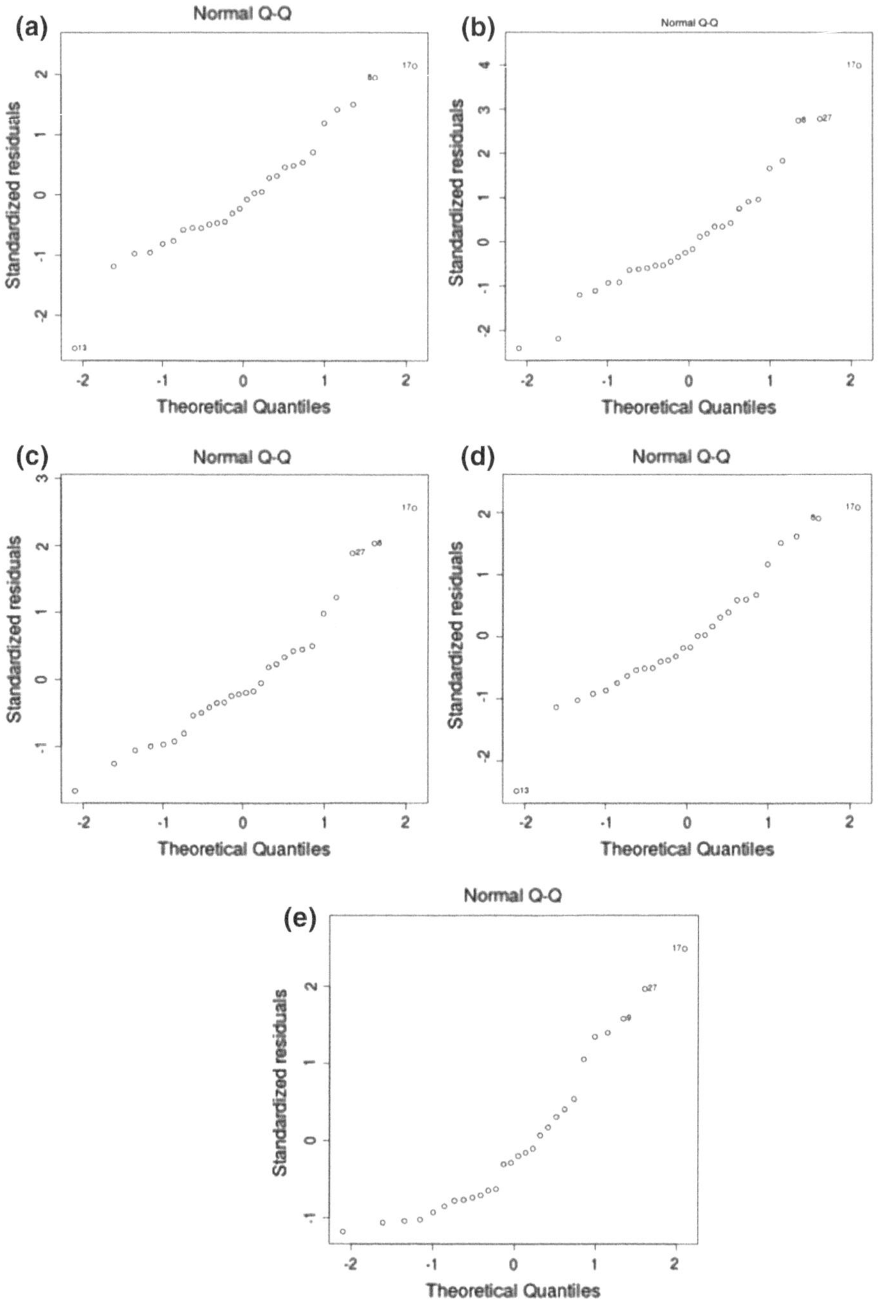

Fig. 5 **a–e** The following graphs include the Q-Q plot of the residuals of the main simple linear regression models

Whereas electricity and GNI per capita seem to be good predictors of internet usage, urbanization might be a prerequisite for digitization, but it is not the last stage before digitization starts to evolve. While none of the countries with a low urbanization rate has a higher share of internet users, none of the countries with a high share of internet users has a low urbanization rate. However, not all countries with a high urbanization rate have high shares of internet users. There are no countries with a low urbanization rate, which have a high share of Internet users too. No country with a lower urbanization rate than 40% has a higher share of internet users of 30%. Those countries, whose urbanization rate is higher than 40%, can be divided into one group, which has a higher share of internet users, and one group, which remains at the lower level of internet usage. Those countries, in which urbanization coincides with high share of internet users, are Ghana, Namibia, Botswana, Algeria, Egypt, Morocco, Tunisia or Cote d'Ivoire.

A similar case is the literacy rate. It is a prerequisite but not a causal factor. The variable corruption also showed some relationship with the share of Internet users, although considerably weaker than access to electricity or GNI per capita. None of the countries with a high share of internet users have a low literacy rate, but a high literacy rate doesn't automatically mean a high share of internet users. Here too, literacy seems to be a prerequisite, but it is not the final cause that leads to more internet usage.

While the problem of multicollinearity was already identified in the correlation analysis, it could be also detected through the stepwise regression procedure. The redundancy of the independent variables can be seen in the determination coefficients which do not significantly increase with more variables being included. In fact, the adjusted R^2 even decreases in model 8 and 9 with more control variables, which confirms that such factors as access to electricity, economic well-being, urbanization and literacy are strongly related to each other. The question is here, therefore, how these factors interact. Access to electricity and the economic development of a country are important predictors for the general level of digitization. Previous research suggests, on the one hand, that the access to electricity and other technological advancements such as mechanization contribute to economic growth. On the other hand, energy production, electrical infrastructure and other technological infrastructure developments require huge investments—a factor, that explains why several studies found a one-directional impact of economic growth on energy consumption. Digitization is, therefore, affected by both access to electricity and economic growth. Since studies show an impact of economic growth on energy consumption one can also assume that economic growth has some causal impact on Internet usage, whether it is a direct impact on Internet usage or indirect impact through electricity access on Internet usage. As the correlation and regression analysis indicate, the effect of economic strength on Internet usage is lower than the effect of electricity access on Internet usage.

Whereas linear regressions are about the independent effects of predictive variables on a dependent variable, multicollinearity reflects the dependency between predictive variables, which gives us biased parameters. The problem of multicollinearity cannot be solved without a reassessment of the theoretical model. In fact, multicollinearity is the result of model complexity and the inability to narrow down the number of independent variables that can be expressed or summarized by one factor. As a consequence, one has to make a decision of how to simplify the model and how to choose variables that are either independent from each other or that are able to subsume interrelated factors. Considering the correlation and regression analysis, the best variable to predict Internet usage has been electricity access. However, given the strong linkage between electricity and economic well-being (logGNIpC) and the correlation between economic well-being and such variables as urbanization, literacy and corruption, it seems that financial capital plays the most important role in this complex mixture of factors. On the one hand, the economic well-being is affected by a variety of variables including infrastructure, urbanization, literacy or low levels of corruption. On the other hand, economic well-being is an enabler for technological development, infrastructure or literacy at the same time. One factor, that was not been explored, but which might be one of the most important aspects of economic development, is the existence of a free market and the individual freedom of people to do business.

7 Conclusion and Discussion

Since the space domain and the information domain are increasingly intertwined, the digital development has gained growing importance within the context of African space programs. The goal of this analysis was to examine general trends of the digital development in Africa and its driving factors behind it. In doing so, the study conducted a theoretical and empirical analysis, of which the former included a discussion of the main academic literature and the factors, which are considered to have an impact on digitization, and of which the latter consisted of an analysis of time series trends, on the one hand, and a correlation and regression analysis, on the other. The empirical analysis was based on theoretical considerations emphasizing the factors access to electricity, GNI per capita, urbanization, literacy and political factors such as stability, democracy and corruption. At the theoretical level it was assumed that these factors have a causal impact on the share of Internet users, by which the level of digitization was measured.

In contrast to previous academic studies, the trend analysis of time series data revealed a positive trend towards higher shares of mobile subscribers and Internet users. Moreover, the time series illustrated that the positive digital development went hand in hand with positive economic and social developments such as higher growth rates, lower inflation rates and increasing literacy and access to electricity rates. In order to explore digitization in more depth, the correlation and regression analysis was supposed to analyze the relationships between independent variables such as GNI per capita, electricity access, literacy, urbanization and political factors, on the one hand, and the dependent variable share of Internet users, on the other. As the empirical results indicate, there is a strong connection between access to electricity and Internet usage. Furthermore, Internet usage depends on economic well-being and, to some extent, on urbanization, literacy and political management, of which political management/mismanagement was measured by the corruption index. What this study could not answer due to the problem of limited time and space is to specify the relationships between the independent variables, which positively correlate with each other.

African space programs can contribute to the digital infrastructure such as broadband access through satellite systems, which would be a relatively cheap and attractive alternative for remote areas which lack telecommunication lines. Furthermore, space programs have the potential to accelerate technological advancement and to boost scientific research, which can have a positive impact on economic and digital development. What African space programs do not address, however, is the lack of electrical infrastructure which was identified as one of the main predictors of Internet usage. Energy production and electricity infrastructure are considered as the main prerequisites not only for Internet usage but also for economic development in general. Given the lack of financial resources and governmental investments in many African countries, one possible solution to this problem would be to open up energy markets for foreign direct investments, as China did in the late 1970s to overcome energy shortages and to satisfy industry demand. Furthermore, international initiatives and national governments could invest into energy efficiency in order to decouple electricity consumption and economic growth.

Christopher Yoon M.Sc. works as a researcher and data scientist at the Institute for Higher Military Leadership at the National Defense Academy in Vienna, where his academic attention aims to digitization, data science, cyberspace and AI in military applications at the operational level of war. After he graduated in political science and social science he switched to economics and completed his master's in international relations at the University of Edinburgh. Before that Christopher gained professional experience as a research assistant at the University of Innsbruck, the Institute for Federalism and as an intern at the Austrian embassy in New Delhi, India. Furthermore, Christopher gained experience as an entrepreneur and founded a business focusing on data analytics and the development of machine learning algorithms and applications for different sectors and industries.

Health from Above: Space-Based Healthcare Services in Africa

Julia Selman Ayetey and Harold Ayetey

Abstract
Despite recent improvements, health indicators such as child and maternal mortality and life expectancy in Sub-Saharan Africa remain amongst the worst in the world. The meagre healthcare budgets, poor transportation links, unpredictable energy supplies and unreliable internet services that characterise large swathes of the continent exacerbate the problem. Further, the significant mismatch between disease burden and qualified healthcare personnel continues to stifle efforts to reverse these trends. Space-based healthcare systems may be the key to substantial change in healthcare while stimulating economic growth within and beyond the healthcare sector. Indeed the Pan-African e-Network on Tele-education and Tele-medicine ('PAeN'), an African-wide platform with satellite-based connectivity, was established to help address these problems. Despite its sub-optimal outcomes to date, its potentially transformative role remains clear. In this chapter, a case is made for the urgent revival and refinancing of PAeN specifically and wider implementation of space-based healthcare services generally. An explanation of space-based healthcare services is followed by detailed discussion of African health indicators which make its use particularly vital to the continent. The arrested development of PAeN is then discussed and examples of successful space-based healthcare services are highlighted. Barriers to effective and extensive use of space-based healthcare services are outlined and recommendations on how to overcome them made.

J. Selman Ayetey
Faculty of Law, University of Cape Coast and Doctoral Candidate, Institute of Air and Space Law, McGill University, Montreal, Canada

H. Ayetey (✉)
Department of Internal Medicine and Therapeutics, School of Medical Science and Cape Coast Teaching Hospital, Cape Coast, Ghana
e-mail: harold@cantab.net

A. Froehlich (ed.), *Space Fostering African Societies*, Southern Space Studies,
https://doi.org/10.1007/978-3-030-32930-3_10

1 Introduction

The third UN Conference on the Exploration and Peaceful Uses of Outer Space ('UNISPACE III'), held in July 1999, resulted in the adoption of the *Vienna Declaration on Space and Human Development ('Vienna Declaration')*. The *Vienna Declaration* recognised the significance of space science and space applications to health and recommended that steps be taken to '...improve public health services by expanding and coordinating space-based services for telemedicine and for controlling infectious diseases...' in the interest of human security and welfare.[1] This objective is particularly important for many countries in Sub-Saharan Africa, given the paucity of health services and facilities across the continent. In the same year, the Indian initiated and African-Union ('AU') backed Pan-African e-Network on Tele-education and Tele-medicine ('PAeN') was established. PAeN, a multi-modal platform which utilised both fibre optic (ground-based) and satellite (space-based) connectivity from the Regional African Satellite Communication Organisation ('RASCOM') delivered tele-education and tele-medicine services Africa-wide.[2] Its telemedicine operations failed to make the expected impact due to a number of reasons including a lack of awareness of the program and underappreciation of its benefits.

Africa bears one-quarter of the global disease burden, yet, has only 2% of the world's doctors.[3] The World Health Organisation estimates that life expectancy for those born in Sub-Saharan Africa is 61.2 years compared with an average 77.5 years for those born in Europe.[4] Indeed in 2018, Luxembourg which has the highest per capita GDP in the world of $113,594 had a life expectancy of 81 years compared to Guinea-Bissau which had a per capita GDP of $840 and a life expectancy of 57 years.[5] These and other typically low health indicators in many parts of Sub-Saharan Africa are driven and compounded by meagre national healthcare budgets, insufficient and inadequate primary and tertiary healthcare facilities, poor transportation links, unpredictable energy supplies and unreliable internet service. Impoverished levels of health have a negative impact on various socio-economic indicators. Innovative and effective healthcare strategies are key to the economic future of Sub-Saharan Africa. Space-based healthcare systems have the potential to drive transformative change in healthcare across the continent while stimulating economic growth within and beyond the healthcare sector.

---

[1]Para 1(b)(i), *The Space Millennium: Vienna Declaration on Space and Human Development*, Third United Nations Conference on the Exploration and Peaceful Uses of Outer Space ('UNISPACE III') (Vienna, Austria, 1999) http://www.unoosa.org/pdf/reports/unispace/viennadeclE.pdf (accessed 2 May 2019).

[2]African Union, *First Progress Report of the Chairperson of the Commission on the Pan African e-Network on Tele-education and Tele-Medicine*, PRC/2 (ii) (7-2018) (Addis Ababa: Meeting of the Permanent Representatives' Committee, 2018).

[3]World Health Organization, *Global Health Observatory (GHO) Data* (2016).

[4]*Ibid.*

[5]*Ibid.*

This chapter advocates for the urgent reintroduction of a reinvigorated, better-resourced and strategically improved PAeN and an expansion of space-based services to improve access to high quality healthcare across Sub-Saharan Africa. It begins with a brief outline of key terms. This is followed by a discussion of space-based healthcare services with an emphasis on health and broader socio-economic benefits. It then examines PAeN alongside two case studies of successful local telemedicine projects which could be integrated into a future PAeN. Barriers to effective, efficient and widespread use of space-based healthcare services are outlined and recommendations made. The chapter concludes by suggesting that effective and efficient delivery of healthcare services in Africa can be augmented through the comprehensive application of space-based technologies.[6]

2 Definitions

Several terms, such as cybermedicine, mHealth and e-health, are employed to describe the use of digital technology for medical-related purposes.[7] However, the most commonly used terms are 'telemedicine' and 'telehealth'. Although the terms are often used interchangeably, distinct definitions have been noted in the literature.[8] Telemedicine generally refers to 'the use of electronic communications and information technologies to provide clinical services when participants are at different locations'.[9] The term is thus often limited to the prevention, diagnosis and treatment of illness or injury. Telehealth is used more broadly to refer to the 'application of technologies to distance education, consumer outreach, and other applications wherein electronic communications and information technologies are used to support healthcare services'.[10] Telehealth can thus include research, public service broadcasting of health information, electronically delivered medical training, continuing education, artificial intelligence (AI) platforms and integrated health information management systems. Underlying all of these terms is the notion that they represent,

[6] Countries in other regions of the globe have also sought or are in the processes of attempting to utilise space-based telemedicine, Canada is one example, see "McMaster researchers head north to explore space-based telemedicine"21 July 2009 http://fhs.mcmaster.ca/main/////news/news_2009/space_based_telemedicine.html.

[7] Claudia Pagliari et al., "What is eHealth (4): A Scoping Exercise to Map the Field" (2005) 7:1 Journal of Medical Internet Research e9.

[8] Kylie Cox Orme, "Guardians of the Galaxy: How Space Tourism Regulation Will Shape Telehealth" (2017) Air and Space Law 163, online: http://www.kluwerlawonline.com/document.php?id=AILA2017013.

[9] The American Telemedicine Association, *Telemedicine, Telehealth, and Health Information Technology* (2006) see also; David Pratt, "Telehealth and Telemedicine in 2015" (2015) 25 Albany Law Journal of Science & Technology 495; Bonnie Kaplan & Sergio Litewka, "Ethical Challenges of Telemedicine and Telehealth" (2008) 17:4 Cambridge Quarterly of Healthcare Ethics 401.

[10] *Ibid.*

> ...not only a technical development, but also a state-of-mind, a way of thinking, an attitude, and a commitment for networked, global thinking, to improve health care locally, regionally, and worldwide by using information and communication technology.[11]

Specific subcategories of telemedicine include: telemonitoring, the surveillance of disease patterns and outbreaks; teleconsultation, the assessment and treatment of patients using digital communication technologies; teleradiology, diagnosis by experts through digitalised radiological images; telepathology, diagnosis by experts of digitalised images of pathological samples; teleprescribing, prescription and treatment of diseases as well as teleintervention, surgical intervention conducted through robotics.[12] As the prefix 'tele' indicates, these services are undertaken remotely.[13] Many of these services may be provided through wired ground-based infrastructure, however, the focus of this chapter is on services provided through space-based systems.

3 Spaced-Based Healthcare Services

Space-based healthcare services are provided through the use of satellites which are stationed in outer space.[14] They are used in conjunction with ground-based equipment which control and monitor the satellites and facilitate the transfer of data. Although interruptions to services due to adverse weather conditions or satellite malfunction may occur, the advantages of space-based healthcare services over those delivered through traditional ground-based infrastructure are clear. First, some services cannot be provided by ground-based services. Second, several space-based services can be delivered in the absence of ground-based infrastructure making it a vital tool in Sub-Saharan Africa where ground-based infrastructure is often sub-standard or absent.

There are various types of satellites, each developed for a distinct purpose. Communication satellites cover vast geographical areas and may thus provide a reliable means of connecting medical personnel with patients or other health experts in villages or other hard-to-reach areas which lack adequate or any wired telephone and internet services. They facilitate synchronous or asynchronous communication, meaning that transmission either occurs instantaneously in real-time or is recorded

[11] Gunther Eysenbach, "What is e-health?" (2001) 3:2 Journal of Medical Internet Research E20.
[12] N Ateriya et al., "Telemedicine and Virtual Consultation: The Indian Perspective" (2018) 31:4 National Medical Journal of India 215.
[13] The Oxford English Dictionary defines 'tele', when used as a prefix, as 'action, observation, or communication at, over, or across a distance, or denoting devices used for this'.
[14] For the sake of brevity, 'space-based healthcare services' will be used in this chapter as an umbrella term to refer to all medically related services, education and information provided remotely via satellite in orbit.

and exchanged later through "store-and forward" communication.[15] Earth observation ('EO') satellites, for instance, enable the production of images which can be utilised to assess light and radiation patterns from the earth's surface.[16] These features are useful for a variety of medical purposes including the monitoring of migration patterns of vector (disease-carrying organisms), such as mosquitoes carrying malaria parasites for example, as well as analysis of data relating to the wind or rain which may reveal the probability of communicable disease outbreaks. Imagery from EO satellites may also help predict crop failure due to floods or droughts which often herald mass starvation and malnutrition and assist authorities to better manage the risk of water-borne diseases.[17] Global positioning system ('GPS') services provide users with precise locations of specified objects on Earth and have obvious applications for the delivery of healthcare including the tracking of infected persons and populations as well as determining the source of disease outbreaks.[18]

Afarikumah and Kwankam rightly note that 'Improving the health of individuals and communities, and strengthening health systems, disease detection and prevention are crucial to development and poverty reduction'.[19] Good healthcare leads to better participation in local economies and results in improved levels of educational attainment, employment, productivity and GDP. International organisations and civil society have published reports and guidelines, stimulated dialogue and promoted regional cooperation with regard to the use of outer space for the development of Africa. The biannual African Leadership Conference on Space Science and Technology for Sustainable Development has been held since 2005. For over a decade the United Nations Committee on the Peaceful Uses of Outer Space ('COPUOS') has pushed the agenda of using space science and technology for global health.[20] This has included formulating an Expert Group on Space and Global Health who in turn recommended the establishment of a longer-term

---

[15]Richard Wootton, John Menzies & Paula Ferguson, "Follow-up Data for Patients Managed by Store and Forward Telemedicine in Developing Countries" (2009) 15:2 Journal of Telemedicine and Telecare 83.

[16]Ramesh S Krishnamurthy & Jason Hatton, "Space Science and Technologies to Advance Health-Related Sustainable Development Goals" (2018) 96:3 Bulletin of the World Health Organization 3.

[17]PricewaterhouseCoopers, *Cost Benefit Analysis of Satellite-Enhanced Telemedicine and eHealth Services in Sub-Saharan Africa* (2008).

[18]*Ibid.*

[19]E Afarikumah & SY Kwankam, "Deploying actor-network theory to analyse telemedicine implementation in Ghana" (2013) 1:2 Science Journal of Public Health 77 at 77.

[20]United Nations Committee on the Peaceful Uses of Outer Space, *Report of the Working Group on Space and Global Health*, Annex III, Report of the Scientific and Technical Subcommittee on its fifty-sixth session A/AC.105/1202 (Vienna, Austria, 2019); United Nations Committee on the Peaceful Uses of Outer Space, *Fourth Meeting of the Expert Group on Space and Global Health held from 31 January to 1 February 2018: Progress Report*, Scientific and Technical Subcommittee Fifty-fifth session A/AC.105/C.1/2018/CRP.17 (Vienna, Austria, 2018); United Nations Committee on the Peaceful Uses of Outer Space, *Final report of the Action Team on Public Health: The Use of Space Technology to Improve Public Health - Note by the Secretariat*, A/AC.105/C.1/L.305 (Vienna, Austria, 2011).

working group on the subject. The Working Group on Space and Global Health was duly formed and their multi-year agenda has be endorsed under the auspices of the Scientific and Technical Subcommittee.[21] In specific reference to Africa, COPUOS and the Economic Commission for Africa, in consultation with members of the United Nations Inter-Agency Meeting on Outer Space Activities, published a report in 2009 entitled, *Space Benefits for Africa: Contribution of the United Nations System*. It detailed ways in which the UN was employing space technology in, *inter alia*, its health activities in Africa.[22] More recently, the Continental Africa Telemedicine Alliance was established in February 2019 to engage with regulatory agencies in an effort to increase and support telemedicine initiatives.[23] Whilst these conferences, reports and policy initiatives are useful, the actual use of space science and technology for health purposes remains relatively limited.[24] Despite these challenges, the global telemedicine market, just one form of service provided by space-based systems is predicted to surpass 65 billion USD by 2021.[25] The African space industry is currently valued at $7 billion and expected to surpass $10 billion by 2024.[26] More needs to be done in order to reasonably exploit the benefits of space-based healthcare services globally. In Sub-Saharan Africa, implementation should be viewed as an urgent humanitarian effort which could transform the health of a continent of 1.2 billion with massive spin-off economic benefits across sectors from tourism through manufacturing to transport and finance to mention a few.

4 Burden of Disease in Africa

Considerable improvements have been attained in a number of health and socio-economic indicators in Africa over the last decade. There have been reductions in child, maternal and adult mortality rates and some evidence that the medical field and governments are better coping with the HIV epidemic.[27] However, poverty, conflict, weak institutions, lack of infrastructure, human resource

[21]United Nations Committee on the Peaceful Uses of Outer Space, *supra* note 20.

[22]United Nations, *Space Benefits for Africa: contribution of the United Nations System*, Committee on the Peaceful Uses of Outer Space, A/AC.105/941,20 August 2009 http://www.unoosa.org/pdf/reports/ac105/AC105_941E.pdf.

[23]Eugen Davies, "Continental Africa Telemedicine Alliance inaugurated", *Business& Financial Times Online*, 25 February 2019 https://thebftonline.com/2019/business/health/continental-africa-telemedicine-alliance-inaugurated/ (accessed 9 May 2019).

[24]This is also the case in Europe and North America, see Richard Wootton et al., eds, *Telehealth in the Developing World* (Ottawa: Royal Society of Medicine Press and International Development Research Centre, 2009).

[25]Mordor Intelligence, *Telemedicine Market - Growth, Trends, and Forecast (2019 - 2024)*, 2018. https://www.mordorintelligence.com/industry-reports/global-telemedicine-market-industry.

[26]Space in Africa, *African Space Industry Annual Report: Executive Summary* (Lagos, 2019). https://africanews.space/report/.

[27]World Health Organization, *The Health of the People: What Works - The African Regional Health Report 2014* (Brazzaville, Republic of Congo: WHO Regional Office for Africa, 2014).

shortages and low educational attainment remain considerable problems.[28] The UN estimates the population of Africa in 2017 was approximately 1.2 billion people.[29] In that year 8.8 million deaths were registered. The top causes of death included lower respiratory tract infections (10.4%), HIV/Aids (8.1%), Diarrhoea (7.4%), Malaria (4.6%), Ischaemic Heart Disease (5.8%) and Stroke (4.2%).[30] Significantly, deaths due to infectious diseases remain high (56%), while mortality resulting from non-communicable disease continues to increase accounting for 3 million deaths (34%) in 2016.[31] These mortality figures significantly outstrip those from other regions.

Doctor patient ratios in Sub-Saharan Africa are amongst the worst in the world. Germany and Austria, for example, have 4.2 and 5 physicians per 1000 people respectively compared to 0.1 and 0.4 for Ghana and Nigeria.[32] Despite increasing levels of urban-dwellers, Africa remains the least urbanised continent. Paradoxically, medical personnel, especially those with specialist training, are concentrated in urban areas.[33] This leaves a significant gap in care in many countries across the continent often exacerbated by poor transport networks. This renders access to healthcare, particularly emergency medical services, difficult. There is consensus that the application of space-based technologies broadly and telemedicine in particular, including the use of AI, could increase access to services, make some healthcare services more affordable to patients, provide savings to the national healthcare budget and lower the number of preventable deaths.

Until recently, policies and initiatives aimed at improving public health in Africa have largely focused on infectious diseases. Research demonstrates that '[S]urgery is the neglected stepchild of global health' in developing countries.[34] Many patients requiring surgery are unable to receive it due to lack of specialists, unaffordability or inability to travel to the city. The Lancet Commission on Global Surgery estimated that there was a deficit of 143 million surgeries in low-income and middle-income countries.[35] In other words, there is a need to add that many surgeries to the current number carried out in order to avoid deformation, disability and hundreds of thousands of preventable deaths. Where surgeries are performed,

---

[28]World Bank, *Poverty and Shared Prosperity Report 2018*; World Health Organization, *supra* note 27; World Bank, "Year in Review 2018 in 14 Charts".

[29]United Nations, Department of Economic and Social Affairs, *World Population Prospects 2017 – Data Booklet*, Data Booklet ST/ESA/SER.A/401 (Population Division, 2017).

[30]World Health Organization, *Global Health Estimates 2016: Estimated Deaths by Cause, Age and Sex*. (World Bank Income Group, 2018).

[31]*Ibid.*

[32]The World Bank: Data, *Physicians per 1000 People, 2016*. World Health Organization's Global Health Workforce Statistics, OECD, supplemented by country data. https://data.worldbank.org/indicator/SH.MED.PHYS.ZS.

[33]United Nations Economic Commission for Africa, *The Demographic Profile of African Countries*, March 2016; Centre for Strategic and International Studies, *Urbanisation in Sub-Saharan Africa*, 2018.

[34]Alvin Powell, "Where Surgery is Lacking", *The Harvard Gazette*, 8 November 2010 https://news.harvard.edu/gazette/story/2010/11/where-surgery-is-lacking/.

[35]John G Meara et al., "Global Surgery 2030: Evidence and Solutions for Achieving Health, Welfare, and Economic Development" (2015) 386:9993 The Lancet 569.

they are sometimes unsafe due to lack of experience of the surgeon, absence of necessary equipment, intermittent electricity or are performed by medically unqualified "operators". Together with Asia, Sub-Saharan Africa has the greatest deficit in needed surgeries.[36] Teleintervention—either with the assistance of a medically trained surgeon or through the use of robotic equipment supported by space-based technologies—could drastically reduce these negative surgery-related outcomes.

Another overlooked area of where space-based healthcare services can be useful is for the health of prisoners. Inmates residing in close proximity to each other make the spread of diseases such as tuberculosis, HIV and syphilis common. These conditions are often left untreated due to sporadic access to community hospitals, a lack of prison doctors or unavailability of prescription medication within prison infirmaries.[37] The potential for these diseases to be transferred to the wider society upon the release of infected inmates from custody make it a significant public health concern.[38] The provision of health care via space-based applications could be of immense benefit in this regard.[39]

5 PAeN and Possibilities

PAeN was initiated and funded by India to provide teleeducation and telemedicine to students and patients in 48 African countries through 169 centres.[40] During its lifetime over 6,000 continuing medical education sessions for doctors and nurses were offered through the platform and 22,000 students obtained various degrees from Indian universities. However less than 800 teleconsultations and teleexpertise meetings were undertaken annually.[41] This pales in comparison to the more than

[36]*Ibid.*

[37]Manop Kanato, "Drug Use and Health Among Prison Inmates" (2008) 21:3 Current Opinion in Psychiatry 252.

[38]Whilst most of the literature in this area are based on findings in North America or Europe, evidence suggests that similar or worse situations exist in African countries see, Richard Smith, "Prisoners: an end to second class health care?" (1999) 318:7189 British Medical Journal 954; Department of Health, *Joint Prison Service and National Health Service Executive Working Group. The Future Organisation of Prison Health Care* (London, 1999); Victor Kwawukume, "Prisons Council to Construct Modern Hospital", *The Daily Graphic* (3 June 2016), online: https://www.graphic.com.gh/news/general-news/prisons-council-to-construct-modern-hospital.html; A A Sarpong et al., "An Assessment of Female Prisoners' Perception of the Accessibility of Quality Healthcare: A Survey in the Kumasi Central Prisons, Ghana" (2015) 5:3 Annals of Medical and Health Sciences Research 179.

[39]Buddhika Senanayake et al., "Telemedicine in the Correctional Setting: A Scoping Review" (2018) 24:10 Journal of Telemedicine and Telecare 669; World Health Organization, *Telemedicine: Opportunities and Developments in Member States: Report on the Second Global Survey on eHealth* (Geneva, Switzerland, 2009).

[40]African Union, *supra* note 2.;Ministry of External Affairs, Government of India, http://www.mea.gov.in/Portal/ForeignRelation/Pan-Project-March-2012.pdf.

[41]African Union, *supra* note 2.

60,000 teleconsultations provided by a telemedicine network in France to French Guyana in 2018.[42] Although the use of PAeN had benefits, the teleeducation component dominated its operations. Clearly, the telemedicine component needs to be scaled-up.

Having operated for eight years, the programme was temporarily suspended in 2017 when operational and financial responsibility was transferred from India to the AU. Whilst it is unfortunate that the programme has come to a standstill primarily due to funding shortfalls, the AU 'now has full ownership of the PAeN's infrastructure and management'.[43] This provides the opportunity to learn lessons and implement corrective action. 'Restoring the operations of the network will therefore be a clear demonstration of the Union's capacity and will to manage and sustain it'.[44] Programmes such as PAeN can make a huge contribution to healthcare delivery in Africa. Indeed a number of promising initiatives have also made a very positive impact. Two of these are discussed here.

The first example was revealed in a 2008 PricewaterhouseCoopers report commissioned by the European Space Agency. It provides a valuable cost-benefit analysis of investment in satellite-based telemedicine and e-health services for public health purposes in Sub-Saharan Africa.[45] Using a Ugandan telemedicine programme aimed at providing services to rural areas as a model, the report found, that continent-wide implementation of the programme could save 151,800 lives and $259 million per year.[46] The combined capital and operating costs of the scaled-up program for one year was estimated at $65 million thus demonstrating an enormous return on investment. Additional health and economic benefits included reduced impact of illness, healthy years of life preserved and improved well-being. This programme alone demonstrated the potential to be gained by an even wider implementation of satellite-based health services.[47]

Another example of the value of space-based healthcare services may be seen in a Nigerian polio project. In 2010, affordable portable GPS devices were used in Nigeria to track the delivery of polio vaccinations.[48] Reporting on the project, Gammino et al., highlight that it not only provided synchronous monitoring of medical delivery but also improved the planning of the operation, both of which contributed to a substantial boost in the numbers of children vaccinated. Traditional methods of vaccine delivery often utilise hand-drawn maps which lack sufficient detail to provide adequate guidance for efficient delivery of vaccines to all

[42]Eurisy, 'Satellite-based healthcare solutions: bringing services closer to patients', 24 April 2019 https://www.eurisy.org/article-satellitebased-healthcare-solutions_40 (accessed 28 May 2019).
[43]African Union, *supra* note 2.
[44]*Ibid* at 7.
[45]PricewaterhouseCoopers, *supra* note 17
[46]*Ibid* at 5.
[47]*Ibid* at 4.
[48]Victoria M Gammino et al., "Using Geographic Information Systems to Track Polio Vaccination Team Performance: Pilot Project Report" (2014) 210 The Journal of Infectious Diseases S98

households'.[49] The study also found that some vaccination teams missed or ignored many households, with some personnel even switching off GPS devices or departing from preapproved locations in an effort to avoid remote or unpleasant neighbourhoods.[50] This revelation would be difficult to unearth without space-based technology. The benefits of modern GPS systems for tracking, auditing and maximising the performance of vaccination teams are clear.

6 Hindrances

There remain significant obstacles to the implementation of sustainable, efficient and effective continental space-based healthcare services on the continent of Africa. The AU and its future partners in this endeavour should take the following concerns into consideration.

6.1 Legal, Regulatory and Policy Framework

Although the International Bar Association submitted a *Draft International Convention on Telemedicine and Telehealth* to the UN in 1999, there is no international treaty which specifically governs the global use of telemedicine.[51] King asserts that,

> [E]ven modest legislative and regulatory reforms create the opportunity for telemedicine as the emerging standard of care for a variety of medical needs, with the right specialists able to serve patients in the right place at the right time, and at the right price-achieving long-term savings and health improvements...[52]

Thus, there is a need to put in place the necessary laws, regulations and policies to govern the use of telemedicine. Patient data may be transmitted via space-based healthcare platforms and shared with various companies, many of which may not be primarily medical in nature. What standards of confidentiality and secure storage of patient information are to apply? Will a medical practice licence held by a Kenyan doctor permit them to provide telemedicine services to a patient in Senegal? Will liability for medical negligence arising out of space-based health services be determined by the laws of the country where the patient was treated or those of the country where the medical expert is based? Who should be responsible in cases of

[49]*Ibid* at S100 For the prevalence of hand-drawn maps see also, Ed Yong, "Most Maps of the New Ebola Outbreak Are Wrong", *The Atlantic* (21 May 2018), online: https://www.theatlantic.com/health/archive/2018/05/most-maps-of-the-new-ebola-outbreak-are-wrong/560777/.

[50]Gammino et al., *supra* note 49 at S100.

[51]Marlene M Maheu et al., *The Mental Health Professional and the New Technologies: A Handbook for Practice Today* (Mahwah, New Jersey: Lawrence Erlbaum Associates, 2011) at xviii and 451.

[52]Michael W King, "Telemedicine: Game Changer or Costly Gimmick" (2018) Denver Law Review 289.

defective hardware or loss of satellite connection during robotic surgery?[53] Video/audio consultations may be recorded, for example, by medical personnel to be used as a teaching tool in medical schools or by patients for litigation purposes. Can consent to being recorded be implied or must it be express? Should telemedicine services provided by a health professional in one country qualify for reimbursement by the national health service of another? Faced with these questions some countries have adopted guidelines. In South Africa, for example, the regulatory body for health professionals created and adopted the *General Ethical Guidelines for Good Practice in Telemedicine* in August 2014.[54] Whilst this is useful for services provided entirely within national borders, it is less so when the services involve doctors or patients in different countries which may have different laws and regulations or none at all.

6.2 Technical Expertise, Research & Development

Of the 54 countries in Africa, only 8 have satellites in space.[55] Moreover, only 40% of the 35 African satellites have been built by African engineers, and not all of them in Africa.[56] There are also insufficient numbers of professionals with the knowledge and skills needed to develop software for delivering space-based telemedicine or the necessary information management systems. This reveals the anticipated low participation of African engineers and technical experts in Africa's burgeoning space industry. This is particularly worrying given the growing numbers of cooperation agreements concluded between African and foreign countries or private companies located therein. Nkengasong et al., note that, '[T]he continent's ability to generate, warehouse, and use quality public health data in real time is a challenge'.[57] Thus, the AU objective to embark upon '...an education and skills revolution emphasizing science and technology...' is encouraging.[58] The promotion, utilisation and funding of national institutions such as the Space Science and Technology Institutes of Ghana and Ethiopia, Nigeria's National Space Research and Development Agency.[59] These are critical to the development of the knowledge, skills and

[53]A Le Roux, "Telemedicine: A South African Legal Perspective" (2008) Journal of South African Law.

[54]Health Professions Council of South Africa, *General Ethical Guidelines for Good Practice in Telemedicine*, Pretoria, South Africa, August 2014 https://www.hpcsa.co.za/Uploads/editor/UserFiles/downloads/conduct_ethics/Booklet%2010.pdf.

[55]Space in Africa, "Ivory Coast to Become the 9th African Country with Satellite in Space", 26 July 2018 https://africanews.space/ivory-coast-to-become-the-9th-african-country-with-satellite-in-space/.

[56]Space in Africa, *supra* note 26.

[57]John Nkengasong, Benjamin Djoudalbaye, & Olawale Maiyegun, "A New Public Health Order for Africa's Health Security" (2017) 5 The Lancet e1064.

[58]African Union, *Agenda 2063: Framework Document*, pg.v, September 2015.

[59]Ethiopian Space Science and Technology Institute http://etssti.org/establishment/; Ghana Space Science and Technology Institute https://gssti.gaecgh.org; National Space Research and Development Agency https://nasrda.gov.ng/en/tenders-eoi/.

attitudes necessary for the successful roll-out of a space technology-driven digital health system continent-wide.

6.3 Financial

With the start-up costs of PAeN having been paid for by India, the AU is left with financial responsibility for the operation and day to day maintenance of the network. The annual operating costs of a revived PAeN are estimated at just over $4 million for which the AU has requested the 48 participating Member States to each contribute $90,000.[60] This amount should be affordable for all African nations, though it is unclear whether guarantees of payment have been given by governments. Beyond these meagre costs however are those expected for sporadic serious refurbishments and upgrades. The ability of the AU to generate funds perhaps by raising taxes (on cigarettes, alcohol and other unhealthy food products and behaviours which contribute to bad health outcomes) and engaging the international community for funding agreements will be key to ensuring that there are no further suspensions, should PAeN get up and running again.

6.4 Awareness and Actual Utilisation

A recent study by Dietrich et al., found that widespread implementation of telemedicine services was hampered by a deficiency in awareness by both those in the space and medical fields and by a lack of 'space-associated skills and knowledge' by health workers.[61] Services through PAeN were however delivered by Indian service providers very aware of the value of the technology. Whilst many African doctors are aware of space-based healthcare services, the number of nurses and/or allied health professionals with knowledge of or an interest in such services, especially in rural areas, is unclear. Indeed, some studies have found that a significant number of healthcare personnel are either unaware of the use of space science and technology for medical purposes or unfamiliar with its benefits.[62] For example, in the Eastern Cape Province of South Africa, 'Despite large investments from the National Department of Health, only a third of telemedicine sites in the province are operational.'[63] Research has also shown that some healthcare workers whilst viewing telemedicine as positive were nevertheless reluctant to use it for a

[60]African Union, *supra* note 2.

[61]Damien Dietrich et al., "Applications of Space Technologies to Global Health: Scoping Review" (2018) 20:6 Journal of Medical Internet Research e230.

[62]Maurice Mars, "Telemedicine and Advances in Urban and Rural Healthcare Delivery in Africa" (2013) 56:3 Progress in Cardiovascular Diseases 326.

[63]Liezel Cilliers & Stephen Flowerday, "User Acceptance of Telemedicine by Health Care Workers A Case of the Eastern Cape Province, South Africa" (2014) 65:1 The Electronic Journal of Information Systems in Developing Countries 1 at 1.

variety of reasons, including legal and other concerns.[64] Importantly, with respect to PAeN, research conducted in Ghana revealed that staff did '…not even know that the eNetwork exists'.[65]

6.5 Ethical and Other Unintended Consequences

Borgetti et al., rightly point out that 'physicians providing care via technology have the same ethical responsibilities as those providing care in brick and mortar settings'.[66] Yet, concerns remain that telemedicine could result in more negligent, irresponsible or outright criminal medical practices in comparison to traditional face-to-face doctor-patient interactions.[67] Further, some studies imply that telemedicine could inadvertently result in groups such as the elderly or the poor being excluded from accessing much needed services due to an aversion to technology or the inability to afford internet services.[68] There have even been suggestions that telemedicine paves the way for uncontrollable sale of dangerous prescription drugs.[69] An important advantage of telemedicine is the use of allied health personnel and technicians to perform some of the remote assessments and data collection for transmission to experts for interpretation.[70] However, these creative and flexible practices may create opportunities for charlatans to evade detection or sanction in a telemedicine environment. Thus, the use of space-based health services raise regulatory issues and ethical dilemmas which must be seriously considered.

7 Influencing the Future

Article 16 of the *African Charter on Human and Peoples' Rights* ('Banjul Charter') places an obligation on State Parties to 'take the necessary measures to protect the health of their people and to ensure that they receive medical attention when they

[64]Cilliers & Flowerday, *supra* note 64; S Nwabueze et al., *The Effects of Culture of Adoption of Telemedicine in Medically Underserved Communities* (Hawaii: International Conference on System Sciences., 2009); Mars, *supra* note 63.
[65]Afarikumah & Kwankam, *supra* note 19 at 83.
[66]Scott A Borgetti, Philip J Clapham & Jeremy D Young, "Telehealth: Exploring the Ethical Issues" (2017) DePaul Journal of Health Care Law.
[67]Ateriya et al., *supra* note 12.
[68]Borgetti, Clapham & Young, *supra* note 67 cf. Nola M Ries, Briony Johnston & Shaun McCarthy, "Technology-Enabled Legal Service Delivery for Older Adults: What Can Law Learn from Telehealth: Findings from an International Review of Literature" (2016) Elder Law Review
[69]Madeleine Rosuck, "Telemedicine Is the New Narcotics Candy Store: Teladoc Opens the Floodgates for the Unrestricted Sale of Dangerous Drugs Case Notes" (2018) SMU Science and Technology Law Review 89.
[70]Caroline Klimek, "Using Outer Space in the Fight Against Ebola", Faculty of Medicine, University of Toronto, 20 August 2015 https://medicine.utoronto.ca/news/using-outer-space-fight-against-ebola (accessed 3 May 2019).

are sick'. Several pan-African policies have made efforts in this regard. The *African Health Strategy 2016–2030*, for example, recommends standardisation of various aspects of medical care and harmonisation of public health legislation. It notes improvements in several health indicators and outcomes across the continent but highlights that Africa 'still faces an urgent need to accelerate progress' and specifically refers to the need to establish 'effective systems for disease surveillance and disaster management'.[71] Yet, there is only a passing reference to telemedicine and no detailed analysis of how outer space may assist in this regard.

The use of outer space to drive development in Africa is well-captured in the AU's *Agenda 2063* and consolidated in the *African Space Strategy*.[72] Whilst both policy documents mention several benefits of telemedicine, neither set out a comprehensive plan to incorporate space-based healthcare services at the regional level. To move space-based health services forward, substantial political will is required. Governments will have to be driven to prioritise the funding and implementation of the many regional and international policies relating to the subject. Strategies to ensure this happens are paramount.

8 Recommendations

In its efforts to re-establish PAeN, the AU, its partners and stakeholders must be cognizant of areas for improvement. The following six recommendations will contribute to an "indigenous space sector". This should form part of "an integrated, prosperous and peaceful Africa, driven by its own citizens and representing a dynamic force in the global arena".[73]

First, where necessary, existing regulations should be amended to take into account the particularities of space-based healthcare services. It is important to review pre-existing laws throughout the continent to ensure they do not impede the operation of space-based telemedicine programmes as was, for example, the case in France.[74] Further, laws regulating telemedicine need to appropriately interconnect with regulations imposed by national medical regulatory councils as well as laws pertaining to traditional African medical practices, consumer protection, commerce, data protection and use as well as electronic communications for example. The African Union has previously sought to provide a harmonised framework in the

[71]African Union, *African Health Strategy: 2016-2030*, pp. 7 and 12 https://au.int/sites/default/files/documents/30357-doc-final_ahs_strategy_formatted.pdf

[72]African Union, *African Space Strategy: Towards Social, Political and Economic Integration*, HRST/STC-EST/Exp./16 (II) (Cairo, Egypt: Second Ordinary Session for the Specialized Technical Committee Meeting on Education, Science and Technology, 2017).

[73]African Union, *African Space Strategy: Towards Social, Political and Economic Integration*, HRST/STC-EST/Exp./16 (II) (Cairo, Egypt: Second Ordinary Session for the Specialized Technical Committee Meeting on Education, Science and Technology, 2017).

[74]Eurisy, 'Satellite-based healthcare solutions: bringing services closer to patients', 24 April 2019 https://www.eurisy.org/article-satellitebased-healthcare-solutions_40 (accessed 28 May 2019).

medical field through its adoption of the *AU Model Law on Medicines Regulation*. It is suggested that the AU similarly seek to secure harmonisation of legislation, regulations and practice guidelines pertaining to telemedicine across the continent. It is also imperative that any such AU-wide policy or treaty pertaining to space-based healthcare services address questions regarding medical practice licences, liability for medical negligence, consent, practice standards, security breaches, national health insurance coverage, confidentiality and privacy. It is also crucial to avoid a two-tier system in which local medical personnel are subjected to a different set of practice standards and legal requirements from those medical personnel providing services outside of the respective country. Issues relating to safety, misleading or false claims and deceptive practices by non-qualified persons on telemedicine platforms need to be legally prohibited. Relatedly, legal safeguards for patients who have been harmed by doctors and other telemedicine practitioners should be robust and applicable beyond national boundaries to ensure fast and effective access to justice for both patients and service providers irrespective of their geographical location.[75]

Second, the continent requires an injection of a variety of professionals including physicians and researchers, allied health practitioners, aerospace engineers, computer and AI scientists, lawyers with expertise in data protection and medical and space law, information management specialists and others to envision, develop, execute, maintain and advance space-based health services. The recent establishment of offices in Africa by a number of tech giants such as Google, Microsoft, Cisco and others will hopefully contribute to a reduction in the technology and skills gap. This must be augmented by African-owned and African-based tech-hubs to spur local innovation.[76] Medical education itself must be adapted to include telemedicine practice principles. This will ensure that the modern medical graduate is better equipped for a career or practice in the telemedicine environment. In particular, Africa-wide recognition of medical qualifications and training should be seriously considered as a central part of the drive to improve telemedicine services across the continent.

Third, a strategy must be developed to multiply African ownership - whether by the AU, African states individually or African companies - of African-built satellites. This would have the important objective of reducing dependence on foreign states or corporate entities, improve cost-benefit ratios and enhance national/regional security. Relatedly, the AU could seek to capitalise on the advantageous equatorial locations of several African countries for space launches. The development of an African spaceport could generate significant revenue, a portion of which

[75]Richard FitzGerald, "Telemedicine doctors abroad don't have to register with the GMC" (2012) 344 British Medical Journal e873.

[76]Julia Selman Ayetey, "Ghana must use space for national development", *The Daily Graphic*, 20 February 2019 https://www.graphic.com.gh/features/features/ghananews-ghana-must-use-outer-space-for-national-development-2.html (accessed 4 May 2019).

could be ringfenced for PAeN and other space-based healthcare initiatives as well as to fund research and development efforts in the fields of STEM and medicine.[77]

Fourth, the success of any telemedicine initiative, especially one as ambitious as "PAeN 2.0", if it may be called that, requires stakeholders to be persuaded to utilise and develop space-based healthcare services. Thus, campaigns targeted at raising awareness of space-based telemedicine applications amongst patients, medical personnel and the general public as well as space science, AI and technology researchers should be established. This can help eliminate barriers to adoption and acceptance of telemedicine by medical personnel and patients.[78] Events such as the annual Innovation Science Technology Africa conference and Ghana Tech Summit may be used as a conduit to raise awareness of using outer space for health and to provide networking opportunities for collaboration.[79] PAeN should also be integrated within national health services as a regular means of service delivery where appropriate.

Fifth, whilst it has been suggested that the AU be the centre of a continental space-based healthcare initiative, the use of telemedicine should neither be limited to PAeN nor should its use be limited to the public sector. The maintenance, advancement and success of such a huge endeavour requires the skills, knowledge and support of the private sector, particularly space tech start-ups, app developers, and non-governmental organisations both domestic and international. The involvement of international governmental and non-governmental organisations such as COPUOS, the International Telecommunication Union, the World Health Organisation, the World Bank and the International Law Association would be advantageous. Civil society including patient advocacy groups, medical associations, academia and representatives and trade union groups for doctors, nurses and allied health professionals all have a meaningful role to play in the development and maintenance of telemedicine services across the continent.

Sixth, the interoperability of both software and hardware across borders is a necessary objective of an AU telemedicine network. It is important that future African partners, whether private healthcare facilities or district hospitals for example, are equipped with infrastructure that is compatible with that used by PAeN. New systems are always being developed, some specifically for telemedicine purposes. Where possible, hardware and software developed in the future for use by both private and public sector in Africa should be capable of integrating with PAeN and other telemedicine platforms.

[77]Though Algeria and Kenya have spacesports they have been inactive for more than 10 years, see Thomas G Roberts, *Spaceports of the World*, CSIS Aerospace Security Project (Centre for Strategic & International Studies, 2019).

[78]Afarikumah & Kwankam, *supra* note 19; Paul J Hu et al., "Examining the Technology Acceptance Model using Physician Acceptance of Telemedicine Technology" (1999) 16:2 Journal of Management Information Systems 91.

[79]Innovation Science Technology Africa http://www.ist-africa.org; Ghana Tech Summit http://ghanatechsummit.com.

9 Conclusion

The prosperity of Africa depends to a large degree on achieving massive improvements in the health status of its citizenry. To date this has proved a massive challenge. Governments and peoples of AU Member States must begin to see space-based healthcare as a critical component of overcoming this challenge. It is acknowledged that space science and technology will not eliminate Africa's health problems. However, space-based healthcare services can be a substantial part of the solution, particularly if used in conjunction with traditional healthcare delivery systems such as clinics and hospitals. Opportunities exist to foster an integrated (space-based and traditional) healthcare system which has the ability to enhance the quality of care received by patients while generating economic growth across sectors.

The efficient and effective use of a continental telemedicine regime depends considerably on surmounting legal and regulatory challenges arising from this web of disciplines. Harmonisation of medical and telemedicine laws, regulations and policies across AU Member States would provide safeguards to ensure that the practice of telemedicine has patient protection at its core. Space-based healthcare pose technical, physical, and ethical dilemmas and may also result in unintended consequences, but most agree the benefits outweigh the risks - most of which can be minimized. An increase in STEM education, support for technology start-ups and the creation of innovation hubs in an effort to design, build, launch and track African satellites would also enhance Africa's position on the international space stage and make it less reliant on foreign-owned companies or other states.

Africa must entrench and enhance the development and implementation of space technologies for healthcare. The successful re-introduction of PAeN requires the political will and financial support of AU Member States who should see PAeN as an indispensable strategic pathway to the achievement of the ambitious health and other socio-economic objectives laid out in various African policies.

Julia Selman Ayetey is a Barrister called to the Bar of England and Wales (*Middle Temple*) and is a Solicitor and Barrister of the Supreme Court of Ghana. She has an M.Phil from the University of Cambridge and is a Senior Lecturer at the Faculty of Law, University of Cape Coast. Julia has a longstanding interest in the intersection between law, ethics, science and technology and has been an advisor to the government of England & Wales as a former member of the *National DNA Database Ethics Group*. Currently she is pursuing her doctorate at the Institute of Air and Space Law, McGill University where her research examines the relationship between non-state actors and international space law and governance.

Harold Ayetey holds a Medical degree and a Ph.D. from the University of Cambridge. He trained in General Internal Medicine in London and Oxford before returning to Cambridge as a Wellcome Trust Clinical Research Fellow and Honorary Registrar in Cardiology. He subsequently completed subspecialty training in Advanced Cardiac Imaging at the Royal Brompton Hospital in London. He is a Senior Lecturer and the immediate past Head of the Department of Medicine and Therapeutics at the University of Cape Coast in Ghana and a Consultant Physician and Cardiologist at the Cape Coast Teaching Hospital. His clinical and research interests include cardiovascular imaging, digital medicine and artificial intelligence in healthcare.

Addressing the Un-Addressed: Opportunities for Rural-Africa

Christoffel Kotze

Abstract
In the developed world ownership of a residential address is almost taken for granted, people grow up with the concept, it is used daily in many processes in an almost unconscious way. Mail depends on it to get delivered, so does the courier service delivering e-commerce procured and it provides a pick-up location for a ride sharing service etc. It forms the foundation for legal identification in most developed nations, however for a large portion of the global population it is not the case. Currently development in both terrestrial and space-based information and communications technology (ICT) provides opportunity for potential solutions to provide alternative approaches to mitigate the "address problem". This work aims to provide a short historical overview of the concept with reference to the postal system, investigates technology options, reviews a number of currently available solutions and proposes another conceptual solution—with a special focus on the rural un-addressed in sub-Saharan Africa. It also aims to create awareness of the highly important role played by physical address systems in modern society by highlighting its economic and societal impact against the background of its contribution to progressing the 2030 UN Sustainable Development Goals.

C. Kotze (✉)
NOEZ Consulting & Design, Yzerfontein, South Africa
e-mail: chris@noez.co.za

A. Froehlich (ed.), *Space Fostering African Societies*, Southern Space Studies,
https://doi.org/10.1007/978-3-030-32930-3_11

1 Introduction

Address systems have been around in one form or the other since ancient times, as postal systems and address systems are as the idiom says "joined at the hip". It was handy to see what the Universal Postal Union (UPU)[1] had to say about what constitutes an address. The UPU, one of the oldest surviving international organisations in the world (founded in 1874) is based in Switzerland and serves as a cooperation forum to set the rules for international mail exchanges. Though the UPU is mostly concerned with the physical aspect of what is seen as traditional mail, it acknowledges the broad nature of what can considered to be an address, describing it as "the underlying threads connecting all different actors and their activities, effectively functioning as a network of networks."[2] Borrowing from the aforementioned the following definition for "address" was defined for the purposes of this paper:

> Any mechanism with the basic function of allowing a sender to effectively deliver a physical object to the intended recipient.

In 2009 the UPU launched an initiative—"Addressing the world – An address for everyone"[2] to create awareness of the true value address systems add to the creation of urban and rural infrastructure of countries. Address systems enable effective communication networks between different entities constituting social and economic systems; individuals, commerce, governmental and non-governmental organizations and are essential to the integrity of such systems. At the level of the individual, a residential address forms part of the identification system, which, in turn allows participation in the formalised economy. A residential address is quite often is a prerequisite to vote and the conduit to a variety of other social services as shown in Fig. 1. Importantly it also serves as the anchor between the physical and virtual worlds—a quintessential requirement of retail e-commerce activities.

In 2018 the retail e-commerce market topped USD 2.8 trillion—a figure which is projected to exceed USD 4.8 trillion by the year 2021.[2] The lack of a residential address system therefore is bound to have an impact on the social and economic development on individual and organisational levels as it will limit the ability to interact with any formalised structures that requires a legal address. It is estimated that more than four billion people may be impaired in one way or the other directly as the result of lack of identity and subsequently the lack of a formalised address.[3] The purpose of this work was to investigate how emerging technology, and most notably space based systems, could be harnessed to "address the un-addressed."

---

[1]Universal Postal Union, 'Addressing the world initiative,' http://www.upu.int/en/activities/addressing/addressing-the-world-initiativehtml accessed 2 4 2019.

[2]Statista, 'E-commerce worldwide - Statistics & Facts 2018' https://www.statista.com/topics/871/online-shopping/ accessed 22 February 2019.

[3]Commission on legal empowerment of the poor, 'Making the Law Work for Everyone: Working Group Reports (Vol 1),' United Nations Development Programme, New York, 2008.

Fig. 1 Importance of address infrastructure (Universal Postal Union. 'Addressing the world initiative,' http://www.upu.int/en/activities/addressing/addressing-the-world-initiative.html accessed 22 February 2019)

2 The Postal System

One cannot look at address systems without invariably including one of the oldest known practical applications of an address, namely the postal system. The purpose of a postal service has not changed at all from ancient times to the current; transport of an object i.e. an envelope between sending and receiving parties by means of a trusted third-party service. This service also introduced the concept of the Universal Service Obligation (USO) to the world, credited to Rowland Hill, a British inventor who proposed a global mail service in 1840, which resulted in the introduction of the Penny Post.[4] This section aims to serve as a brief introduction of address systems viewed against the background of the postal system's historical development.

[4]The Economist, 'Amazon is not the only threat to legacy post offices,' https://www.economist.com/business/2018/04/19/amazon-is-not-the-only-threat-to-legacy-post-offices accessed 21 April 2019.

2.1 Postal System—Historical Overview

The postal system in essence owes its existence the invention and adoption of the written language, believed to have developed independently, in three general areas, China, Mesoamerica and the Near-East. The Mesopotamian cuneiform script invented circa 3200 BCE (Before Common Era) is considered to be the oldest writing system.[5] As empires developed and grew, it became important for the ruling centre to regularly communicate with the regions to ensure effective administration —enter the postal system. Using the Nile as primary transportation to deliver papyrus letters, the first postal system was established in Egypt (circa 2000 BCE) for the exclusive use of the pharaohs.[6] China introduced the posthouse relay system (circa 1000 BCE, during the Zhou dynasty), to exchange mail delivered by couriers between pre-determined points located between the sending and the receiving locations.[7] The Roman contribution to the mail system was less revolutionary than evolutionary, borrowing from existing systems but taking it to a much larger scale. One of the key contributions of the Roman empire was indeed its road system, a number of which still exist today. It is estimated that more than 80,000 km of Roman highways were built in all parts of the empire: all the way from Britain, throughout North-African and extending to the East Roman territories.[8] The primary purpose was supporting the military and trade, but it also provided the ideal infrastructure for effective administration via a postal system introduced by Emperor Augustus circa 30 BCE. It was only used to carry mail for official purposes using horses and carriages on the established Roman road network, in a relay fashion.[9] The Roman road network included named roads and milestones detailing information such as distance to the next important destination; it did not feature a formal street address system. Rather than the numbering system in use today, streets in large cities and towns were arranged in a gridiron pattern around important buildings and key landmarks used as reference markers used to locate a building. This concept is still used in many places lacking a street address system.

The "Age of Enlightenment" (1685–1815)[10] in Europe was characterised by the questioning of "the traditional" and a quest to improve the human condition through change leading to a number of ground-breaking inventions, house numbering being one Though the idea of a formal centralised addressing system was not necessarily

[5]Denise Schmandt-Besserat, 'University of Texas, Austin' 2014 https://sites.utexas.edu/dsb/tokens/the-evolution-of-writing accessed 12 March 2019.

[6]Nour Eltigani, 'The Egyptian Postal System: Oldest Postal System in the World', Egyptian Streets, https://egyptianstreets.com/2018/10/01/the-egyptian-postal-system-oldest-postal-system-in-the-world accessed 22 March 2019.

[7]Andrew Brix, 'Postal system' Encyclopedia Britannica https://www.britannica.com/topic/postal-system#ref367051 accessed 22 March 2019.

[8]Alison Eldridge, 'Roman road system' Encyclopaedia Britannica, https://www.britannica.com/technology/Roman-road-system accessed 22 January 2019.

[9]British postal museum, '500BC - Roman Postal System' Bath Postal Museum, https://bathpostalmuseum.org.uk/500bc-roman-postal-systemhtml accessed 22 March 2019.

[10]History, 'Enlightenment' History Channel https://www.history.com/topics/british-history/enlightenment accessed 27 April 2019.

motivated by the desire to improve the lives of the general populace, but rather as a tool of control for the state, it did make it easier for all to find a location. The mail system benefited greatly from this development and addressing is still core to any mail system today. Though the basics remain in place, delivery technology has changed to enable efficiency (postal systems have been early adopters of new technology on a number of occasions). From carrier pigeon to the current highly converged state-of-the-art technology, the postal system has long since demonstrated its ability to adopt new technology (mostly developed for other industries and purposes) to its advantage to help it fulfil its mandate better. The first air-mail occurred in 1911 when thousands of mail pieces were transported using a powered aeroplane over the river Yamuna in India.[11] As new delivery methods became available, it allowed the postal system to introduce innovation and choice to its customers i.e. Depending on the willingness of the customer to pay for an expedited service, a letter could be sent by airmail or by slower means. On the operational side too it also benefited from the adoption of new technology to improve sorting and routing efficiency; barcoding, optical character recognition and radio frequency identification to name a few. The constant pace of technological improvement is increasingly blurring the lines between the physical and the virtual Increasingly the virtual and physical addressing has been merging to present new opportunities not possible a decade or two ago. Hybrid mail and virtual addressing is perhaps the most well-known examples of the blending of the physical with the virtual. Hybrid mail allows an e-mail user the ability to upload content to a provider, where it is converted to a physical version for delivery to the recipient. The use of unmanned aerial vehicles (UAVs) commonly known as drones for mail delivery are being investigated by a number of postal services around the world including France, Japan and South Africa.[12] Autonomous vehicles, the "sharing" economy, 3D printing and cryptocurrencies are all developing areas which are expected to be adopted in one way or the other by global postal systems. Though the traditional postal system is under threat today in the views of some observers, it still remains remarkably efficient in developed countries. Approximately 700 million items were handled and delivered per day in the USA by the US Postal Services in 2011[13] versus just under 485 million pieces of mail in 2019 representing 234 million USD of revenue per day.[14] The postal system is still well positioned to serve as integrator of an increasingly complex society, however to stay relevant though it must be able to cater for the demands of a rapidly changing socio-economic stakeholder ecosystem including the needs of the un-connected poor including the un-addressed.

[11]Cheryl R Ganz, 'India and the World's First Official Air Mail by Airplane' Smithsonian National Postal Museum, https://postalmuseum.si.edu/collections/object-spotlight/india-air-mailhtml accessed 14 April 2019.

[12]Ray Mahlaka, 'SA Post Office turnaround staked on e-commerce and drone deliveries' https://www.moneyweb.co.za/news/companies-and-deals/sa-post-office-turnaround-staked-on-e-commerce-and-drone-deliveries/ accessed 22 February 2019.

[13]Adam Hartung, 'Why the Postal Service Is Going Out of Business' https://www.forbes.com/sites/adamhartung/2011/12/06/why-the-postal-service-is-going-out-of-business/#3ce635314317 accessed 13 February 2019.

[14]USPS, 'Postal Facts - One Day in the Life of The US Postal Service' https://facts.usps.com/one-day/ accessed 24 April 2019.

Table 1 Istanbul World Postal Strategies (UPU, 'Strategic Direction and Analysis' Istanbul World Postal Strategy 2017–2020, Berne, 2016)

Istanbul world postal strategy goals		
IWPS Goal1	Improve the interoperability of network infrastructure	
	Ensure quality of service	Efficient and secure supply chains
	Develop standards for ICT's	Streamline process and services
IWPS Goal 2	Ensure sustainable and modern products	
	Modernize and integrate products and remuneration systems	Accelerate e-commerce product development
	Diversification of products and service portfolio	Support trade facilitation via the postal network
IWPS Goal 3	Foster market and sector functioning	
	Contribute to the definition and development of the USO (Pape G. Toure, 'Basic Principles of Universal Service' Costs and Tariffs for Member Countries of the TAF Group, 23–25 April 2001, Niamey (NIGER), 2001)	Initiating/implementing relevant development cooperation measures
	Encourage market sector information	Improve policy and regulation efficiency
	Mobilize environmental sustainable development	

2.2 UN Sustainable Development Goals

The UN 2030 SDG's is an action plan promoting "people, planet and prosperity", primarily through the eradication of poverty, to place the world onto a more "sustainable and resilient path", through the implementation of 17 Sustainable Development Goals guided by a set of measurable targets. As the postal sector is dependent on a functioning address system, the role it plays as a vehicle to reach the United Nations 2030 Agenda for Sustainable Development (UN 2030 SDG)[15] can thus also be seen as that of an address system. The global postal system has positioned itself to support the UN SDG's via its own internal strategic goals. Three strategies were adopted at the 2016 UPU congress in Istanbul, known as the Istanbul World Postal Strategies (IWPS), to address the way in which a well-coordinated global system could contribute to economic and societal development. Table 1 provides a summary of the aforementioned goals, all of which will be dependent on a ***functional address system*** in one way or the other.

[15]UN SDG knowledge platform, 'Transforming our world: the 2030 Agenda for Sustainable Development' https://sustainabledevelopment.un.org/post2015/transformingourworld accessed 21 December 2018.

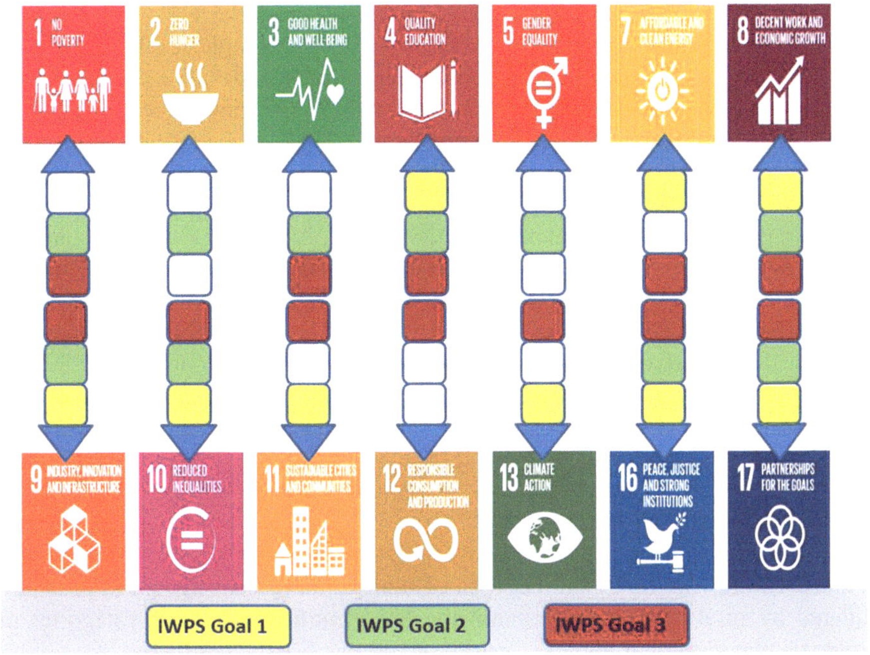

Fig. 2 IWPS Contribution to UN SDG 2030 (UPU, 'The United Nations 2030 Sustainable Development Agenda—Contributions of the Postal Sector and the UPU,' 2015)

Figure 2 provides a graphical representation of how the IWPS goals are positioned in relation to the UN 2030 Agenda for Sustainable Development, associated with support of fourteen out of the seventeen sustainable development goals.

3 The "Un-Addressed"

The most familiar form of a location address system is the concept of a street address. According to the World Bank[16] it enables four basic activities, without which local government becomes challenging to say the least:

I. Provides a basic ability to collect information about cities/towns which in turn can be used to update basic urban planning
II. Planning of investments in the city or town
III. Maintaining facilities and infrastructure
IV. Effective use and mobilisation of local resources.

[16]Catherine Farvacque-Vitkovic, Lucien Godin, Hugues Leroux, Florence Verdet, Roberto Chavez, 'Street addressing and the management of cities' The World Bank Washington DC, 2005.

The quite often quoted figure of 4 billion people from the UPU's white paper "Addressing the world – An address for everyone" refer to people "excluded from the rule of law" as a consequence of the "lack of proper addressing". For people who are unaddressed in the sense that they do not have a formal location address no accurate figures exist, it is estimated that this number are likely to be in the "billions."[17] According to a 2017 United Nations Department of Economic and Social Affairs (UNDESA) report, the global population is projected to reach 86 billion[18] by 2030—the target year for implementation of the UN SDG's. A point to note is a significant growth expected in the population of Africa, with the population of 26 countries projected to double their population by 2050, a significant number of these also representing some of the poorest countries globally. Currently Africa also represents the most "un-connected" society globally. Technological development can be harnessed to drive innovation in the postal sector to enable it to adapt to a market with changing demands characterised by a changing socio-economic environment.

3.1 Challenges of Rural Areas

A study commissioned by the European Commission to investigate the problems suffered by rural communities within the EU identified four main categories of problems plaguing rural areas[19]:

I. Demography—rural areas are typically inhabited by a population over-represented by older people with a diminishing young populace due to urban migration, often leading to underperforming local economic performance.
II. Remoteness—which makes it more difficult to provide and maintain good infrastructure, in turn leading to underperforming economic activity and motivating urban migration, consequently serving as a counter incentive to improve said infrastructure.
III. Education—is typically of a lower level amongst the rural populace, a causative factor in a number of the problems experienced by such areas, lower employment, economic opportunity and increased poverty. Due to a lack of infrastructure, the chance to obtain a better education is diminished, adding another incentive to leave the area for those who can.
IV. Labour market—is limited for the inhabitants of rural areas, the confluence of the other three factors described above. Qualified people leave the area as there is no opportunity, which prevents investment in the area due to a lack of a capable available labour force.

[17]Karen Hao, 'MIT," MIT Technology Review' https://www.technologyreview.com/s/612492/four-billion-people-lack-an-address-machine-learning-could-change-that accessed 11 March 2019.
[18]UNDESA, 'World Population Prospects' UN DESA, New York, 2017.
[19]Paola Bertolini, Marco Montanari and Vito Peragine, 'Poverty and social exclusion in rural areas' 2008.

Though this study was done on developed countries of the EU and EEA[20] and might not be directly applicable to developing countries, it does provide a valuable point of departure to approach challenges affecting the developing world's rural development.[21] One of the many challenges does include the lack of availability of formalised address systems. This is most likely the product of a number of causative factors, including large distances that need to be covered in order to reach remote rural areas.

4 The Technology Toolbox

Just as the postal system continually demonstrates its ability to adopt new technology to fulfil its mandate, the current unprecedented level of simultaneous development on many fronts is presenting opportunity to resolve the address problem. This section briefly looks at these areas of interest which has enabled development of address solutions, it is by no means exclusive but represents key enablers.

4.1 Augmented Reality

Augmented reality (AR) uses a bouquet of converged technologies and can facilitate the display of useful data "mapped" onto the user's physical environment in real-time. The technology is seen as a paradigm shifting technology with a potential impact magnitude analogous to the introduction of the internet and the smartphone.[22] One of the more well-known examples of augmented reality (AR) is the Pokémon GO[23] smartphone game, launched in 2016 for Android and iOS. Encouraging users to progress through the game by finding virtual objects mapped to "real-world" locations the game requires a user to have active internet connectivity and a handset supporting satellite location technology to function optimally. Though the game has emphasized some of the negative consequences of virtual

[20]European Parliament, 'The European Economic Area (EEA), Switzerland and the North.' http://www.europarl.europa.eu/factsheets/en/sheet/169/the-european-economic-area-eea-switzerland-and-the-north accessed 22 April 2019.

[21]Paolo Bertolini, 'Overview of Income And Non-Income Rural Poverty In Developed Countries' in Expert Group Meeting on Eradicating Rural Poverty to implement the Agenda for the 2030 Agenda for Sustainable Development, Addis Ababa, 2019.

[22]Gregory Curtin, '6 ways augmented reality can help governments see more clearly' https://www.weforum.org/agenda/2017/02/augmented-reality-smart-government accessed 22 April 2019.

[23]Pokemongo, 'Gt U And Go' https://www.pokemongo.com/en-us accessed 25 February 2019.

addressing, resulting in a number of legal disputes in US,[24] it did prove how effective the use of the AR technology could potentially be Just how well the game morphed the virtual with the physical can be found in one particularly interesting statistic, it is estimated that in the year the game was launch, players walked a collective 8.7 billion kilometres in pursuit of the virtual tokens.[25] Not surprisingly, augmented reality technology leans itself to address applications in rural areas to aid existing landmark-based systems.

4.2 Machine Learning

Digital transformation is often defined as "The application of digital capabilities to processes, products, and assets to improve efficiency, enhance customer value, manage risk, and uncover new monetization opportunities."[26] Digital capabilities represented by the so-called "SMACT" group of technologies; Social Media, Mobile Tech, AI & Analytics, Cloud & IoT, is currently considered to be the main enabler of digital transformation. Within this space, AI is staring to become a key player as it facilitates the data analytics process crucial to identifying new opportunities. Professor John McCarthy considered to be one of the founding fathers of the science of Artificial Intelligence (AI) described it as "It is the science and engineering of making intelligent machines, especially intelligent computer programs."[27] The field of AI is generally considered to be a nested concept, with AI spanning the fields of Machine Learning and Deep Learning in an inclusive nested fashion. Machine learning is an important tool from the point of view that it allows a high degree of automation in the process enabling the machine to "learn" on its own with very little intervention. There are many definitions for machine learning, but it all involves in one way or the other the ability of machines to analyse many different types of data with the goal to determine patterns. These patterns could be applied to provide solutions to real-world problems. Machine learning is already widely deployed by many multinational corporations in the form of virtual assistant and "chat bots."[28]

---

[24]A.J. Dellinger, 'Pokémon Go' settlement promises action on nuisance Pokéstops,' https://www.engadget.com/2019/02/15/niantic-pokemon-go-trespassing-lawsuit-settlement accessed 29 March 2019.

[25]Iqbal Mansoor, 'Pokémon GO Revenue and Usage Statistics' http://www.businessofapps.com/data/pokemon-go-statistics accessed 30 April 2019.

[26]Bill Schmarzo, 'What is Digital Transformation' https://www.cio.com/article/3199030/what-is-digital-transformation.html accessed 29 April 2019.

[27]John McCarthy, 'What is AI?' http://jmc.stanford.edu/articles/whatisai/whatisaipdf accessed 22 March 2019.

[28]Pal Taylor, 'Machine Learning In The Real World' https://www.digitalistmag.com/digital-economy/2018/05/09/machine-learning-in-real-world-06166003 accessed 23 April 2019.

4.3 Space Technology

Space based technology that can be utilised to provide solutions to addressing falls into three categories, namely communication technology in the form of broadband, spatial data acquisition using earth observation and location and navigations services using Global Navigation Satellite System (GNSS). This section provides a brief positioning of the technologies.

4.3.1 Satellite Broadband

Broadband satellite is currently positioning itself for significant capacity growth with a number of mega-constellations announced for delivery in the next couple of years driven by significant technological innovation both in the form of satellite hardware and availability of launch systems. High throughput satellites (HTS) in lower MEO (Medium Earth Orbit) and LEO (Low Earth Orbit) orbits have the ability to deliver increased bandwidth whilst reducing latency at lower costs. Not confined by last-mile infrastructure challenges, it is the ideal choice to deliver communication in hard-to-reach areas not viable using traditional terrestrial means, is making satellite broadband the "go-to" technology to connect the "un-connected". One of the challenges to broadband delivery is the ability to make it available to the unconnected billions of humans globally. One of the keys to satellite broadband adoption, in addition to adding the significant capacity represented by the planned systems, is to provide an affordable user terminal.[29] Traditionally mass adoption of flat panel antenna (FPA) systems have been prevented by cost and performance issues, though driven by new demand and technological innovation these systems can open up new opportunity for satellite broadband delivery. With a number of companies entering the market using new technology[30] companies are promising systems with the ability to access high-throughput satellites in different orbits using no movable parts. If these virtually maintenance free systems can be provided to the market at affordable prices, it will aid adoption of satellite broadband in hard-to-reach places and where parabolic antennae will not be ideal.[31]

4.3.2 Remote Sensing

The word remote sensing refers to "science/art of identifying, observing, and measuring an object without coming into direct contact with it."[32] Tough first used by Eveyln Pruitt in 1958, it was not necessarily a new concept—the first aerial

---

[29]Debra Werner, 'Cheap satellite terminals key to bridging digital divide, execs say' Space News, 10 March 2017.

[30]Alcan, 'Alcan Systems' https://www.alcansystems.com/technology accessed 1 December 2018.

[31]Caleb Henry, 'Space News' https://spacenews.com/german-startup-takes-kymeta-like-lcd-approach-to-flat-panel-antenna-manufacturing accessed 29 July 2018.

[32]Steve Graham, 'Remote Sensing - Introduction and History' https://earthobservatory.nasa.gov/features/RemoteSensing accessed 21 April 2019.

photograph was taken from a gas balloon in France as far back as 1858.[33] With the progression of photographic technology[34] it became possible to employ a variety of different aerial platforms for the task of aerial photography: kites, rockets, even carrier pigeons (in the form of the famed Bavarian pigeon corps) and naturally airplanes.[35] By the 1930s aerial photography was a well-established practice and apart from military and mapping applications it was being used in the USA for basic crop inventory purposes.[36] The satellite era of remote sensing was announced by the launch of Sputnik on 4 October 1957[37] opening the door to a number of former "cold war era" military programs such as Corona.[38] According to the satellite database of the Union of Concerned Scientists (UCS), there were just under 2000 operational satellites in orbit of which around 36% represented some form of remote sensing purpose at the end of 2018[39] indicating the importance of the sector. Whereas satellite imagery used to be the domain of specialised individuals as little as a decade ago, it is now a common feature available to billions of people globally Though currently there many different options available to the individual or organisations to access satellite imagery for free[40] it was the arrival of Google Earth, more than any other service, that introduced the bulk of the global population to the world of satellite imagery. Launched by Google in 2004, it presented users with an easy to use interface to view the globe, using a variety of sources of satellite imagery stitched together. Its rapid adoption was confirmed when by 2011 it recorded its one billionth download,[41] subsequent availability for smartphone platforms have increased the usage footprint significantly.

[33]Paul R. Baumann, 'History of remote sensing, aerial photography' http://employees.oneonta.edu/baumanpr/geosat2/RS%20History%20I/RS-History-Part-1htm accessed 21 March 2019.

[34]Gerald K. Moore, 'What is a picture worth? A history of remote sensing/Quelle est la valeur d'une image? Un tour d'horizon de télédétection,' Hydrological Sciences Bulletin, vol 24, no 4, pp 477–485, 1979.

[35]Dan Schlenoff, 'Aerial Spying, 100 Years before Drones' https://blogs.scientificamerican.com/anecdotes-from-the-archive/aerial-spying-100-years-before-drones/ accessed 25 April 2019.

[36]Duane Nellis, Kevin Price, Donald Rundquist, 'Remote sensing of cropland agriculture' The SAGE handbook of remote sensing, 1, pp 368–380, 2009.

[37]Andrew Tatem, Scott Goetz and Aimon Hay, 'Fifty years of earth observation satellites: Views from above have lead to countless advances on the ground in both scientific knowledge and daily life' American Scientist, 96(5), p 390, 2009.

[38]CIA, 'CORONA: Declassified' Central Intelligence Agency, https://www.cia.gov/news-information/featured-story-archive/2015-featured-story-archive/corona-declassifiedhtml accessed 27 April 2019.

[39]Union of Concerned Scientists, 'UCS Satellite Database' https://www.ucsusa.org/nuclear-weapons/space-weapons/satellite-database#VRsK4fnF8wo accessed 3 March 2019.

[40]GISGeography, '15 Free Satellite Imagery Data Sources' https://gisgeography.com/free-satellite-imagery-data-list accessed 29 April 2019.

[41]Gill Press, 'The Origins Of Google Earth, Microsoft, Barcodes, And The World's Most Valuable Company—Apple' https://www.forbes.com/sites/gilpress/2016/12/11/the-origins-of-google-earth-microsoft-barcodes-and-the-worlds-most-valuable-company-apple/#4719ec116b90 accessed 13 March 2019.

4.3.3 GNSS

Throughout history the ability to navigate accurately has played an important role in maintaining the balance of power and wealth. The British Empire—an island nation highly dependent on maritime trade and therefore naval power—was very aware of the strategic importance of accurate navigation. Realising that a method to accurately determine longitude was a key factor to naval competitive advantage, the British parliament offered a reward of £20000 (a princely fee for the time) for a method to determine longitude to within half a degree (the 1714 Longitude Act).[42] Today, a smartphone costing less than USD 25,[43] puts the aforementioned ability into anyone's pocket, thanks to Global Navigation Satellite System (GNSS). The system is the product of a cold war era military request to develop an all-weather system with the capability to provide accurate information about military assets and resources in a common global reference framework, in terms of position, speed and time.[44] Based on the principle of trilateration,[45] the position of a receiver unit on the earth's surface is determined using calculations involving the propagation of a signal relative to three component segments namely; space, ground and user. In addition to the truly global systems GPS[46] (USA), GLONASS[47] (Russian Federation) and Galileo[48] (EU), there are also a number of regional systems in use. This technology is arguably the space technology which had the most economic impact recently, spawning whole new industries e.g. Disruptive business concepts such as ride-sharing is very dependent on location meta-data. GNSS technology is logically positioned to support addressing the un-addressed by the nature of being a truly global system not excluding any geography, and it is available in many platforms such as smartphones driving user friendly navigating applications e.g. Google Maps.

[42]Richard Dunne 'The Longitude Act, 1714' https://longitudeprize.org/about-us/history accessed 27 April 2019.

[43]Helena Wasserman, 'MTN's super-cheap R300 smartphone has two cameras and a long battery life – here's what it looks like' https://www.businessinsider.co.za/mtns-super-cheap-r300-smartphone-has-two-cameras-and-a-long-battery-life-heres-what-it-looks-like-2018-11 accessed 25 April 2019.

[44]DARPA, 'Transit Satellite: Space-based Navigation' https://www.darpa.mil/about-us/timeline/transit-satellite accessed 22 March 2019.

[45]GISGeopgraphy, 'Trilateration vs Triangulation – How GPS Receivers Work' https://gisgeography.com/trilateration-triangulation-gps accessed 24 April 2019.

[46]GPS, 'GPS: The Global Positioning System' https://www.gps.gov accessed 24 April 2019.

[47]GLONASS, 'Information and Analysis Centre for Positioning, Navigation and Timing' https://www.glonass-iacru/en accessed 24 April 2019.

[48]GSA, 'Galileo is the European global satellite-based navigation system' https://www.gsa.europa.eu/european-gnss/galileo/galileo-european-global-satellite-based-navigation-system accessed 28 April 2019.

5 Addressing the Un-Addressed

There are numerous ways in which technological application could be used to provide solutions to the problem of the un-addressed and, in particular, to the rural un-addressed. In recent years technological development has started to accentuate the emerging dominant nature of "virtual addressing" and the potential advantages it has to create address systems that, though not necessarily familiar in the traditional sense of the word, still can provide the basic function. Relying heavily on the availability of space-based systems and terrestrial ICT services, there are solutions emerging to address the address problem This section looks at possible solutions to the problem; firstly it presents a number of current initiatives.

5.1 Current Initiatives

This section briefly investigates three different interpretations of tackling the problem, all involving some aspect of space technology. These are by no means the only initiatives but are used as a point of departure in the approach proposed in this work in the Sect. 6.

5.1.1 "What3words"

"What3Words"[49] is an attempt to take the complexity of out of referring to physical location though the use of a very intuitive way of description. A private start-up company based in the UK and focussed on geo-mapping, What3Words, have divided the world up into 3 m × 3 m squares each linked to its own unique coordinates (longitude and latitude). Forgoing the traditional numerical coordinate format in favour of one consisting out of a unique description consisting out of a compound "word" made up from three unique words concatenated with two full stop punctuation marks e.g. "word1.word2.word3."[50] Currently it is available in a variety of languages with the ability to present the address in the language of choice for the location involved, has a number of products on the market including smartphone apps for Android and iOS, API's and a variety of GIS Plugins. The company presents the solution as a way to solve the address infrastructure problem globally, a particularly interesting application being the use of the application to find locations of Airbnb's provided by nomads in Mongolia (Fig. 3).[51]

[49]What3Words, 'About, "What3Word"' https://what3words.com accessed 22 April 2019.
[50]Liane Yvkoff, 'This Geo-Mapping Startup Could Save Your Life In A Crisis' https://www.forbes.com/sites/lianeyvkoff/2018/09/18/what3words-could-save-your-life/#890720712d02 accessed 22 April 2019.
[51]Stephen Shankland, 'Airbnb uses What3words locations so you'll find your host in a Mongolian forest' https://www.cnet.com/news/airbnb-uses-what3words-locations-to-find-host-in-a-mongolian-forest accessed 25 April 2019.

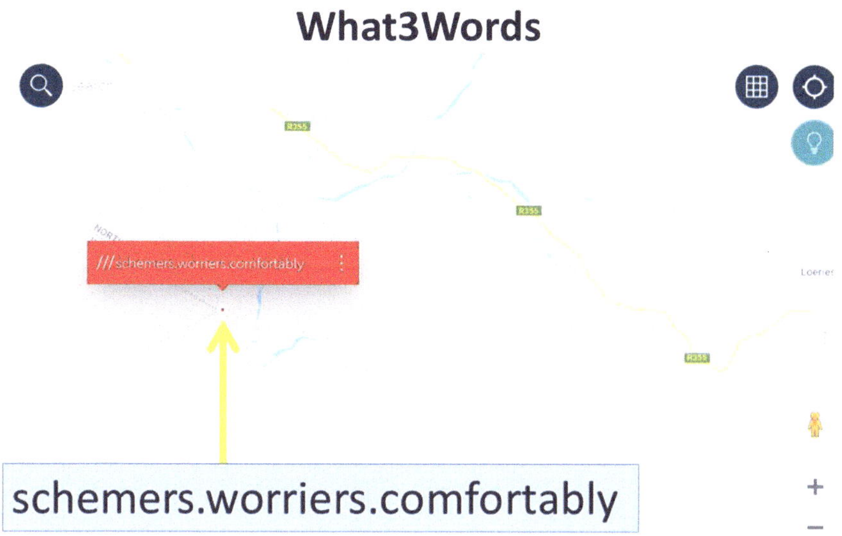

Fig. 3 What2Words code example (What3Words. https://map.what3words.com/schemers.worriers.comfortably?utm_source=w3w&utm_medium=owned&utm_campaign=B2C_4561_W3_Pt_Map-Site_Org_WO_en_Main-Site-Explore-Map accessed 25 April 2019)

5.1.2 MIT Media Lab and Facebook

A collaborative project involving the MIT Media Lab and Facebook proposes a different way of solving the address problem, especially for infrastructure lacking rural areas.[52] Using machine learning as the primary tool in a very innovative way, the project uses two algorithms to identify road networks from satellite data. The first process involved an algorithm that was trained using deep learning to be able to identify pixels from remote observation satellites imagery that could be used to identify a road network for a particular area. A secondary algorithm used the road topology to determine communities within the road network, ultimately identifying "city centres" based on density surrounding the road networks. These city centres were used as centres points to orient direction (N, S, E and W) quadrants around which can then be divided into street names and numbers. A perceived advantage of the system is that is based on relative relationships of objects familiar to the local population and naming conventions could be used to reflect local culture and custom. Compared to the "What3Words" system which could be interpreted as very abstract and lacking local relevance to some communities, this system could be perceived as potentially being more user-friendly. The system is aimed at the rural "un-addressed" market.

[52]Karen Hao, 'MIT Technology Review' https://www.technologyreview.com/s/612492/four-billion-people-lack-an-address-machine-learning-could-change-that accessed 11 March 2019.

5.1.3 Plus Codes

"Plus Codes"[53] is a system developed by Google also based on latitude and longitude coordinate system, but using a simpler format. The code consists out of two parts to form a 10 character string, the first part identifies the greater region and the second a more precise local code. The initial four characters denotes the region representing an area of approximately 100 km × 100 km. The second segment consists out of six charters representing the local code, an area of 14 m × 14 m. Depending on the area where the target location is, the code could be used in different ways. The area code could be discarded if the target location is within 25 km of a town, where the name of the town is used in lieu of the area code, an optional additional code is available to increase the accuracy to 3 m × 3 m. The service is available as smartphone apps, API and on-line version, importantly it could also be deployed off-line as a printed map. The system is open source, free, and aimed at both the "addressed" and un-addressed market (Fig. 4).

6 Solution Proposal: Addressing for Rural Sub-Saharan Africa

What if it was possible to identify within a rural community an object of common utility which can be "sensorized" to the extent that in can be used as location reference point for the area against which a local address system could be implemented using available technology as described (refer Sect. 4).

Though much emphasis has been on the rapid urbanisation of the African continent, the reality is that there is still a very large number of people in Sub-Saharan Africa living in rural areas. In Sub-Sharan Africa it is estimated that almost two thirds of the population currently resides in rural areas as compared to roughly a quarter of the population in the European Union.[54] Challenged by a lack of formal infrastructure in many aspects electricity, telecommunication including address systems, leaves rural areas at a disadvantage to partake in a developing economy dependent on the availability of these services. Though services such as the aforementioned What3Words (refer Sect. 5.1.1) and Plus Codes (refer Sect. 5.1.3) are practical systems presenting it in relevant way could be challenging as it seeks to introduce an abstract concept into an environment plagued by illiteracy. More than a quarter (27%) of the global functionally illiterate adults reside in sub-Saharan Africa[55] with 17 countries with literacy rates of less than 50%, most of these countries are also plagued by poverty and adverse civil conditions with the

[53]PlusCodes, 'Plus Codes how it works' https://plus.codes/howitworks accessed 22 April 2019.

[54]GlobalGrowing, 'Fact 1: The majority of Sub-Saharan Africans live in rural areas, Europeans predominantly in cities' http://global-growing.org/en/content/fact-1-majority-sub-saharan-africans-live-rural-areas-europeans-predominantly-cities accessed 12 January 2018.

[55]UNESCO Fact sheet no 45, 'Literacy Rates Continue to Rise from One Generation to the Next' UNESCO, Paris, 2017.

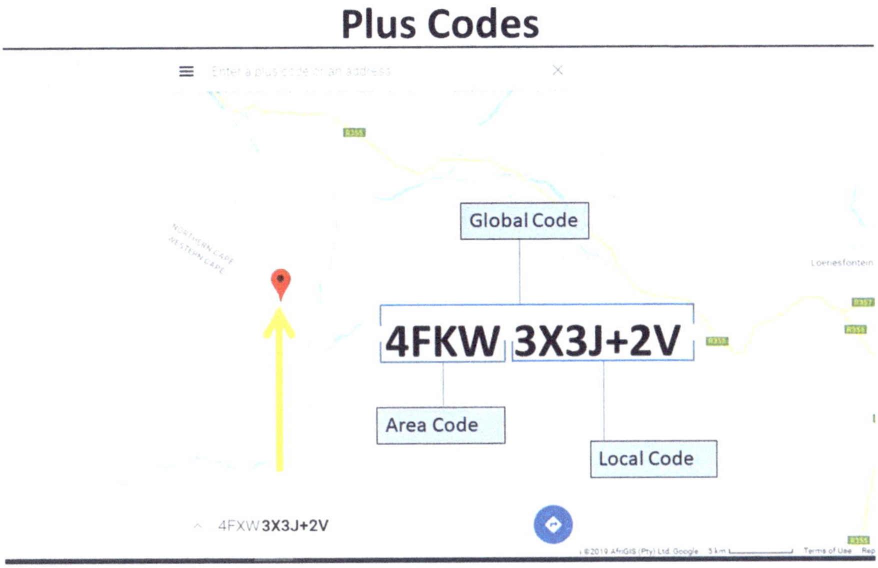

Fig. 4 Plus Codes example (PlusCodes https://plus.codes/4FXW3X3J+2V accessed 23 April 2019)

remote rural areas most affected.[56] Taking the aforementioned into account, we would like to present our interpretation of a simple system to provide an address infrastructure for remote rural areas the form of what we would like to call CUPSS (Common Utility-Point Sensor System).

6.1 Common Utility Point Sensor System—CUPSS

Taking a leaf out of the ancient Roman system and recently the Facebook/MIT collaborative approach (refer Sect. 5.1.2) this system also revolves around identification of a central point of reference. This point is equipped with a sensor enabled appliance which could be used to determine geolocation with a fair degree of accuracy and provide a facility to create local addresses based on geolocation and user data, which could be mapped to existing systems such as Plus Codes and What3Words.

The operator uses the CUPSS App on a tablet to get the geolocation of the user (resident) location relative to the CUPSS device, this is a representation of the address using the Plus Code protocol which is then interpreted into a symbol-based code (Fig. 5).

[56]UNESCO, 'UNESCO e-Atlas of Literacy' https://tellmaps.com/uis/literacy/#!/tellmap/-1003531175 accessed 28 March 2019.

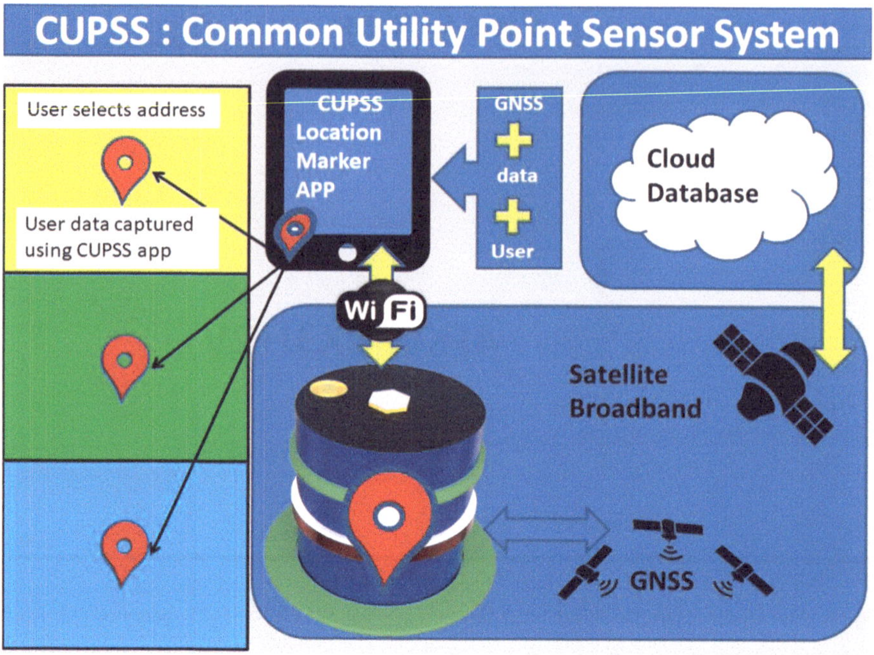

Fig. 5 Common utility point sensor system (Graphic courtesy of the author)

The system can be operated by an external party or preferably by a trained member of the community.

6.2 CUPSS Address

The idea with the address is to keep it simple by focusing on the use of symbols and colours which are relatively easy to recognise and easy to be identified by technology as well. It consists out of five elements arranged in a rectangle divided into four quadrants and a central colour code.

- Quadrant 1—Resident Symbol
  Consisting out of two symbols allocated to the user on the creation of the address it is linked to the colour code. The user can be presented with choices by the system, to encourage user participation.
- Quadrant 2—Direction Symbol
  Used to indicate the direction to the CUPSS location from the residential location.

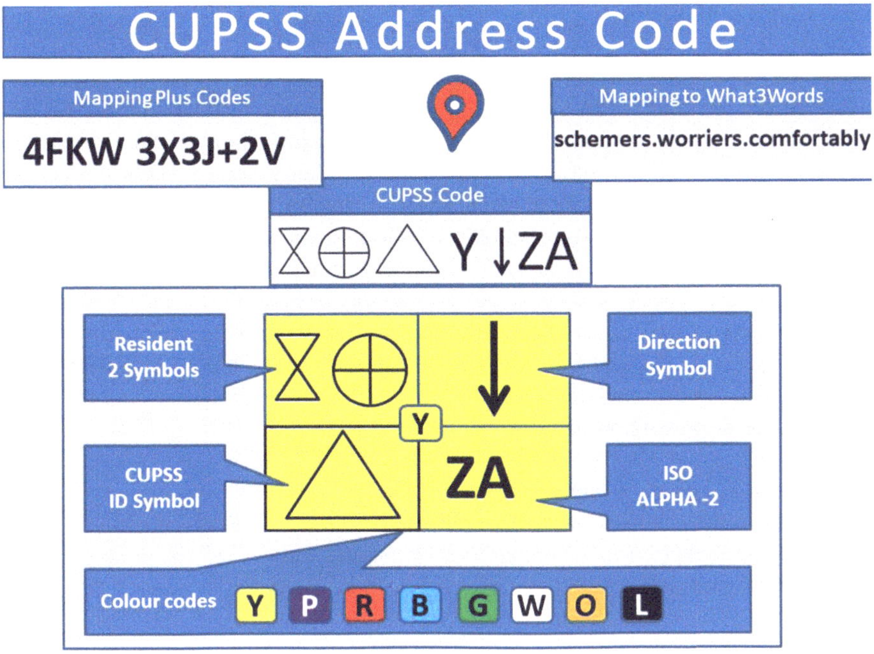

Fig. 6 CUPSS address code (Graphic courtesy of the author)

- Quadrant 3—ISO ALPHA-2 Code.[57]
  Used to identify the country hosting the CUPSS unit.
- Quadrant 4—CUPPS ID
  A unique symbol linked to the area the CUPSS unit is deployed, it is automatically allocated.
- Colour code—One of eight colours to represent eight square zones arranged in a gridiron pattern surrounding the CUPPS.
  All symbols apart from the country and "direction" codes must be researched to ensure it is culturally acceptable and relevant to the community. Two symbols and a color code would be able to accommodate quite a number of people, though the system is designed for small remote communities (Fig. 6).

[57]ISO, 'Country Codes - ISO 3166' https://www.iso.org/iso-3166-country-codeshtml accessed 15 April 2019.

6.3 CUPSS Module

The module is a multifunction unit that is physically presented in a single enclosure, envisioned to be mounted on top of a common utility point, the example used in this work a water tank (Fig. 7) though it could be any other fixed utility that is central to providing to the needs of all members of the community. Functionally the system provides three services, firstly a mechanism to determine the geolocation of the utility used as mounting point, secondly as communication enabler and finally the means to capture geolocation data of community residents into a cloud repository. The module is also responsible to provide a standardised time-stamp and geolocation meta-data to all data collected by the sensors using data obtained from the GNSS sensors. All raw data collected by the sensors are collected in a data buffer and "stamped" with meta-data before being sent to a centralised cloud-based storage container, available for retrieval.

- Location
  This sub-module consists out of a satellite positioning sensor capable of transmission and reception to accurately determine and "advertise" the geolocation of the CUPSS unit.

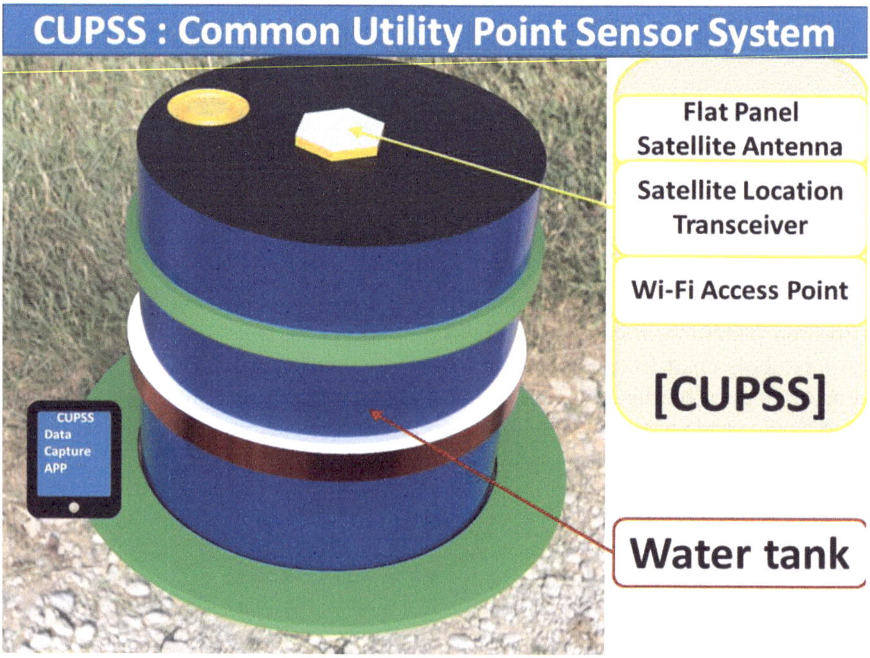

Fig. 7 Water tank based CUPSS

- Communication
  This module is responsible for the communication function of the CUPSS, it has two basic primary functions, firstly it needs to resolve the "last mile" problem and secondly to provide local communication to data capture devices used to record user data. Satellite broadband is used to provide internet connectivity to the location, for this purpose compact flat panel satellite antenna technology is deployed. The antenna is integrated on the top of the enclosure housing to obtain a broadband signal which is then in turn redistributed locally using Wi-Fi.
- Data Capture
  The user's data is captured ensuring compliance to the regulatory ecosystem, using a tablet device to host the CUPSS application software, available symbols are suggested for the user to select. This software application is dependent on the availability of internet though to send the user data to a cloud repository where it will be available to be mapped to existing systems.

7 Conclusion

The aim of this work was to shortly present the relative importance for a person to "belong" to some form of an address that is geographically locatable and the technology enabling a number of solutions including our own take on a possible solution for sub-Saharan Africa. Implementation of such a solution could present useful location meta-data to the benefit of any number of parties such as Governmental and commercial organizations, to aid interaction with the communities in question. The community in turn can participate in activities where a delivery address is a prerequisite e.g. e-commerce transactions. Simultaneous development in drone technology as well as the regulations to govern their use is increasingly allowing the technology to provide unique solutions, especially remote rural areas, Zipline,[58] a drone company operating in Ghana and Rwanda, serving millions[59] of people by delivering medical supplies such as vaccines and blood products to remote health centres can make use of CUPSS as a location beacon for the package drop-zone.

Creating the address system is however not the only challenge that needs to be overcome, many different diverse factors play into whether an address system will be accepted. Lessons[60] learned from the telecommunication industry to fulfil its

[58]Zipline, 'Lifesaving deliveries by drone' https://www.flyzipline.com accessed 25 April 2019.
[59]Riley de Leon, 'Zipline takes flight in Ghana, making it the world's largest drone-delivery network,' https://www.cnbc.com/2019/04/24/with-ghana-expansion-ziplines-medical-drones-now-reach-22m-people.html accessed 27 April 2019.
[60]Pape Toure, 'Basic Principles of Universal Service', Seminar on costs and tariffs for member countries of the TAF Group, 23–25 April 2001, Niamey (NIGER), 2001.

USO of providing communications services to all have identified at least three barriers that needs to be overcome:

- Availability—infrastructure should be available everywhere
- Accessibility——to all without discrimination useable by all social groups and in particular by the disabled
- Affordability—services must be affordable for the general public.

In addition to the above any new service should not be imposed on the community and should be not only culturally acceptable by the community, but also relevant to such an extent that the community "wants" to use it.

To conclude, we believe that rather than to serve purely as an address facilitator, CUPSS represents rather a bigger opportunity that should be further investigated namely; how can it be expanded to assist in bridging the digital divide by providing broadband internet to the un-connected which could not only provide address systems but provide a gateway to a plethora of additional services including governance, health and education.

Christoffel (Chris) Kotze established a boutique technology strategic advisory company in 2012 after a successful corporate career spanning two decades. This company specialises in providing assistance to Digital Transformation projects within organizations, with a special interest in the use of technology resources to support sustainable development. Current research interests include space technology, dematerialisation through digital transformation and solutions to the "digital divide". M.Phil. (Space Science) candidate at the University of Cape Town. Other qualifications include a Bachelor of Commerce Honours (Information Systems)—University of Cape Town, Bachelor of Science (Physiology & Microbiology)—University of Pretoria, Diploma in DataMetrics (Computer Science) University of South Africa, a number of strategy focussed executive management courses at the Graduate School of Business from the University of Cape Town.

MENASat—Proposal for a Space-Based Refugee Assistance Programme

Nicolas Ringas

Abstract
This article investigates the Syrian refugee crisis and explores how services derived from remote sensing and earth observation data could assist relief organizations in aiding the millions of people displaced throughout Africa and the Middle East. Three core services are described: (i) planning refugee camps based on high density routes, informal settlement locations, border post locations, environmental considerations and water resource data, (ii) identification and monitoring of key areas in the Mediterranean Sea to reduce fatalities associated with illegal refugee smuggling, and (iii) identifying minefields remotely.

1 Introduction

"Syria is the biggest humanitarian and refugee crisis of our time, a continuing cause of suffering from millions which should be garnering a groundswell of support around the world". Filippo Grandi—United Nations High Commissioner for Refugees (UNHCR).

In March 2011, the Arab Spring movement saw anti-government demonstrations sweep through middle-eastern countries. In Syria, the government responded with force which resulted in numerous armed opposition groups being formed. By July

N. Ringas (✉)
University of Cape Town, Rondebosch, South Africa
e-mail: RNGNIC002@myuct.ac.za

A. Froehlich (ed.), *Space Fostering African Societies*, Southern Space Studies,
https://doi.org/10.1007/978-3-030-32930-3_12

Table 1 Number of Syrian refugees in asylum countries (UNHCR, Operational portal: refugee situations, accessed 8 June 2019, https://data2.unhcr.org/en/situations/syria#_ga=2.39024227.1555627567.1559675991-1903133917.1559675991, European University Institute, Syrian refugees: a snapshot of the crisis, accessed 20 May 2018, http://syrianrefugees.eu/inflows-recognition/)

Country	Number of Syrian Refugees	Date of data
Turkey	3 606 737	May 2019
Lebanon	938 531	April 2019
Jordan	664 330	June 2019
Iraq	253 371	April 2019
Egypt	132 473	April 2019
Greece	547 270	Jan 2016
Hungary	189 135	Jan 2016
Germany	158 015	Jan 2016

of the same year, the situation had spiralled into civil war between the government and the Free Syria Army, exacerbated by fighting between armed secular and religious groups.[1]

The conflict resulted in one of the worst refugee crises in modern history, with a total of 5.6 million people having fled Syria since 2011 and 6.6 million people displaced within their own country.[2] The UNHCR, also referred to as the UN Refugee Agency, states that as of April 2018, there are 13.1 million people in need in Syria. In 2017, more than 1.8 million Syrians were newly displaced—approximately 6 550 people displaced every day, suffering horrors such as besiegement, hunger and disease.[3] The majority of the refugees that fled Syria entered Syria's neighbouring countries—Turkey, Lebanon, Jordan and Iraq. Table 1, shows the number of refugees taking asylum in each of these countries, as well as other European countries.

For many Syrian refugees seeking asylum, Europe is an attractive option. Many Syrians have successfully applied for asylum within European countries, however due to the large number of applications and increasing restrictions on asylum grants, an increasing number of refugees attempt to illegally enter Europe—relying on smugglers to transport them across the Mediterranean Sea. The most common points of arrival in Europe are Spain, Italy and Greece, due to their relatively short distances from Morocco, Egypt and Turkey, respectively. Figure 1 shows the number of refugees arriving at each of these countries in 2015 and 2016.

[1]Mercy Corps, Quick facts: what you need to know about the Syria crisis, accessed 16 May 2018, https://www.mercycorps.org/articles/iraq-jordan-lebanon-syria-turkey/quick-facts-what-you-need-know-about-syria-crisis.
[2]UNHCR, Syria emergency, http://www.unhcr.org/syria-emergency.html, accessed 17 May 2018.
[3]Mercy Corps, Quick facts: what you need to know about the Syria crisis.

Fig. 1 Number of refugees arriving by sea in Spain, Italy and Greece (2015 and 2016) (UNHCR Bureau for Europe, Refugees and Migrants Sea Arrivals in Europe, Monthly Data Update, Dec 2016)

Table 2 UNHCR data on refugees travelling to the EU by sea (UNHCR Refugee Situations, Operational Portal, http://data2.unhcr.org/en/situations/mediterranean)

Year	Sea Arrivals in EU	Dead/Missing
2014	216 054	3 538
2015	1 015 078	3 771
2016	362 753	5 096
2017	172 301	3 139
2018	141 472	2 277
2019 (of 7 June 2019)	21 591	530

Refugee smugglers often operate in unsafe conditions, using unseaworthy vessels and/or exceeding the safe capacity of their boats. This has resulted in a considerable number of boating accidents resulting in numerous fatalities—in April 2015 alone, more than 800 refugees died while trying to cross the Mediterranean Sea.[4] Table 2 shows the number of refugees arriving in Europe by sea, as well as the number of registered deaths and missing persons from 2014 to 2019.

It is clear from these figures, that the Syrian refugee crisis is a humanitarian tragedy which requires the assistance and help of the world. Furthermore, the plight of refugees is not restricted to Syria alone. In Myanmar, over 713 000 refugees have

[4]UNHCR Bureau for Europe, Refugees and Migrants Sea Arrivals in Europe, Monthly Data Update, Dec 2016.

fled to Bangladesh since August 2017.[5] Violent conflicts in South Sudan displaced more than 4 million people from their homes and resulted in over 2.4 million fleeing to neighbouring countries.[6] Similarly in Iraq, conflict and brutal mass executions displaced more than 3 million Iraqis and resulted in over 260 000 seeking refuge in other countries.[7] The UNHCR global trend report stated that at the end of 2016, 65.6 million people worldwide had been forcibly displaced due to conflict and human rights violations.[8]

2 Proposed Space-Based Solution—MENASat

This article suggests a hypothetical non-profit company to assist relief efforts aimed at helping refugees using remote sensing and earth observation services for humanitarian applications. The aim of the company is to utilize the opportunities afforded by high spatial resolution commercial satellites through data analysis and processing to provide relief organizations with assistance in their work. The specific focus of the South African-based company is to aid the millions of refugees displaced throughout Africa and the middle-East. However, the proposed methodologies can be applied in all parts of the world with some minor modifications. The proposed hypothetical non-profit company will be referred to as MENASat within this document, after Saint Menas—a saint born in Egypt and venerated as the protector of pilgrims and travellers by the Coptic Christians of Northern Africa.[9,10]

Remote sensing services do not require a large capital investment when compared with establishing ground and space-based infrastructure. It is advantageous in that the analysis and interpretation of the data can be performed remotely, and hence a ground presence is not required. Furthermore, partnerships and agreements can be negotiated with commercial satellite operators due to the humanitarian aspect of the work, which may offer reduced rates for data acquisition, or even possibly free data transfer.

MENASat could seek to establish partnerships and contracts with organizations such as the United Nations High Commission Refugee Agency (UNHCR), the United Nations Children's Fund (UNICEF) and Islamic Relief USA, and could work closely with these organizations while developing and optimizing their products. The company could seek to establish local technical partnerships with the South African National Space Agency (SANSA), the Council for Scientific and

[5]UNHCR, Rohingya Emergency, http://www.unhcr.org/afr/rohingya-emergency.html (20/05/2018).

[6]UNHCR, South Sudan Emergency, http://www.unhcr.org/afr/south-sudan-emergency.html (20/05/2018).

[7]UNHCR, Iraq Emergency, http://www.unhcr.org/afr/iraq-emergency.html (20/05/2018).

[8]UNHCR, Global Trends: Forced Displacement in 2016.

[9]Catholic Online, Saint Menas, https://www.catholic.org/saints/saint.php?saint_id=5084 (9/05/2018).

[10]https://www.johnsanidopoulos.com/2009/11/saint-menas-great-martyr-and-miracle.html.

Industrial Research (CSIR) and universities within South Africa, as well as engage with the global relief aid and earth observation communities.

The suggested company MENASat is a practical example of how space-based assets can be leveraged to aid refugees within Africa and the world. MENASat will provide services derived from the analysis of remote sensing data to compliment relief support for refugees in the following means:

1. Planning location of refugee camps based on identified high density routes, informal settlement locations, border post locations, environmental considerations and availability of water resources.
2. Monitoring of key areas in the Mediterranean Sea to co-ordinate search and rescue missions and reduce fatalities of refugees.
3. Identify mine fields in conflict zones and provide this information to United Nations Mine Action Services (UNMAS) to support their decision-making procedures.

The activities of MENASat will assist in realizing the following UN sustainable development goals:

- Peace, justice and strong institutions
- Good health and well-being
- Quality education
- Gender equality
- Clean water and sanitation

3 Technical Description of Services

A technical overview of each of the services of MENASat is offered below. The output from the various services can be displayed as layers in a GIS system and distributed to relevant organizations along with reports to aid decision makers.

3.1 Service 1—Assisting Refugees Travelling by Land

The aim of this service is to provide meaningful information to key UNHCR decision makers when planning, locating and sizing new refugee camps and relief zones, as well as to identify urban areas where additional assistance and/or resources may be required. The service consists of five core areas, listed below:

- Identifying common travel routes and determining the risk profile of each route based on number of people, environmental aspects, armed threats and total distance,

- Monitor weather data to provide storm and severe weather warnings to organizations and identify travelling refugees that may be at risk,
- Observing informal settlements, population growth and population densities in neighbouring countries,
- Identification and mapping of open surface water bodies and using remote sensing data to help locate underground water sources where possible to assist with providing refugee camps sufficient water,
- Integration of above information with existing Geographic Information Systems (GIS) to allow for easy interpretation of data and support decision making.

Satellite remote sensing data and telemetry systems have been used extensively in the past for tracking global animal migration patterns. These systems usually rely on GPS transmitters or ARGOS transmitters located on tagged animals to track their movements.[11,12] However, due to the availability of high resolution satellite imagery, satellite remote sensing is being increasingly used to track the movements of land-based animals without the use of transmitters.[13]

MENASat will make use of data obtained from Digital Globe's GeoEye-1 satellite which has a 0.5 m spatial resolution to track the movements of Syrian refugees to their neighbouring countries Turkey, Lebanon, Iraq and Jordan. Artificial intelligence algorithms will be developed which will automate the tracking process. The required temporal resolution will vary depending on numerous factors such as the local terrain, the total distance of the route, the number of people traversing the route and weather conditions.

To determine which areas to monitor, extensive discussions will be held with local Syrian refugees and UN officials with valuable knowledge of the local conditions. Another method of identifying high traffic routes is to monitor busy border posts using satellites which have a high spatial and temporal resolution of those areas. For example, if one were to monitor the Jordan border post (see Fig. 2) with a high temporal resolution it would be possible to go back in time to determine where the refugees originated, thereby determining the routes they travelled to get to the border post.

Another approach MENASat will investigate is to use the thermal infrared bands of Landsat 8 data to try and track the refugees. If one acquired images at night time, any refugees travelling by foot will have high heat signatures when compared to the desert sand surrounding them. Currently the spatial resolution of this thermal imagery is not precise enough to track individual people; however groups or clusters of travellers could be identified.

The information obtained from monitoring the refugees and their travel routes will be integrated into a GIS system. The other layers in the GIS system will be created from data obtained from the UNHCR and similar organizations currently

[11]G. Viljoen et al., Animal migration tracking methods, Springer, 2016.
[12]J. Pierre, Satellite tracking and avian conservation in Asia, Keio University, 2005.
[13]Satellite Imaging Corporation, Wildlife Monitoring, 20 May 2018, https://www.satimagingcorp.com/applications/environmental-impact-studies/wildlife-and-marine-conservation/wildlife-monitoring/.

Fig. 2 Syrian refugees at a border post waiting to cross into Jordan (Khalil Mazraawi, Getty Images, 14 Jan 2016)

operating in Syria. The layers will include: vegetation indices, transport networks, known areas of militant activity, digital elevation models (DEMs), areas where landmines are thought to be present (see Sect. 3.3), border post locations, known patrol paths and areas of operation, location of refugee camps and the location of other important infrastructure (such as hospitals, clinics and water sources).

Once the most dense routes have been identified and mapped, they will each be assigned a risk factor calculated from the following factors: total distance, number of people using route, presence to known militant operations, availability of water sources along route, proximity to hospitals, proximity to refugee camps and proximity to minefields. This information will be used to increase monitoring of high risk routes and to plan relief and support missions in correspondence with the UN. Weather information can also be incorporated into the risk matrix and may be obtained through the American National Oceanic and Atmospheric Administration (NOAA) using their Geostationary Operational Environmental Satellite (GOES). The information will be used to issue alerts and warnings of adverse weather events which may threaten the safety of refugees travelling by foot.

Open surface water bodies will be monitored using optical remote sensing data, using Landsat imagery. The Landsat imagery is available for free, and its 30 m spatial resolution is sufficient for open surface water body monitoring. However, since Syria is an arid country, there are not many open surface water bodies. As such, MENASat will use remote sensing data coupled with geological information to

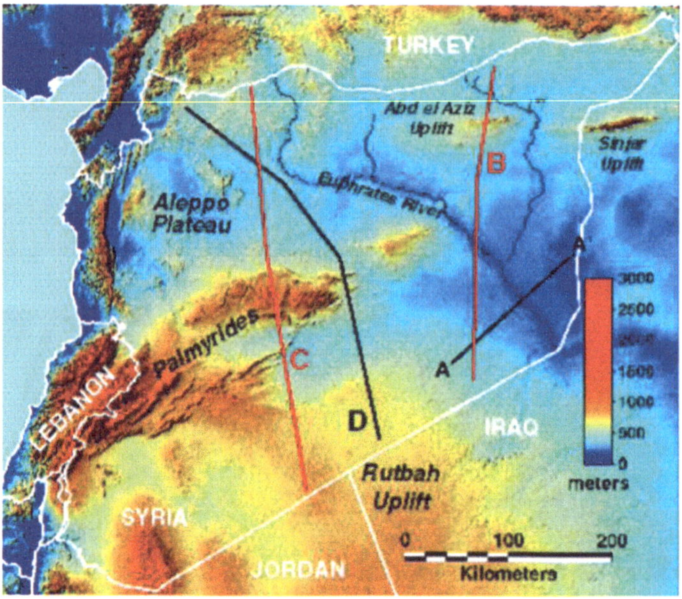

Fig. 3 Topographical map of Syria indicating the Dead Sea fault system (G. Brew et al., Summary of the geological evolution of Syria through geophysical interpretation, Cornell University, New York, 1997)

identify areas with a high likelihood of possessing groundwater. This will be achieved by determining a ground water potential index (GWPI) based on work done by K. Khodaei et al.[14] The GWPI value indicates the likeliness of groundwater in a specific area based on weighted values assigned to various groundwater indicators. Landsat and SPOT data will be analysed and co-registered, along with DEMs and local geological information of the common rock types in the area to allow for calculation of the GWPI values using the method discussed later on in this section.

Figure 3 shows a topographical map of Syria and surrounding areas and indicates the Dead Sea fault system. Geological maps of Syria are available, but will need to be purchased from European Soil Data Centre (ESDAC) and will form a critical input to this process. The Palmyride fold (shown in Fig. 3) is comprised of hard bed rock extrusions. As such, MENASat will focus on identifying hard rock and karstic aquifers. Due to the low porosity of hard rock and karstic terrains, groundwater occurrence is limited to areas exhibiting increased secondary porosity zones. These zones are most often located along bedding plains, fractures and lineaments. This allows for certain surface features to be identified using remote sensing data to determine the GWPI.[15]

[14]K Khodaei et al., Groundwater exploration using remote sensing and geographic information systems in a semi-arid area, Shahid Behesti University, Iran, 2011.
[15]Ibid.

The groundwater indicators used in calculating the GWPI are: lineaments, vegetation, rock type classifications, drainage density and slope. Lineaments are linear geological features along the Earth's surface such as faults, fractures or anticline fold lines.[16] These areas exhibit increased porosity and permeability and hence facilitate the presence of groundwater. To identify lineaments, the middle infrared band of the Landsat and SPOT images are employed. Due to the linear nature of these features, directional filters are applied to the remote sensing data to enhance the identification of the lineaments. Distance zones are then defined as follows: 0–100 m, 100–300 m, 300–500 m, and greater than 500 m. The closer a location is to a lineament, the more likely the occurrence of groundwater.[17]

The next indicator is vegetation. Since Syria is an arid area, the presence of green vegetation in the dry season is a good indicator of shallow groundwater. A vegetation map can be easily prepared using Landsat data acquired in the hot and dry summer season (May–October).

The local rock type plays a crucial role in determining the secondary porosity at a site. With hard rock terrains, the secondary porosity decreases with time as the rock weathers to clay or feldspar which hardens in the dry seasons. Conversely, secondary porosity of carbonate rock types increases over time due to the increase in pore spaces resulting from the solution of minerals. In order to determine the rock type, false colour composites are created using the Landsat data. Specifically, RGB 7, 4, 1 is used to enhance tonal differences of rock types. Band rationing and principle component analysis are also incorporated to enhance the mapping of the rock types in the area along with geological data and lithological maps.[18]

Drainage density is calculated using the lineament information and DEM data. The ratio of drainage density to lineament density offers an indication of groundwater recharge potential and permeability conditions. The last indicator is topographic slope, which is calculated using DEM data—gently sloped lands increase the percolation of groundwater.[19]

These indicators are all assigned numerical numbers from 1 to 10, with 10 being a strong indicator for the presence of groundwater. They are weighted according to their importance relating to groundwater occurrence and then summed to obtain the GWPI. The GWPI is then used to rank the areas into six groups: excellent, very likely, likely, moderate, not likely and no potential. This information will be incorporated into the GIS system and areas exhibiting "excellent" and "very likely" ratings of groundwater presence will be recommended for further geophysical investigation using appropriate in situ groundwater location techniques (resistive, electromagnetic or other) to confirm the presence of groundwater.[20]

[16]S Gay, Joints, linears and lineaments—the basement connection, AAPG Rocky Mountain Section Meeting, Colorado, September 2012.

[17]K Khodaei et al., Groundwater exploration using remote sensing and geographic information systems in a semi-arid area, Shahid Behesti University, Iran, 2011.

[18]Ibid.

[19]Ibid.

[20]Ibid.

The last aspect relating to this service will involve the monitoring and growth of informal settlements in and around Syria, in order to better plan refugee camps and relief work. Previous work has been done showing that fairly accurate modelling of informal settlements can be achieved using lower spatial resolution Landsat imagery, which is freely available.[21] Additional research performed at Ruhr University in Germany highlights how high spatial resolution imagery from Digital Globe's IKONOS and QuickBird satellites can offer improved mapping and classification of informal settlements and the growth thereof.[22]

MENASat will engage with Digital Globe to acquire high resolution optical imagery of key areas in Syria and its neighbouring countries to allow for this analysis. The areas will be evaluated and ranked with regard to their population, their densities, the availability of critical infrastructure (such as water, electricity and sanitation) and their proximity to conflict zones, hospitals and schools. The analysis will highlight areas that are under stress as a result of overpopulation or rapid population growth caused by refugees. These areas will be passed on to the UNHCR for further investigation.

3.2 MENASat Service 2: Assisting Refugees Travelling Be Sea

In 2016 alone, more than 5 000 people were reported missing or dead due to illegal smugglers traversing the Mediterranean Sea (see Table 2 in Sect. 1). This is due to overcrowding of boats (an example of which is shown in Fig. 4), as well as operating unsafe sea vessels. The second service of MENASat involves monitoring key areas of the Mediterranean Sea to locate and identify illegal smugglers in an effort to reduce the number of accidents and fatalities associated with these trips, as well as reduce the overall number of illegal immigrants entering Europe.

This will be achieved using data obtained from the COSMO-SkyMed (Constellation of Small Satellites for Mediterranean-basin Observation) satellite constellation, consisting of four identical radar satellites. This constellation was conceived by the Italian Space Agency (ASI) and is capable of operating in all visibility conditions at real-time. The satellites possess different operational modes, depending on the application, each with related spatial resolutions, swath widths and temporal resolutions, as shown in Table 3.

COSMO-SkyMed offers the lowest target revisiting on the synthetic aperture radar (SAR) market, with at least one acquisition every twelve hours. Furthermore, since the satellites are tasked twice a day, in the case of an emergency, COSMO-SkyMed offers the best response time for new acquisitions. It is also possible to have a Commercial User Terminal which, if negotiated, would allow MENASat to receive COSMO-SkyMed acquisitions in near real-time.[23] It is for

[21]Ibid.

[22]M Netzban et al., Urban and suburban areas as a research topic for remote sensing, Ruhr University, 2010.

[23]Ibid.

Fig. 4 A boat filled with migrants is rescued by the Italian Police (UNHCR, The sea route to Europe, p. 8, July 2015)

Table 3 Operational modes of the COSMO-SkyMed constellation (Agenzia Spaziale Italiana, COSMO-SkyMed SAR Products Handbook, Rev 02, 2009)

Name	Type	Spatial resolution	Swath width
Spotlight	Spot Obs. Area	≤ 1 m	10×10 km grid
HIMAGE	Stripmap	3–15 m	40 km
WideRegion	ScanSAR	30 m	100 km
HugeRegion	ScanSAR	100 m	200 km
Ping Pong	Stripmap	15 m	30 km

these reasons that COSMO-SkyMed was chosen over Airbus's TerraSAR-X and TanDEM-X satellites.

e-GEOS is an earth observation and geo-spatial information business owned by ASI and Telespazio. They are the exclusive distributor of COSMO-SkyMed data. As such MENASat will need to purchase the required data from e-GEOS after deciding its acquisition needs in order to task the satellites correctly. e-GEOS has released a new maritime surveillance platform named SEonSE (Smart Eyes on the Sea) which integrates maritime traffic data with COSMO-SkyMed satellite data and delivers applications in near real-time. The software also contains customizable early warning notifications and includes the e-GEOS toolkit. The software allows for regular and continual missions, instantaneous synoptic maritime pictures, characterization of vessel behaviours, threat assessments and other features relating to vessel identification.[24]

As shown in Fig. 4, the vessels transporting the refugees are often dinghy-type boats. Due to their small size, increased spatial resolution is required. The data MENASat will request for this application is HH polarized data, as this polarization type is sensitive to backscattering effects and is hence well-suited to maritime domain awareness applications. Ship detection is often performed using the L-band ($\lambda = \pm$ 23 cm) while oil spill detection and bilge dumping is performed using the X-band ($\lambda = \pm$ 3 cm). Due to the size of the vessels used to transport the refugees, it is recommended that the C-band is used for this application, which has a wavelength of approximately 5 cm.

MENASat will make use of the *Spotlight* and *WideRegion* operational modes of COSMO-SkyMed. The *WideRegion* mode will be used to continuously monitor target zones, such as ports in Libya, Egypt and Turkey up to 4 times a day (see Fig. 5). Once target vessels have been identified, the *Spotlight* mode will be used to identify and confirm specific targets. The SEonSE software will be used to compare this with known registered vessels in the area and evaluate the likelihood that the vessel is smuggling refugees across the Mediterranean Sea. If this is confirmed, then the details of the vessel will be sent to the relevant authorities for action.

It is also recommended that refugee camps located in close proximity to ports and harbours disseminate information regarding the Cospas-Sarsat beacons to help raise awareness about these life-saving devices. There are different types of Cospas-Sarsat beacons, also known as distress radio beacons or emergency beacons, which, when activated send a distress signal at 406 MHz to a network of satellites. The Cospas-Sarsat Programme comprises 43 countries and agencies and directs the alerts to the proper authorities in more than 200 territories.[25] The UNHCR could partner with Cospas-Sarsat to obtain Personal Locator Beacons (PLBs), which can then be issued to registered refugees within the refugee camps before they begin a journey to a new destination. The UNHCR staff would need to train the refugees on the PLBs and register them with the Cospas-Sarsat Programme. This would ensure that if a ship carrying refugees were to capsize and a PLB was activated, the refugees would receive help in the shortest amount of time possible, thereby possibly reducing the number of fatalities associated with such an incident.

[24]e-GEOS, SEonSE Smart Eyes on the Sea, https://www.e-geos.it/#/hub/hubPlatforms/platform/platform-sense.

[25]International Cospas-Sarsat Programme, Cospas-Sarsat System, https://cospas-sarsat.int/en/system-overview/cospas-sarsat-system.

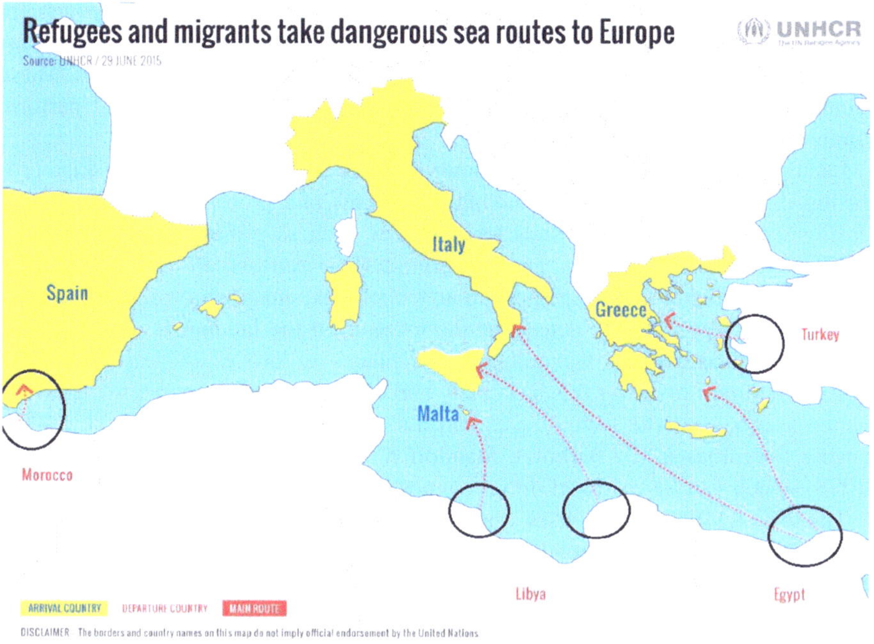

Fig. 5 Popular sea routes to Europe (UNHCR, The sea route to Europe. p. 9, July 2015) and areas to be monitored using *WideRegion* COSMO-SkyMed data highlighted in black circles

3.3 MENASat Service 3: Support Demining Activities Using Remote Sensing

The Landmine & Cluster Munition Monitor states that in Syria, there have been at least 3 075 casualties since 2012 arising from landmines and improvised explosive devices (IEDs), but note that the actual number of fatalities is expected to be significantly higher than the recorded tally.[26] It estimates that the deaths in Libya over the same time period exceed many thousands, with a total of 1 610 recorded casualties in 2016 alone.[27] The Egyptian Ministry of Foreign Affairs reported in 2006 that there had been a total number of 8 313 mine casualties in Egypt since 1982.[28]

It is estimated that there are currently millions of anti-tank and anti-personal landmines located in more than seventy countries. The majority of these mines are

[26]ICBL, Syria Casualties, July 2017. http://the-monitor.org/en-gb/reports/2018/syria/casualties.aspx.

[27]ICBL, Libya Casualties, July 2017. http://the-monitor.org/en-gb/reports/2018/libya/casualties.aspx.

[28]Jano Charbel, "Egypt continues to suffer from WWII landmines," 4 April 2017; and Ministry of Foreign Affairs of Egypt, "A paper on the problem of Landmines in Egypt," 27 July 2006.

still active and account for an average of ten casualties per day, four of which are children.[29] Existing landmines pose a severe threat to everyday life of local citizens, relief and aid operations, agricultural operations, and also military operations. These numbers highlight the urgent need to identify and locate minefields and perform demining operations.

Landmine detection techniques comprise three sections: ground sensing, processing of signals and decision-making algorithms and procedures.[30] The ground sensing segment employs various technologies such as ground penetrating radar (GPR), infrared (IR) sensors, and ultrasound (US) sensors. Processing the data involves noise reduction using appropriate techniques, image processing techniques and applying algorithms to determine the location of the landmines.[31]

Many ground sensing technologies use planes or drones to perform remote sensing of mines, however this is a costly and time consuming operation due to the limited swath areas the technologies can assess. A successful example of remote sensing for mines is the Airborne Standoff Minefield Detection System (ASTAMIDS) which can be mounted on unmanned aerial vehicles (UAV).[32]

The ASTAMIDS device uses optical imagery to perform change detection to sense where the soil or ground has been disturbed by placement of a mine. The change detection is done with complex algorithms to minimize false positives but still ensure sufficient sensitivity to identify small roadside IEDs. The optical change detection is used in conjunction with SAR and multispectral images (including near and mid-wave infrared) to penetrate the ground and identify possible mines. By combing the two technologies, the number of false positives is reduced. Once a target is identified using these two technologies, its location is sent to a laser rangefinder which then emits pulses to confirm if it is indeed a mine.[33]

MENASat could use multispectral remote sensing imagery to help identify and prioritize airborne sensing missions and demining operations. The service can use digital elevation models (DEMs), and other GIS information as inputs to identify areas which are likely to contain landmines. This information will be supplemented by additional analysis techniques applied to the remote sensing data. In order to perform change detection, multi-temporal data are required. The service will comprise of three core technologies:

1. Zone probabilistic classification,
2. Multispectral change detection algorithms,
3. Thermal infrared analysis

[29]ICBL, Country profile 2015 (www.the-monitor.org).

[30]S. Elkazaz et al., Towards landmine detection using ubiquitous satellite imaging, Egypt-Japan University of Science and Technology, 2016.

[31]C. Bruschini, B. Gros, A survey of current sensor technology research for the detection of landmines, International Workshop on Sustainable Humanitarian Demining, Volume 6, Citeseer (1997).

[32]E. Chamberlayne, A GIS model for minefield area prediction: The minefield likelihood procedure, Virginia Polytechnic Institute, Nov 2002.

[33]Global Security, Airborne Standoff Minefield Detection System (ASTAMIDS), 20 May 2018, https://www.globalsecurity.org/military/systems/ground/astamids.htm.

The first technology takes existing information such as DEMs, vegetation maps, soil maps, hydrography maps and transport network information and processes the data to determine areas most likely to possess mines and is based on previous work performed by the Virginia Polytechnic Institute.[34] The algorithm determines the trafficability of each zone, which indicates if an area can be traversed by common military vehicles. This is determined by analysing the vegetation zones (based on vegetation, density and tree spacing) along with the local slope, obstacles, and soil type and moisture—and can be obtained from the American National Imagery and Mapping Agency (NIMA) as part of their Vector Interim Terrain Data. Other key indicators such as roads, tracks, fencing and land use patterns are analysed in relation to known mine laying techniques. Each zone is then ranked according to its likelihood of containing mines based on the following assumptions:

- Trafficable areas are more likely to be mined,
- Areas that can be viewed from defensive positions are more likely to be mined,
- High speed roads are more likely to be mined than other roads,
- Areas not covered by surface water are more likely to be mined,
- High slope areas are less likely to be mined than flat areas.

The land areas are classified as: highly likely, likely, possible and not likely. The remote sensing data is then acquired for the highly likely zones for further identification and location of mines.

The second technology involves multispectral satellite data to perform change detection in an attempt to locate newly placed mines. This has its limitations in that it can only be performed for time periods where there is multispectral data available. However, work has been done to show that identifying older minefields may be possible using imagery from the Russian satellite KFA-1000, which has high spatial resolution (up to 5 m) film images from 1974 to 1993.[35] The analysis of the data seeks to identify and locate soil disturbances resulting from the construction of minefields.

To analyse recently installed landmines, SAR data will be obtained from the Canadian Space Agency's RADARSAT-2 satellite, which has a 1 × 3 m spatial resolution and operates in the C-band (centre frequency 5.4 GHz) and has data from 2007, or the previously mentioned COSMO-SkyMed satellites, which have higher spatial resolution. For older minefields, the relevant data can be acquired from ESA's ERS-1 or ERS-2 satellites which also operate in the C-band.

Analysis in the optical spectrum and near infrared wavelengths will be achieved using high resolution data from Digital Globe's GeoEye-1 satellite which was launched in 2008 and has a resolution of up to 0.46 m. For minefields thought to be installed in the 2000s, data will be obtained from the QuickBird satellite, which has

[34]E. Chamberlayne, A GIS model for minefield area prediction: The minefield likelihood procedure, Virginia Polytechnic Institute, Nov 2002.

[35]B. Maathuis, Remote sensing based detection of landmine suspect areas and minefields, International Institute for Aerospace Survey and Earth Sciences, 2001.

a spatial resolution of up to 0.65 m but is now decommissioned. For older minefields, the relevant LandSat data will be used from the multispectral scanner (MSS) and thematic mapper (TM) allowing for analysis as far back as 1972 (but with poorer spatial resolution).

The change detection requires numerous images of the same area captured at different dates to be analysed to identify soil disturbances which may indicate the installation of a landmine. In order to avoid false positives, one must obtain information of the local agricultural practices, cropping patterns, natural vegetation, construction projects, land usage and hydrological conditions. Furthermore, to be able to compare data from different satellites, look angles and orbits, a substantial amount of pre-processing is required. The images need to be normalized and co-registered using a nearest neighbourhood procedure as the first step. In addition to this, comparing data with different pixel sizes requires pixel resampling.[36] To compare satellite data containing multispectral data, the pixels must first be normalized to the same total energy to correct for look angles and topographic effects. Once this is performed, the internal relative reflectance of each pixel is calculated by dividing each pixel spectrum by the overall scene average spectrum. This, in conjunction with absorption analysis corrects for atmospheric absorption and instrument variations.[37] The change detection will be performed using existing techniques such as image differencing, image rationing, vegetation index differencing and Tassled Cap differencing.[38,39]

The third technique MENASat will use to locate landmines involves thermal infrared images. The premise of this technique is that the metal mines exhibit different thermal properties to the surrounding soil and as such can be identified from images captured at dusk and/or dawn.[40] This requires a relatively high spatial resolution and currently there are not many satellite imagers which have high resolutions in the thermal infrared bands. Since this approach does not use change detection, temporal resolution is not a factor and the latest satellite data can be used to locate landmines. Currently, the best options for thermal imagery data are provided by the Enhanced Thematic Mapper Plus (ETM+) on the Landsat 7 satellite and the two Thermal Infrared Sensors (TIRS) aboard Landsat 8. The ETM + acquires thermal images with a 60 m resolution, but this is resampled to obtain 30 m pixels. Similarly, the TIRS acquire thermal imagery at 100 m resolution which is resampled to provide 30 m resolution.[41] Clearly this resolution is too coarse for locating anti-personal mines. However, a resolution enhancement algorithm has

[36]Ibid.

[37]Erdas Inc, Erdas filed guide 5th edition, Erdas Inc., USA, 1991.

[38]B. Maathuis, Remote sensing based detection of landmine suspect areas and minefields, International Institute for Aerospace Survey and Earth Sciences, 2001.

[39]R. Lea et al., Using the Tasseled cap transformation to identify change in the Missouri Ozark Forests, University of Missouri, 2004.

[40]S. Elkazaz et al., Towards landmine detection using ubiquitous satellite imaging, Egypt-Japan University of Science and Technology, 2016.

[41]U. S. Geological Survey, Department of the Interior, "Landsat 7 (L7) Data Users Handbook", LSDS-1927, Version 1.0, South Dakota, USA, June 2018.

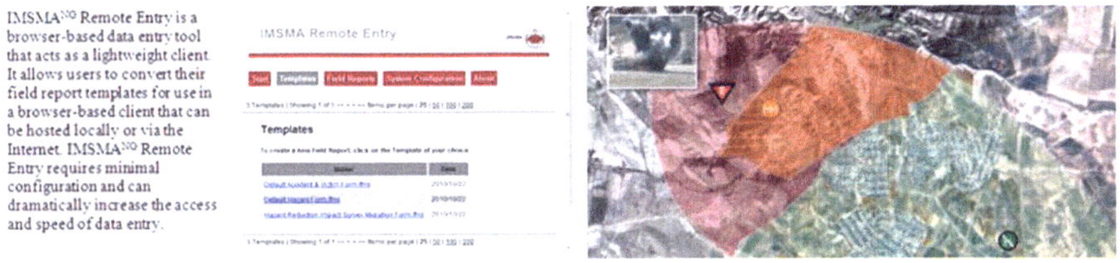

Fig. 6 Overview of the IMSMA software developed by GICHD (Ibid.)

been developed which uses multiple images of the same region to greatly improve the resolution of thermal images. This algorithm has been successfully employed to use 500 Landsat images, with a spatial resolution of 100 m to reconstruct a single image with a spatial resolution of 5 m per pixel. With this increased resolution, good detection rates for anti-tank mines have been achieved.[42] This algorithm will be used, along with Landsat images to help identify anti-tank mines in the zones classified as "highly likely".

The targets identified using the thermal analysis will be compared with those identified from the change detection technique to reduce the number of false positives obtained by the system.

MENASat may engage with ESA regarding their MIDAS programme, which seeks to "develop an integrated suite of mine clearance support tools to provide a comprehensive solution from enhanced operator training, planning of clearance campaigns and real-time monitoring of clearance in progress".[43] The MIDAS programme is ongoing and has secured financing through ESA and as such will prove to be an important partner, both technically and strategically.

MENASat can look to partner with the Geneva International Centre for Humanitarian Demining (GICHD) which is a well-established organization focussed on reducing the impact of mines in partnership with mine action and human security organizations. MENASat would hold discussions with GICHD to ensure the information derived from its minefield identification service can be incorporated into the GICHD's Information Management System for Mine Action (IMSMA, see Fig. 6) and the Mine Action INTelligence Tool.[44]

---

[42]S. Elkazaz et al., Towards landmine detection using ubiquitous satellite imaging, Egypt-Japan University of Science and Technology, 2016.

[43]ESA Business Applications, MIDAS Mine & IED Detection Augmented by Satellite, 20 May 2018, https://business.esa.int/projects/midas.

[44]GICHD, Information management system for mine action (IMSMA), 21 May 2018, https://www.gichd.org/topics/information-management/information-management-system-for-mine-action-imsma/#.WwGIQu6FOUl.

4 Future Expansion

With all of the aforementioned remote sensing applications, MENASat could investigate using neural networks (artificial intelligence), to help automate the data processing. MENASat could make use of the Google Earth Engine to process the large amounts of remote sensing data quickly and efficiently.

The core focus of MENASat revolves around remote sensing and earth observation. However, once the company is well-established and has learnt valuable lessons about operating in Africa and the middle-East, established partnerships, and fostered successful relationships with UN organizations, the company could look to invest in two types of ground terminals in their portfolio. The first type would involve establishing a local centre of operations with its own terminals, infrastructure and servers to allow for near real-time acquisition and processing of satellite data.

The other ground terminals that the company may wish to acquire are EUMETCast reception stations, which connect to EUMETCast—a multi-service dissemination system that uses telecommunication geostationary satellites with multicast technology. MENASat could purchase and install these ground terminals at selected refugee camp sites around the continent to provide education and tele-medicine services to refugees.

5 Conclusion

Millions of people have been displaced from their homes as a result of violent conflicts and are suffering atrocious human rights violations on an ongoing basis. This article proposes a remote sensing company called MENASat as an example to highlight how remote sensing and EO data can be employed to assist. MENASat would provide invaluable information to assist relief aid organizations such as UNHCR by providing the following services:

1. Planning location of refugee camps based on identified busy routes, informal settlement locations, border post locations, environmental considerations and availability of water resources.
2. Monitoring of key areas in the Mediterranean Sea to co-ordinate search and rescue missions and reduce fatalities of refugees.
3. Identify mine fields in conflict zones and provide this information to UNMAS to support their decision-making procedures.

This article has highlighted the vital need for these services and detailed how they may be actioned in an attempt to help with the largest refugee crisis of all time.

Nicolas Ringas graduated with a Bachelor of Science degree in Electrical Engineering from the University of the Witwatersrand, in Johannesburg, South Africa, in 2012. Since graduating Nicolas has been working at an engineering consultancy firm in the water, oil and natural gas sector, specializing in electromagnetic interference issues with AC power lines, rail ways and cables. He is currently completing a Master's of Philosophy in Space Sciences at the University of Cape Town in South Africa.

Satellites and Their Potential Role in Supporting the African Union's Continental Early Warning System

David Lindgren

Abstract
This article outlines the role of satellites in supporting the African Union's Continental Early Warning System to detect and report on outbreaks of conflict that occur on the continent. The AU and its regional counterparts currently rely on open source information, such as media reports, to develop and inform their early warning systems. As Earth observation data has become more readily available and accessible over the past decade, including with its use in academic and public policy research, it holds potential for complementing and strengthening the reports developed by early warning centers for policymakers to make decisions. Sustainable development and the fostering of African societies will not be possible without addressing the significant levels of conflict occurring on the continent, thus an emboldened and effect early warning system that incorporates satellite imagery and data in its efforts to prevent and mitigate conflict will help the continent achieve these goals.

1 Introduction

Many within the international development sector argue that development cannot be achieved without security, thus making security an important prerequisite for any successful development interventions. Research by academics and leading research organizations, such as Brookings Institution and the Stockholm International Peace Research Institute (SIPRI), have explored and determined the importance of the relationship between security and development, including the powerful reciprocal

D. Lindgren (✉)
SpaceLab, University of Cape Town, Rondebosch, South Africa
e-mail: davidlindgrenj@outlook.com

A. Froehlich (ed.), *Space Fostering African Societies*, Southern Space Studies,
https://doi.org/10.1007/978-3-030-32930-3_13

effects that they create for each other. Security facilitates development, while development reinforces security and reduces risk factors.

1.1 The Security and Development Relationship

Citing various cases, it remains evident that security affects development. For example, the Center for Global Development, another leading research group, has argued that development gains made since the invasion of Afghanistan remain at risk as security deteriorates in that country following the withdrawal of coalition troops. The gains that were made since 2001–2017 included a more than doubling of income per capita, increase to primary health from nine percent to 82% of the population, and the increase in school enrollments from less than one million children in 2001 to nine million presently, including an increase of girl school enrollments from 10 to 40%.[1] The mayor of Baghdad, Thikra Mohammed Jabir Alwash, has also publicly commented on the role of improved security resulting in greater development in her own city in Iraq, specifically citing the clearing of roads of mines and other obstacles that has led to investments which generate jobs and long-term stability.[2] SIPRI supports these findings and observations, particularly as it writes "As conflict increases, development gains in countries affected by the conflict are reversed and new poverty is created."[3] SIPRI notes that a civil war or humanitarian crisis in a developing country can set back a country's economic growth and human development by a generation, and that 30 of the 50 fragile states facing such scenarios exist within Africa.[4]

1.2 Africa's Insecurity and Its Implication for Development

While the security-development nexus remains evident for areas throughout the world, it plays an outsized role in Africa, where recent years have featured the highest number of conflicts on the continent since the end of World War II. Africa experienced 21 state-based conflicts in 2016, which was the highest number of

[1]Pisa, Michael, "Fragile Gains in a Fragile State: Economic Development in Afghanistan," 2017, Center for Global Development, Accessed via: https://www.cgdev.org/blog/fragile-gains-fragile-state-economic-development-afghanistan.
[2]Kharas, Homi and Jones, Bruce, "Improving the relationship between security and development," 2018, Brookings Institution, Accessed via: https://www.brookings.edu/blog/future-development/2018/04/27/improving-the-relationship-between-security-and-development/.
[3]Milante, Gary and Jan, Suyoun and Burt, Alison, "Security and Development," 2015, Stockholm International Peace Research Institute, Accessed via: https://www.sipri.org/sites/default/files/SIPRIYB15c08sI.pdf.
[4]Ibid.

state-based conflicts ever recorded for the continent, and a peak of 50 non-state conflicts in 2017.[5]

It has been estimated that armed conflicts cost African countries at least 15% of gross domestic product (GDP) each year, or nearly the equivalent of $18 billion per year. The case study of Guinea-Bissau demonstrates this during its 1998–1999 conflict in which the country had a negative 10.15% GDP growth rate, whereas it had been projected to receive a positive 5.24% growth rate before the outbreak of conflict.[6] Conflict is not only accompanied by significant economic costs, but also severe loss of human life. In 2017 alone, the continent had more than 60,000 battle deaths resulting from conflict. The intensity and number of conflicts results in varying battle deaths year to year, but have ranged from an estimate of more than 20,000 in 2010 to more than 500,000 in 1994, the latter being the year of the Rwandan genocide.[7] Studies have determined that poor living and health conditions caused as a result of conflict in Africa have led to the death of five million children under the age of five from 1995 to 2015.[8] A report developed by the Food and Agriculture Organization and the World Food Programme found that for 2019 nearly 25 million people across five conflict-affected countries were food insecure, and six million of these people were experiencing emergency or catastrophe levels of food insecurity.[9]

1.3 The Role of Security in the African Union's Agenda 2063

Given the large number of conflicts occurring in Africa and the devastating consequences that they produce, it remains pivotal for the continent to achieve security in order to begin realizing sustainable development results. Viewing security as crucial to a prosperous Africa, the African Union highlighted its importance in the organization's Agenda 2063, which sets out the continental body's objectives and priorities for the next 50 years and is "the continent's strategic framework that aims to deliver on its goal for inclusive and sustainable development."[10] Agenda 2063

[5]Bakken, Indgrid and Rustad, Siri Aas, "Conflict Trends in Africa, 1989–2017", 2018; Accessed via: https://reliefweb.int/sites/reliefweb.int/files/resources/Conflict%20Trends%20in%20Africa%2C%201946%E2%80%932017%2C%20Conflict%20Trends%20Report.pdf.

[6]Al Jazeera, "Africa aid swallowed up by conflict," 2007, Al Jazeera, Accessed via: https://www.aljazeera.com/news/africa/2007/10/200852513936102455.html.

[7]Bakken, Indgrid and Rustad, Siri Aas, "Conflict Trends in Africa, 1989–2017", 2018; Accessed via: https://reliefweb.int/sites/reliefweb.int/files/resources/Conflict%20Trends%20in%20Africa%2C%201946%E2%80%932017%2C%20Conflict%20Trends%20Report.pdf.

[8]Bhalla, Nita, "African wars killed five million sick children in 20 years: study," 2018, Reuters, Accessed via: https://www.reuters.com/article/us-africa-conflict-children/african-wars-killed-five-million-sick-children-in-20-years-study-idUSKCN1LF2R4.

[9]Rolby, Christin, "Conflict expected to deepen Africa's hunger crisis in 2019," 2019, Devex, Accessed via: https://www.devex.com/news/conflict-expected-to-deepen-africa-s-hunger-crisis-in-2019-94195.

[10]African Union, "Agenda 2063: The Africa We Want," 2013, African Union, Accessed via: https://au.int/agenda2063/overview.

was adopted in 2013 to celebrate the founding of the Organization of Africa Unity/African Union 50 years prior. As the AU states, Agenda 2063 is meant to facilitate the achievement of a pan-African vision of "an integrated, prosperous and peaceful Africa, driven by its own citizens, representing a dynamic force in the international arena."[11]

Contained within Agenda 2063, the AU identified an independent security-related "aspiration area" supported by underlying "goals" and "priority areas." The aspiration area "a peaceful and secure Africa" is supported by three goals (peace, security and stability is preserved; a stable and peaceful Africa, and a fully functional and operational APSA) followed by four priority areas, namely maintenance and preservation of peace and security, institutional structure for AU instruments on peace and security, defence and security and peace, and fully operational and functional APSA pillars.[12] Here, APSA refers to the African Peace and Security Architecture.

1.4 The African Union and Its Security Mechanisms

The Peace and Security Council Protocol, which includes the elements of the APSA, was adopted in 2002 and entered into force in 2003. It calls for the creation of a Peace and Security Council (PSC) that should function in the promotion of peace, security, and stability; early warning and preventive diplomacy, peace-making, peace support operations and intervention, peace-building and reconstruction, and humanitarian action and disaster management.[13] As part of fulfilling these various functions, the PSC receives support from the PSC Commission, Panel of the Wise, continental early warning system, the African Standby Force, and the Peace Fund.[14]

The emphasis on the APSA within Agenda 2063 likely results from a 2010 assessment of the APSA conducted by the AU, which found that there was limited coordination of the APSA elements outside of the early warning system and African Standby Force and that there was differing paces of development among the elements themselves.[15] Importantly, the relationship between the AU and many of

[11]Ibid.

[12]The African Union Commission, "First Ten-Year Implementation Plan 2014–2023," 2015, African Union, Accessed via: https://www.un.org/en/africa/osaa/pdf/au/agenda2063-first10yearimplementation.pdf.

[13]African Union, "Protocol Relating to the Establishment of the Peace and Security Council of the African Union," 2002, African Union, Accessed via: https://au.int/sites/default/files/treaties/7781-treaty-0024_-_protocol_relating_to_the_establishment_of_the_peace_and_security_council_of_the_african_union_e.pdf.

[14]African Union, "African Peace and Security Architecture," 2019, African Union, Accessed via: http://www.peaceau.org/uploads/african-peace-and-security-architecture-apsa-final.pdf.

[15]African Union Peace and Security Department, "African Peace and Security Architecture (APSA): 2010 Assessment Study," 2010, African Union, Accessed via: https://www.securitycouncilreport.org/atf/cf/%7B65BFCF9B-6D27-4E9C-8CD3-CF6E4FF96FF9%7D/RO%20African%20Peace%20and%20Security%20Architecture.pdf.

the regional economic communities expected to support the functioning of the ASPA, including on early warning, was unclear, especially around whether the AU should be implementing or only coordinating information between the various regional groupings.[16]

2 Continental Early Warning System

The African Union's Continental Early Warning System (CEWS) was established to help address outbreaks of conflict similar to those as described previously. Following the failure of the Organization of African Unity (OAU) to address and prevent the atrocities that constituted the Rwandan genocide and ongoing civil war in the Democratic Republic of the Congo during the 1990s, the continental grouping in fact disbanded and reformed to become the African Union largely around the issue of whether African states can intervene to prevent conflict in other African states.[17] The AU, which emerged as the successor organization to the OAU, expressly included in its founding constitution that states did have the right to intervene pre-emptively to prevent the failures that occurred under the OAU regime. The pre-emptive policy found within the AU constitution signified one of the first times the concept was applied in international law.[18] "The AU prohibits interference in the internal affairs of one member by another but, however, gives the AU itself the right of intervention 'in respective of grave circumstances namely war crimes, genocide and crimes against humanity.'"[19] As part of this effort, the AU established the CEWS to support the organization in fulfilling this mission. CEWS consists of an observation and monitoring center (i.e. a situation room) that advises the PSC on potential conflicts and threats to security.[20]

Similarly, the AU recognizes that it does not operate independently within the continent on peace and security issues, and relies on and closely collaborates with regional economic communities (RECS), including the Arab Maghreb Union (AMU), Economic Community of West African States (ECOWAS), East African Community (EAC), Intergovernmental Authority on Development (IGAD), Southern African Development Community (SADC), Common Market for Eastern and Southern Africa (COMESA), Economic Community of Central African States (ECCAS), and Community of Sahel-Saharan States (CENSAD). As such, the

[16]Ibid.
[17]Kufuor, Kofi Oteng, "The Collapse of the Organization of African Unity: Lessons from Economics and History," 2005, Journal of African Law Vol. 49 No. 2, Accessed via: https://www-jstor-org.proxyau.wrlc.org/stable/27607944?seq=3#metadata_info_tab_contents.
[18]Ibid.
[19]Ibid.
[20]African Union Peace and Security Department, "African Peace and Security Architecture (APSA): 2010 Assessment Study," 2010, African Union, Accessed via: https://www.securitycouncilreport.org/atf/cf/%7B65BFCF9B-6D27-4E9C-8CD3-CF6E4FF96FF9%7D/RO%20African%20Peace%20and%20Security%20Architecture.pdf.

RECS named above are expected to play a similar and complementary role in supporting and collaborating on the continent-wide early warning system.

2.1 Challenges Affecting the Continental Early Warning System

Although as mentioned above, a 2010 assessment found a lack of clarity on how these bodies should practically be engaging each other and the AU on CEWS. The RECs themselves as well were found to have been slow in developing their own region-specific early warning systems. The assessment determined that automated data collection and reporting were advanced at the AU CEWS level in addition to the ECOWAS-specific and IGAD-specific early warning systems, ECOWAS Early Warning and Response Network (ECOWARN) and Conflict Early Warning and Response Mechanism (CEWARN), respectively. However, the other RECs beyond ECOWAS and IGAD have not yet developed effective data collection and reporting for their regional early warning systems, particularly among CEN-SAD, EAC, and COMESA.[21] In addition to the region-specific challenges associated with early warning and the pace and emphasis on setting up such systems, the report also identified funding, staffing, and technical capabilities and hardware limitations affected the effective operation of CEWS.[22]

In terms of the practical operation of the early warning system, it relies upon open-source information collected from government, civil society, and media reports. For example, the European Union's Joint Research Centre (JRC) developed the Africa Media Monitor tool that supports CEWS to produce daily news publications used by the CEWS situation room and governmental decision-makers.[23]

2.2 ECOWARN as a Regional Early Warning System

Similarly, the ECOWAS-specific early warning system, ECOWARN, operates by collecting reports from government and civil society representatives. ECOWARN relies upon country-level offices that include representation from the national government and the West Africa Network for Peacebuilding (WANEP), which serves as the civil society voice in reporting. On a weekly basis, a designated contact person within each country prepares a risk indicator form for the country, and transmits this to the ECOWAS department responsible for early warning, which is then used by national and regional officials to make decisions based on the

[21]Ibid.
[22]Ibid.
[23]Joint Research Centre, "The African Union and the Joint Research Centre (JRC)," 2019, European Union, Accessed via: https://ec.europa.eu/jrc/sites/jrcsh/files/jrc_leaflet_african_union_en.pdf.

reports.[24] ECOWAS also has developed a geographic information systems (GIS)-based tool that maps security incident reports, but this has not been updated since 2017 thus limiting its efficacy.[25] Nonetheless, ECOWARN does serve as a positive regional model for other RECs to emulate, particularly in regard to the political support and attention the early system receives, such as was expressed by the Nigerian Permanent Representative to ECOWAS during 2017 in which he remarked on the importance of fully developing the capacity of the system to effectively prevent and mitigate conflict.[26]

Unlike ECOWAS which includes civil society representation in the compilation of its risk reports, other RECs do not include civil society as observed by the AU 2010 assessment of CEWS, which could prove problematic in arriving at a real-time and accurate understanding of security risks as often times civil society maintains extensive networks within countries to verify and augment official government information. The 2010 report supported this when it stated, "Such stakeholders can provide valuable information, expertise and increase awareness of CEWS and the work carried out by AUC [African Union Commission] and the RECs."[27]

3 Satellites and the Continental Early Warning System

Satellites can serve as powerful tools in support of CEWS and its regional counterparts. Satellites as space-based assets are useful across a wide range of areas, and importantly in carrying out earth observation missions. Earth observation entails the collection of information about the physical, chemical, and biological systems on Earth, including the monitoring of the status of these systems and the changes that occur within in them whether they are nature or man-made systems.[28] The Group on Earth Observations comprises over 100 national governments and over 100 organizations that currently work to collaborate on realizing the benefits of Earth observations by making decisions and actions based on the collected information.[29] Separately, the Committee on Earth Observation Satellites (CEOS), a committee supported by 28 space agencies and 20 other organizations, also works to

[24] Sagna, Augustin, "The ECOWAS Early Warning and Response Network," 2009, Sahel and West Africa Club – OECD, Accessed via: http://www.oecd.org/swac/theecowasearlywarningandresponsenetwork.htm.

[25] ECOWAS, "ECOWARN Incidents," 2017, ECOWAS, Accessed via: https://ecowas.maps.arcgis.com/apps/View/index.html?appid=9c6878d7d90c42d592aa5957c768ce1d.

[26] ECOWAS, "ECOWAS strengthens capacity of its Early Warning Directorate on governance and human rights," 2017, ECOWAS, Accessed via: https://www.ecowas.int/ecowas-strengthens-capacity-of-its-early-warning-directorate-on-governance-and-human-rights/.

[27] African Union Peace and Security Department, "African Peace and Security Architecture (APSA): 2010 Assessment Study," 2010, African Union, Accessed via: https://www.securitycouncilreport.org/atf/cf/%7B65BFCF9B-6D27-4E9C-8CD3-CF6E4FF96FF9%7D/RO%20African%20Peace%20and%20Security%20Architecture.pdf.

[28] Group on Earth Observations, "FAQ," 2019, Group on Earth Observations, Accessed via: https://www.earthobservations.org/g_faq.html.

[29] Ibid.

coordinate "civil space-borne observations of the Earth."[30] Similar as the Group on Earth Observations, CEOS does this to "enhance international coordination and data exchange and to optimize societal benefit."[31]

3.1 Growth of Earth Observation Users and Uses

Whereas intergovernmental and national level bodies have established efforts dedicated to Earth observation, private actors, including companies and non-profit groups, have increasingly relied on Earth observation to market and sell products and services or to improve and complement their on-the-ground work. Planet, a company established around developing and deploying satellites for collecting satellite imagery, and Earth Observation PBC, which relies on satellite imagery to sell services for managing land and forests, are examples of for-profit entities making use of the explosion and interest in Earth observation. Radiant.Earth Foundation serves to make satellite imagery widely accessible and usable for development practitioners seeking to incorporate Earth observation information in their work. These remain just a few of the many organizations established within the much broader Earth observation industry.

Beyond these uses of Earth observation, it is increasingly being sought by academics, researchers, and development practitioners to build new knowledge, identify trends, and construct new analyses. The Council on Foreign Relations (CFR), the Center for Strategic and International Studies (CSIS), and the Harvard Humanitarian Initiative (HHI) prove as examples of such groups seeking make use of Earth observation data. For example, the CFR Global Conflict Tracker serves as a tracking tool that maps conflicts and collates data from various sources of information to provide in-depth detail on ongoing conflicts around the world. As a United States-based research organization, it categorizes these conflicts into three categories depending on their level of impact on the United States's interests.[32] Separately, CSIS has recently released a report entitled "Space Threat Assessment 2019" that relied heavily on Earth observation data and satellite imagery as evidence to support its findings. For example, in cooperation with DigitalGlobe, another private firm offering satellite imagery, CSIS was able to identify locations in the South China Sea where China has deployed jamming systems. Similarly, CSIS, again working with DigitalGlobe, identified a satellite launching station in North Korea to support the findings within its report.[33]

[30]Committee on Earth Observation Satellites, "CEOS," 2019, Committee on Earth Observation Satellites, Accessed via: http://www.un-spider.org/sites/default/files/CEOS.pdf.
[31]Ibid.
[32]Council on Foreign Relations, "Global Conflict Tracker," 2019, Council on Foreign Relations, Accessed via: https://www.cfr.org/interactive/global-conflict-tracker/?category=us&vm=grid.
[33]Aerospace Security Project, "Space Threat Assessment 2019," 2019, Center for Strategic and International Studies, Accessed via: https://aerospace.csis.org/space-threat-assessment-2019/.

3.2 Satellite-Based Early Warning in Sudan

Outside of these groups, the Harvard Humanitarian Initiative launched a program in 2017 on crisis mapping and early warning to examine the use of information communications technology (ICT) in conflict and disaster settings.[34] In particular, HHI focuses on examining the "impact of crisis mapping, geospatial and crowd sourcing technologies to prepare, mitigate, and respond to emergencies."[35] In particular, HHI has worked with the Satellite Sentinel Project, which has now evolved into the Sentry, to collect evidence of conflict in Sudan, especially on mass atrocities committed by the government. The Satellite Sentinel Project, which was a joint collaborative effort by DigitalGlobe and the Enough Project as an advocacy organization based in the United States, collected satellite imagery, analyzed it for changes to detect bombings and burnings of settlements, and used that information to inform policymakers. The Satellite Sentinel Project used changes in satellite imagery as cues to investigate further to determine bombings and settlement burnings, and further complemented this data with on-the-ground reports from local sources to produce reports that were then delivered to international and national level officials for making policy decisions.[36] In a review and evaluation of the project, researchers identified that the successes made possible with the project were only made possible with the use of Earth observation technologies, namely "improving the speed, agility and tempo with which evidence could be collected and presented for the benefit of vulnerable populations."[37] In particular, the short periods of time that passed between each new satellite capture of an area by DigitalGlobe's satellites allowed for a "tempo of refresh (i.e. collection of new imagery) over key locations that made the observation of micro-changes in a village or a military base possible."[38] To complement this satellite imagery, HHI staff collected and geo-coded other data found outside of the imagery to help make sense of what was observed.[39]

3.3 The AU's CEWS and Making Use of Satellite Imagery

As demonstrated above, satellite imagery and Earth observation data offer significant insights into supporting and expanding upon findings related to conflict situations. Ranging from mapping and tracking conflicts on a location basis and using

[34]Harvard Humanitarian Initiative, "Crisis Mapping and Early Warning," 2019, Harvard University, Accessed via: https://hhi.harvard.edu/resources/crisis-mapping-and-early-warning.
[35]Ibid.
[36]Satellite Sentinel Project, "Our Story," 2015, Satellite Sentinel Project, Accessed via: http://www.satsentinel.org/our-story.
[37]Raymond, Nathaniel and Davies, Benjamin and Card, Brittany and Al Achkar Ziad and Baker, Isaac, "While We Watched: Assessing the Impact of the Satellite Sentinel Project," 2013, Science & Technology, Georgetown Journal of International Affairs, Accessed via: https://reliefweb.int/sites/reliefweb.int/files/resources/185-191-SCI%2BTECH_Raymond.pdf.
[38]Ibid.
[39]Ibid.

satellite imagery to detect changes in military installations to combining Earth observation data with local reports to verify observed changes, satellites hold potential in supporting new ways of doing conflict research, particularly in early warning systems.

While the AU's CEWS suffers from challenges outside of the collection and reporting of information itself, reports that are produced by CEWS and its RECs counterparts can be enhanced and strengthened by the use of satellite imagery and Earth observation data. Prior to a conflict occurring, researchers and analysts could more readily track movements of people or potential armed groups that may be mobilizing for a conflict. Satellite imagery could also be used in a complementary fashion to open source reports that are already being incorporated in the early warning system, such as with the Africa Media Monitor supported by the European Union. Satellite imagery could bolster on-the-ground reports by verifying crowd sizes and the affected areas, whereas local sources could flag areas of interest for further detailed monitoring by satellites which were previously unobserved or closely monitored. The relationship between the different types of sources both open source and satellite-based could be significantly enhanced within CEWS to make its reports more credible and thus more convincing to regional and national level policymakers. Importantly, as noted in the lessons learned from the Satellite Sentinel Project, the tempo and refresh of satellite images could help policymakers stay more directly and regularly informed of often rapid changing conflict dynamics, which otherwise may be difficult to ascertain in remote environments where conflicts occur.

As discussed previously, ECOWARN leads the RECs in its regional early warning system in that it includes both governmental and civil society sources; however, not all RECs follow this model. Thus, satellites could be used to verify information presented within government-only produced reports in cases where a government may not be forthcoming about conflict occurring within its territory for various political or security reasons. Satellite imagery would indirectly support the strengthening of the AU and its regional counterparts as independent institutions capable of not only serving their member states and promoting continental cooperation, but also championing transparency and good governance through an independent verification process made possible by Earth observation data.

Finally, the use of satellite data within the early warning system would build upon efforts by ECOWARN to map conflicts and geo-code information received from ground sources to present a fuller and more comprehensive picture of conflicts as they evolve over time and place.

4 Conclusion

Sustainable development and the fostering of African societies as envisioned by the African Union's Agenda 2063 will not be possible without achieving security on the continent. Agenda 2063 recognizes the high importance of security, and the African Union itself was organized in part in response to its predecessor's failures to provide leadership on security matters for the continent. Thus, the AU established the continental early warning system and envisioned a system of working with its regional counterparts and their own early warning systems to provide a network of information so that relevant officials can make decisions to prevent outbreaks of conflict. In order to promote security, and indirectly the sustainable development and fostering of African societies, the AU could increase its effectiveness on its early warning systems by incorporating the use of satellites and the imagery they produce. Political buy-in and understanding of the information presented to policymakers will be instrumental to ensuring conflict is mitigated and decisions are made in a timely and responsive manner. Early warning reports, supported by Earth observation data, will help achieve this objective and benefit the African continent overall.

David Lindgren is an experienced international development professional with a strong knowledge and practice in program development and management. David serves as the Associate Director of Research at American University's School of International Service, where he works closely with faculty and doctoral students throughout the grant and project development process from idea to submission. Previously, David served as a program officer with Freedom House in Johannesburg, South Africa, where he implemented democracy and human rights programs. Prior to this, he worked in Washington, DC supporting programs across Central, West, and Southern Africa. David graduated from American University's School of International Service in 2011 with a B.A. in International Studies, and is currently a candidate for an M.Phil. in Space Studies at the University of Cape Town.

Study of Fractures Network in the Basement of Socotra Island—Yemen by Using Remote Sensing and GIS Techniques

Khaled Khanbari, Sylvie Leroy, Ahmad Adris, Sami Moheb-Al-Deen and Waheed Al-Sarari

Abstract
Fractured basement rocks have become an increasingly common target for hydrocarbon production in the Republic of Yemen, where the oil is in production from the deeply buried fractured basement. These basement reservoirs underlie a thick sequence of Mesozoic to Tertiary sediments and in order to study the characteristics of this buried basement, indirect methods must be used. The present work deals with the production of structural maps of basement of Socotra Island, and the extraction of lineaments from remote sensing data (satellite imageries and aerial photographs) as well as the analysis of these extracted lineaments by using GIS. The numerous fractures detected in the basement are characterized by consistent variations in trend, length and density when they affect the different lithological units out cropping in the study areas. The granitic basement is expected to have the best fractured reservoir potential. The dominant fracture trends in the basement of Socotra Island are NW-SE and NE-SW. The presence of this fracture system is thought to be due to major tectonic activity which formed new fractures and also reactivated older fractures.

K. Khanbari (✉)
Department of Earth and Environmental Sciences, Sana'a University, Sana'a, Yemen
e-mail: kkhanbari@hotmail.com

S. Leroy
ISTEP UPMC Paris, Paris, France
e-mail: Sylvie.leroy@umpc.fr; Sylvie.leroy@sorbonne-universite.fr

K. Khanbari · A. Adris · S. Moheb-Al-Deen · W. Al-Sarari
Yemen Remote Sensing and GIS Center, Sana'a, Yemen
e-mail: geophysicall@yahoo.com

S. Moheb-Al-Deen
e-mail: samoh2010@gmail.com

W. Al-Sarari
e-mail: walsarari@gmail.com

A. Froehlich (ed.), *Space Fostering African Societies*, Southern Space Studies,
https://doi.org/10.1007/978-3-030-32930-3_14

1 Introduction

The occurrence of naturally fractured basement reservoirs has been known within the hydrocarbon industry for many years. Reservoirs in fractured basements, where the oil and gas in place may be within an extensive fracture network on variety of different scales. Fractured basement rocks have become an increasingly common target for hydrocarbon production in the Republic of Yemen, where the oil is in production from the deeply buried granitic/gneissic fractured basement. These basement reservoirs underlie a thick sequence of Mesozoic to Tertiary sediments and in order to study the characteristics of this buried basement, indirect methods must be used.

Remote Sensing and Geographic Information System (GIS) techniques help to identify the fractures that effect basement outcrops. Generally the fractures of basement outcrops reflect the characteristics of the buried basement. Therefore a greater understanding of fracture distribution within basement reservoirs using Remote Sensing and GIS techniques may prove to be the key tool for improved exploration and management of this hidden resource.

Socotra Island (Yemen) is located in the northwest Indian Ocean near the Gulf of Aden (Fig. 1). Socotra is about 120 by 40 km and covers an area of 3625 km^2. It is composed of a basement complex of igneous and metamorphic rocks of Pre-Cambrian age overlain by sedimentary rocks, mainly limestone and sandstone.

The main aim of this study is to explore the fracture network in the Pre-Cambrian rocks of Socotra Island from remote sensing data (satellite imageries and aerial photographs). Remote Sensing and GIS techniques were used to produce structural maps of the basement areas.

2 Tectonic and Geological Setting

The Socotra Island belongs to the southern continental margin of the Gulf of Aden (Fig. 1) which corresponds to a an active oblique rift system with young margins along the Arabia-Somalia plate boundary connects to the Indian Ocean to the east and to the Red Sea to the west. This rift separates the Somalia plate to the South and the Arabian plate to the North. Rifting history started around 35 Ma and oceanic accretion is recorded since 17.6 Ma to the east.[1,2] The oceanic floor becomes

[1]Leroy, S., Gente, P., Fournier, M., d'Acremont, E., Bellahsen, N., Beslier, M.-O., Patriat, P., Maia, M., Blais, A., Perrot, J., Al-Kathiri, A., Merkouriev, S., Ruellan, P.-Y., Fleury, J.-M., Lepvrier, C., Huchon, P., 2004. From rifting to spreading in the eastern Gulf of Aden: a geophysical survey of a young oceanic basin from margin to margin. Terra Nova, 16, 185–192.
[2]Fournier M., Chamot-Rooke N., Petit C., Huchon P., Al-Kathiri A., Audin L., Beslier M.O., D'Acremont E., Fabbri O., Fleury JM., Khanbari K.,Lepvrier C., Leroy S., Maillot B. and Merkouriev S., 2010, Arabia-Somalia plate kinematics, evolution of the Aden–Owen-Carlsberg triple junction, and opening of the Gulf of Aden, *J. Geophys. Res.*,V. 115, B04102.

Fig. 1 Topographic and bathymetric map of the Gulf of Aden and surrounding regions showing the location of Socotra Island and the direction and rate of motion of Arabia and Somalia with respect to Africa (Nubia and Somalia)

younger in the westernmost part of the Gulf of Aden, where the oldest magnetic anomaly corresponds to 3 Ma.[3,4] This gulf results from an oblique rifting characterized by a direction of extension N20° and a rift oriented N70°.[5,6] The Socotra

[3]Cochran, J.R., 1981. The Gulf of Aden: Structure and Evolution of a Young Ocean Basin and Continental Margin. J. Geophys. Res. 86 (B1), 263–287.

[4]Audin, L., 1999. Penetration de la dorsale d'Aden dans la depression Afar entre 20 et 4 Ma. PhD Thesis, Paris 7 University, France. 278 pp.

[5]Dauteuil, O., Huchon, P., Quemeneur, F., Souriot, T., 2001. Propagation of an oblique spreading center: the western Gulf of Aden. Tectonophysics 332, 423–442.

[6]D'Acremont, E., Leroy, S., Beslier, M.-O., Bellahsen, N., Fournier, M., Robin, C., Maia, M., and Gente, P., 2005. Structure and evolution of the eastern Gulf of Aden conjugate margins from seismic reflection data, Geophys. J. Int., 160, 869–890.

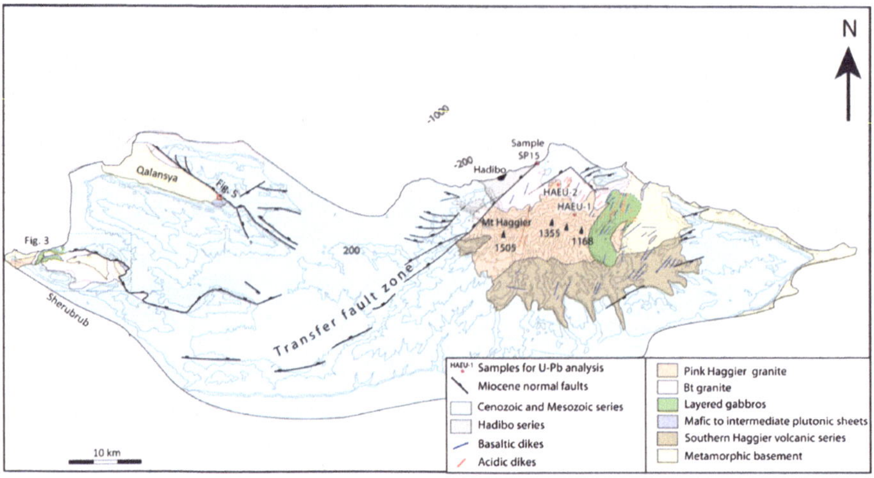

Fig. 2 Geological and topographic map of Socotra Island showing the location of basement exposures in three main areas on Socotra (Denele Y., Leroy S., Pelleter E., Pik R., Talbot JY., Khanbari K., 2012, The Cryogenian arc formation and successive high-K calc-alkaline plutons of Socotra Island (Yemen), Arab J Geosci 5(5):903–924)

Island has occupied, prior to the Oligo-Miocene Gulf of Aden opening, a position between the proto-East Gondwana and the Arabian Nubian Shield.[7]

Socotra Island composed of a basement complex of igneous and metamorphic rocks of Pre-Cambrian age overlain by sedimentary rocks, mainly limestone and sandstone (Fig. 2). It is covered by a carbonate-dominated Cretaceous and Tertiary succession unconformably overlying a Proterozoic and Paleozoic basement. Triassic and Jurassic deposits are locally preserved in a fault bounded area at the eastern end of the island.

Basement is exposed in three main areas of Socotra. The basement mainly includes Pan-African granites, which make up the 1500 m-high Haggier massif (Fig. 2). The oldest basement lithologies, metamorphosed to amphibolites facies, are metasediments, amphibolites and acid orthogneisses. These are intruded by gabbros, acid volcanics and a variety of acid plutons, including granodiorites, adamellites and peralkaline granites.[8,9]

---

[7]Denele Y., Leroy S., Pelleter E., Pik R., Talbot JY., Khanbari K., 2012, The Cryogenian arc formation and successive high-K calc-alkaline plutons of Socotra Island (Yemen), Arab J Geosci 5 (5):903–924.

[8]Beydoun Z.R., Bichan H.R., 1970, The geology of Socotra Island, Gulf of Aden. Quaterly J Soc Lond 25:413–446.

[9]Denele Y., Leroy S., Pelleter E., Pik R., Talbot JY., Khanbari K., 2012, The Cryogenian arc formation and successive high-K calc-alkaline plutons of Socotra Island (Yemen), Arab J Geosci 5 (5):903–924.

On the island of Socotra, the southern passive margin of the Gulf of Aden displays along its strike two different types of asymmetric structures. Western Socotra is made up of a series of southward tilted blocks which are bounded by consistently northward dipping normal faults. Eastern Socotra consists of a broad asymmetric anticline with a steep northern limb and a gently dipping southern limb. A zone of NE–SW striking strike-slip and normal faults separates the two areas. The overall structure is interpreted as representing two rift segments separated by a transfer zone which is called Hadibo transfer zone.[10]

Three main outcrops of basement can be observed in Socotra Island (Fig. 2): Mont-Haggier area, Sherubrub area and Qalantssya area which is small and diffuse.[11] In this study, Mont-Haggier and Sherubrub areas which represent the major outcrops of the basement, have been studied.

3 Study of Fractures Network by Remote Sensing Data

Aerial photo interpretation technique was used for mapping the small fractures which effect the basement of Socotra Island, while Landsat-ETM image is used de identify the main structures such as major faults. Aerial photo mosaic with resolution of one meter and Landsat-ETM image have been interpreted by visual method in order to delineate the structures in the study areas (Mont-Haggier and Sherubrub area).

All structures derived from aerial photos and Landsat-ETM have been implemented in a GIS in order to obtain the related fracture maps, useful for the tectonic interpretation. The recognized structures have been analyzed by their trend. The numerous fractures detected are characterized by consistent variations in trend, length and density when they effect the different lithological units outcropping in the study areas.

3.1 Mont-Haggier Area (Eastern Outcrop)

Haggier mountains are located in eastern–central part Socotra Island (Fig. 2). This basement massif occupies area of 662 Km^2 with the highest altitude in the Island (1500 m). The oldest basement rocks in Mont-Haggier area are the metamorphic basement which comprise Paragneisses and Micaschists, Orthogenesis,

[10]Fournier M., Huchon P., Khanbari K., Leroy S., 2007, Segmentation and along-strike asymmetry of the passive margin in Socotra, eastern Gulf of Aden: Are they controlled by detachment faults?, *G3*, Q03007.

[11]Denele Y., Leroy S., Pelleter E., Pik R., Talbot JY., Khanbari K., 2012, The Cryogenian arc formation and successive high-K calc-alkaline plutons of Socotra Island (Yemen), Arab J Geosci 5(5):903–924.

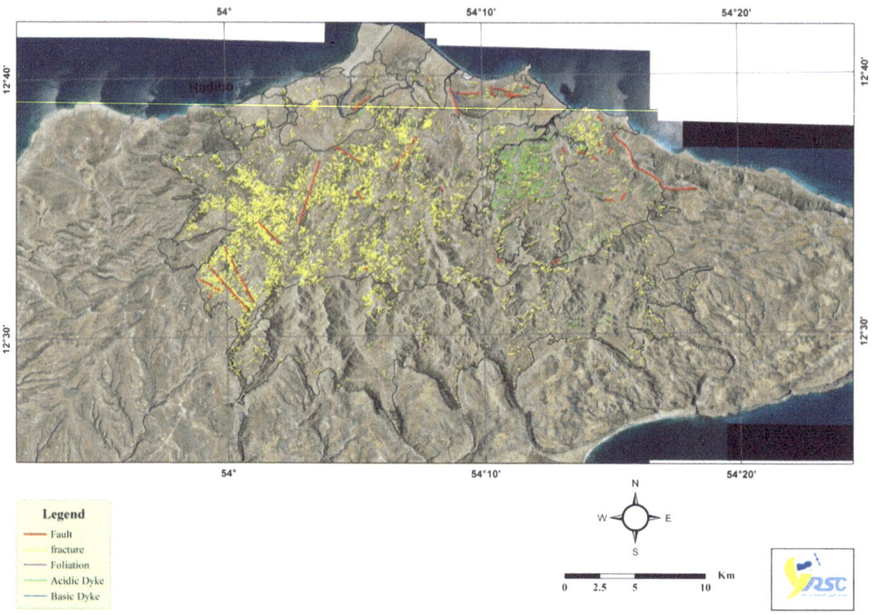

Fig. 3 Structural map of Mont-Haggier basement (the background is aerial photo mosaic with resolution of one meter)

Amphibolite and Quartzites. The Metamorphic basement is overlain by Metavolcanic series which corresponds to an association of pyroclastic and effusive rocks.[12]

Structural map of Mont-Haggier area (Fig. 3) shows that the basement is effected by different type of geologic structures (fractures, faults, dykes) which have different trend, length and density. The rose diagram (Fig. 4) shows that the dominant trends of fractures in Mont-Haggier area are NE-SW and NW-SE. Some other trends: N-S and E-W have been observed. The density map of fractures (Fig. 5) shows that the fracture density is high in the west and central parts of the study area which represent per-alkaline granites, while the other rock units are less affected by fractures.

3.2 Sherubrub Area (Western Outcrop)

Sherubrub basement is located in the western part of Socotra Island (Fig. 2). The basement occupies area about 53 Km^2. Sherubrub area is covered mainly by Granitic pluton which is constituted mainly by a pink Biotite-Hornblende granite.

[12]Denele Y., Leroy S., Pelleter E., Pik R., Talbot JY., Khanbari K., 2012, The Cryogenian arc formation and successive high-K calc-alkaline plutons of Socotra Island (Yemen), Arab J Geosci 5(5):903–924.

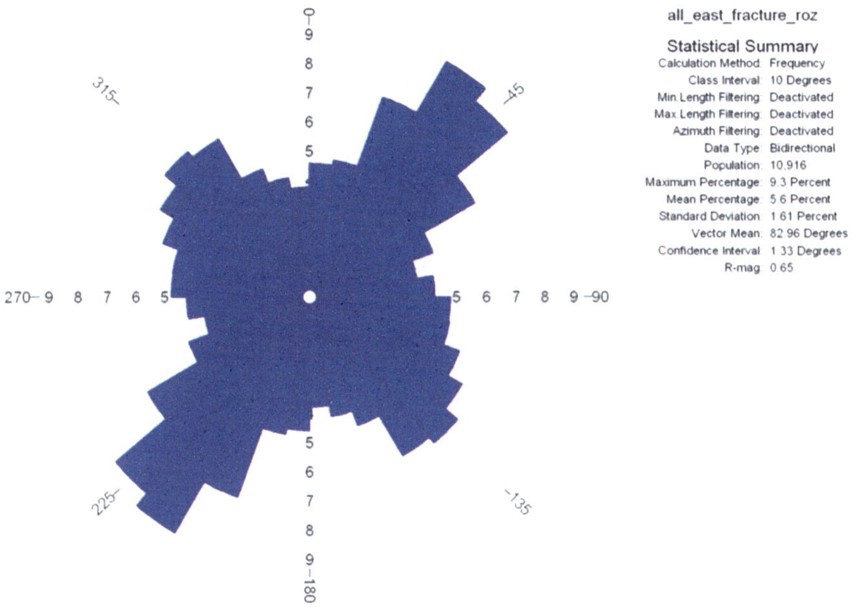

Fig. 4 Rose diagram showing the trends of fractures in Mont-Haggier basement

The pink granite and Gabbro are intended the metamorphic basement (Orthogenesis, Paragneisses, Quartzites). The volcanic dykes cut all petrographic formations in Sherubrub area. These dykes correspond to basaltic dykes with a microganular porphyric texture.[13]

Structural map of Sherubrub area (Fig. 6) shows that the basement is effected by different type of geologic structures (fracture, faults, dykes, floatation) which have different trend, length and density. The rose diagram (Fig. 7) shows that the dominant trends of fractures in Sherubrub area are NW-SE and NE-SW. All other trends such as N-S and E-W have been observed. The density map of fractures (Fig. 8) shows that the fracture density is high in the west and central parts of the study area which represent Paragneisses and Biotite-Hornblende granites, while the other rock units are less affected by fractures.

[13]Denele Y., Leroy S., Pelleter E., Pik R., Talbot JY., Khanbari K., 2012, The Cryogenian arc formation and successive high-K calc-alkaline plutons of Socotra Island (Yemen), Arab J Geosci 5(5):903–924.

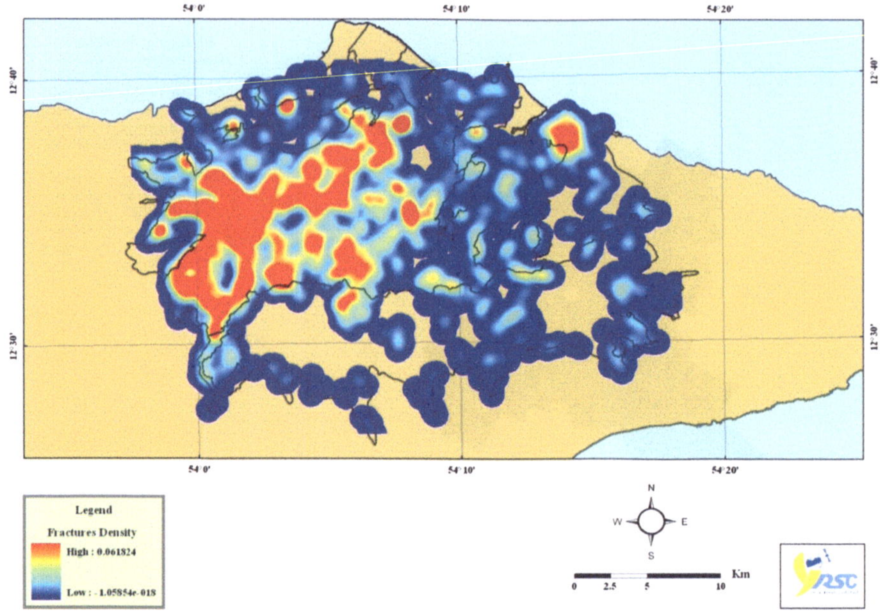

Fig. 5 Density map of fractures in Mont-Haggier area

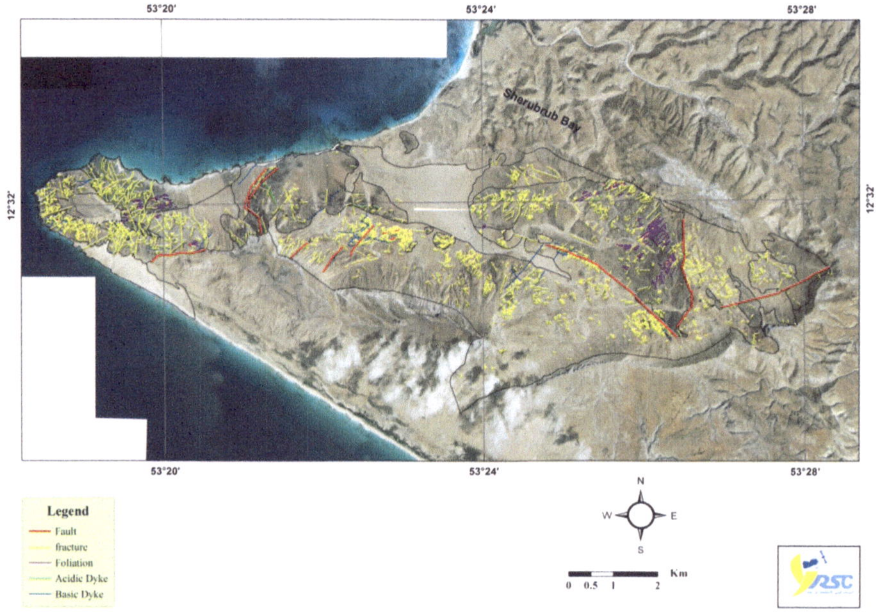

Fig. 6 Structural map of Sherubrub basement (the background is aerial photo mosaic with resolution of one meter)

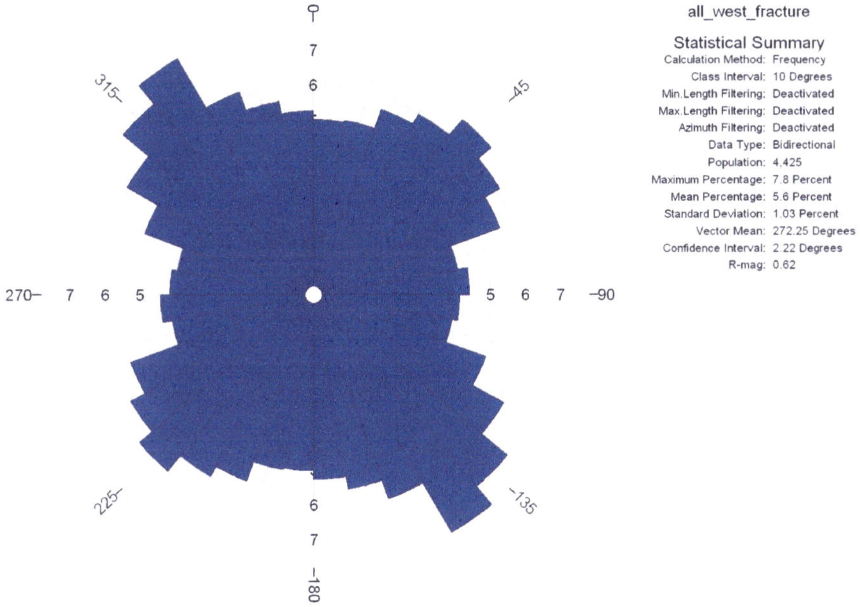

Fig. 7 Rose diagram showing the trends of fractures in Sherubrub basement

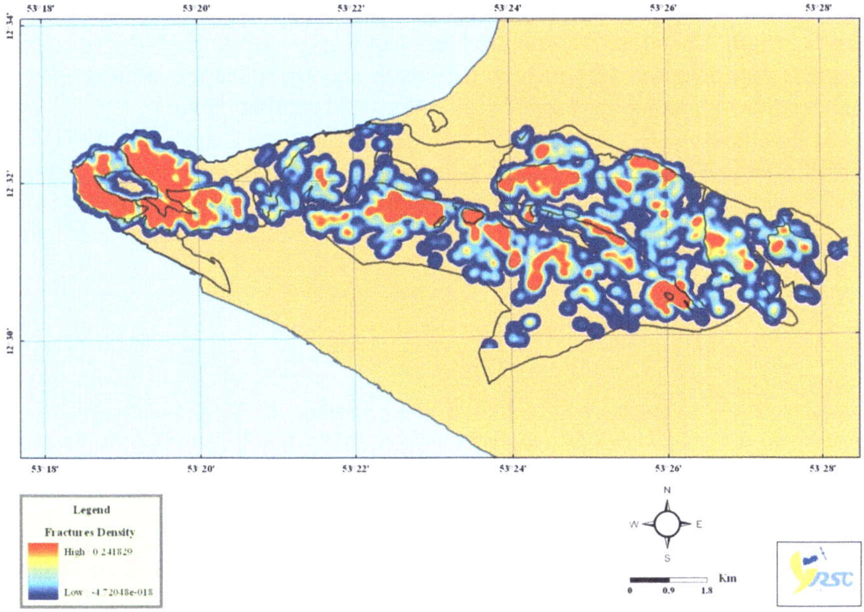

Fig. 8 Density map of fractures in Sherubrub area

4 Discussion and Conclusion

Basement of Socotra is correlated with Proterozoic basement of Arabian plate (especially Yemen and Oman). The construction of the Gulf of Aden shows that the Socotra Island has occupied, prior to the Oligo-Miocene rifting, a position close to the Precambrian Mirbat and Al-Halaaniyat Islands outcrops in Oman.[14,15] The Socotra Island has occupied, prior to the Oligo-Miocene Gulf of Aden opening, a position between the proto-East Gondwana and the Arabian Nubian Shield.[16]

The present work deals with the production of structural map of basement of Socotra Island, and the extraction of fractures from remote sensing data (satellite imageries and aerial photographs) as well as the analysis of these extracted fractures. The basement is exposed mainly in eastern Socotra (Mont-Haggier) and also in several western localities such as Sherubrub area. Basement of Socotra displays a great variety of metamorphic, plutonic and volcanic rocks.

The numerous fractures detected in the basement are characterized by consistent variations in trend, length and density when they affect the different lithological units out cropping in the study areas (Mont-Haggier and Sherubrub area). Fractures networks have been observed in all rock units. They exist over a wide range of scales from micro-fractures to largest faults. The fracture analysis shows two principal structural trends: NW-SE and NE-SW with less expressed trends around N-S and E-W. The average length of the fractures is about 25–100 m.

Basement in Socotra Island is intensely fractured. The nature of fracture network varies with lithology. It appears that more homogeneous lithologies of magmatic origin have been better fracture properties than the layered formations of metamorphic origin. Therefore the granitic basement is expected to have the best fractured reservoir potential. Basement variations in Socotra Island are influenced by a number of factors including basement lithology and tectonic factor.

The dominant fracture trends in the basement of Socotra Island are NW-SE and NE-SW. The presence of this fracture system is thought to be due to major tectonic activity such as Tertiary opening the Gulf of Aden and Mesozoic break-up of Gondwana. These tectonic events formed new fractures and also reactivated older fractures.

[14]Denele Y., Leroy S., Pelleter E., Pik R., Talbot JY., Khanbari K., 2012, The Cryogenian arc formation and successive high-K calc-alkaline plutons of Socotra Island (Yemen), Arab J Geosci 5 (5):903–924.

[15]Leroy, S., Razin, P., Autin, J., Bache, F., d'Acremont, E.,Watremez, L., Robinet, J., Baurion, C., Denele, Y., Bellahsen, N., Lucazeau, F., Rolandone, F., Rouzo, S., Kiel, J.S., Robin, C., Guillocheau, F., Tiberi, C., Basuyau, C., Beslier, M.O., Ebinger, C., Stuart, G., Ahmed, A., Khanbari, K., Al Ganad, I., de Clarens, P., Unternehr, P., Al Toubi, K., Al Lazki, A., 2012. From rifting to oceanic spreading in the Gulf of Aden: a synthesis. Arab. J. Geosci. 5 (5), 859–901.

[16]Denele Y., Leroy S., Pelleter E., Pik R., Talbot JY., Khanbari K., 2012, The Cryogenian arc formation and successive high-K calc-alkaline plutons of Socotra Island (Yemen), Arab J Geosci 5(5):903–924.

Khaled Khanbari has a Ph.D. in Geology, University of South Paris, France. He is Associated Professor at the Department of Earth and Environmental Sciences, Faculty of Sciences, Sana'a University. Dr. Khanbari is also chairman of Yemen Remote Sensing and GIS Center (YRSGISC). He carries out lectures on following courses: Remote sensing and GIS, Structural Geology and Tectonics, Surveying and Field Geology, Volcanic Activities in Yemen. During his work in (YRSGISC), Dr. Khanbari supervised and participated in many projects such as "Satellite Image Atlas of the Republic of Yemen" project and Unified Digital Map project of the Republic of Yemen. Dr. Khanbari has good relationship and experiences on research activities in Geoscience with the French PMCU University. He coordinated one Yemeni-French scientific project YOCMAL which aims to study the geology of Socotra Island and the evolution of the Gulf of Aden. Dr. Khanbari improved his qualification on Geology, GIS and Remote Sensing by attending many of local and regional workshops and conferences. He supervised many of bachelor final projects and master thesis in the fields of Geology GIS, Remote Sensing. He has published many of scientific articles in local and international scientific journals.

Sylvie Leroy has a Ph.D. in Geodynamic, University of Pierre and Marie Curie, France. She is director at the Institute of Earth Sciences of Paris (ISTEP). Dr. Leroy is also mission manager of Marine Geoscience at National Institute for Earth Sciences and Astronomy (INSU). She is a head of the Lithosphere-Structure-Deformation team. Dr. Leroy is manger of many projects such as YOCMAL project which aims to study the geology of Socotra Island and the evolution of the Gulf of Aden. Dr. Leroy supervised many of master and Ph.D. theses. She has published many of scientific articles in international scientific journals.

Ahmad Adris has a Ph.D. in Technical Science. Yerevan State University, Armenia. He is a Researcher in RS, GIS & WRM at the Department of Geology and Water Resources, General Organization of Remote Sensing (GORS), Syria. Dr. Adris is also Scientific Consultant and Founder of Smart Live Yemen Initiative. He worked as expert in RS & GIS at Yemen Remote Sensing and GIS Center (YRSGISC) for 7 years (2010–2017), and at Hydroscope L.T.D. company, Armenia for more 10 years as GIS analyst—Interpreter of Satellite Images—Geophysicist. During his work, Dr. Adris supervised and participated in many projects such as "Determine optimal sites for Dams for Sana'a basin using RS & GIS techniques", "Study the karst phenomena (cavities) in the Ras-Alain district (in favor of the Directorate of Water Resources in Al-Hasakah governorate, Ministry of Irrigation, Syria)". Dr. Adris improved his qualification on RS, GIS and Water Resources Management by attending many of local and regional workshops and conferences. He supervised many of bachelor final projects thesis in the fields of Geology GIS, Remote Sensing. He has published many of scientific articles in local and international scientific journals. He joined: Syrian Geological Society in 1994, Planetearth Forum" Pace for Sustainable Development" in 2014.

Sami Moheb Al-Deen has Postgraduate Degree in Geology (Petroleum & Stratigraphy unit) from Faculty of Science, Sana'a University, Yemen. He was Teaching–Assistant in Geology Sana'a University for one year. Eng. Mohab Al-Deen is the Manager of Remote Sensing Department in Yemen Remote Sensing & GIS Center (YRSGISC). He is a Specialist in Remote Sensing and GIS with +13 years' experience in YRSGISC. Eng. Mohab Al-Deen held a training course on Remote Sensing and GIS at different governmental authorities such as General Telecomm Institute (GTI), Public Electricity Corp & General Authority for the Conservation of Cities and Historical Monuments (GAPHCMY). He is GIS Consultant in Social Fund for Development (SFD) and Emergency Response Unit in the Ministry of Water and Environment.

Waheed Al-Sarari has Postgraduate Degree in Geology from Faculty of Science, Taiz University, Yemen. He is the Manager of Scientific Researches and Environmental Studies Department in Yemen Remote Sensing & GIS Center (YRSGISC). He is a Specialist in Remote Sensing and GIS with +13 years' experience in YRSGISC. Eng. Al-Sarari improved his qualification on GIS and Remote Sensing by attending many of training courses and workshops. During his work in YRSGISC, he participated in many projects such as "Satellite Image Atlas of the Republic of Yemen" project and Unified Digital Map project of the Republic of Yemen.

Signal Coverage of Low-Land Areas Using Geographic Information Systems, Case Study: Kassingar Area, Sudan

M. A. Tajelsir Raoof, Dieter Fritsch and Rifaat Abdalla

Abstract

Telecommunication services of today are one of the most important means of sharing information, which makes the world a small village through the services provided by telecommunication companies. Therefore, there is a continuous interest to provide communication services to facilitate people's life and work on expansions to coverage on a wider scale. In this chapter, we discuss the lack of signals or weaknesses in the Kassingar area, Sudan, as a solution to provide better service for lowland areas. This article aimed to study and analyze the nature (elevation and slope) of the study area, and which effects on signal quality can be derived, using GIS techniques. Through analysis, we found that the study area is within the coverage of just two towers of the total eight towers surround the region. After ensuring that the remaining towers were excluded because of the distance or exist on the other side of the River Nile and because the water surface is a reflection factor of signals. Results are shown in a clearly and directly the strong relationship between the (height and position) of the towers and the lower area under study, that is resulting in a weak signal, or sometimes absences. We propose to build a new tower in the study area at Long

M. A. Tajelsir Raoof (✉)
University of Alfashir, Alfashir, Sudan
e-mail: mohamed89fatih@gmail.com; m.fateh@mail.dlut.edu.cn

M. A. Tajelsir Raoof
School of Computer Science and Technology, Dalian University of Technology, Dalian, China

D. Fritsch
University of Stuttgart, Stuttgart, Germany
e-mail: dieter.fritsch@ifp.uni-stuttgart.de

R. Abdalla
Department of Earth Science, College of Science, Sultan Qaboos University, Muscat, Oman
e-mail: rabdalla@squ.edu.om

A. Froehlich (ed.), *Space Fostering African Societies*, Southern Space Studies,
https://doi.org/10.1007/978-3-030-32930-3_15

31.912523°, Lat 18.596873°, which ensures to provide high-quality signals in the study area and improves the signal coverage in the entire surrounding area.

1 Introduction

Telecommunication services have become a very big challenge in the present time, as it has an impact onto most or all aspects of daily life, for instance on education, health, business management, communication, and banking transactions, to name only a few. Furthermore, it carries a very important economic component, which helps in the development of experienced communication technologies. Initially, communications were limited and restricted to a few consumers in handful countries, but soon became a competitive economic commodity because of the technological progress.[1] In the last 20 years, we have seen an expansion of wired networks and wireless networks to allow for mobility or roaming at any time and any place. Therefore, telecommunication companies have to overcome a big challenge to provide services to wider areas, which allow for having always stable connections while roaming.[2] GIS is part of Geographic Information Science, which is the science of underlying geographic concepts, applications, and systems.[3]

The expansion and provision of telecommunication services depend on many factors, for instance choosing the appropriate location of the towers, which depends on other factors, including the distance between the towers, the expected number of beneficiaries of the services and revenues in the area and so forth.[4] In recent years, the scope of network planning has been expanded using GIS (Geographical Information System) and GNSS/GPS (Global Positioning System) technologies, especially for telecommunication services.[5] These key technologies can help in getting strong and accurate results (Footnote 2). Therefore, telecommunication companies are increasingly becoming very large users of GIS technology. Over the

[1]Simpi B, Chandrashekarappa K N, Nirmala N, Bhargavi P (2011). Mobile Network System of Bhadravathi Town using Remote Sensing. GIS & GPS, Shimoga District, Karnataka, India. Global Journal of Computer Science and Technology, Vol. 11 Issue 14 Version 1.0, pp. 51–56.

[2]Munene E N, Kiema J B K (2014). Optimizing the Location of Base Transceiver Stations in Mobile Communication Network Planning: Case study of the Nairobi Central Business District, Kenya. International Interdisciplinary Journal of Scientific Research, Vol. 1 No. 2, pp. 113–127.

[3]Clarke K C (1986). Advances in geographic information systems. Computers, Environment and Urban Systems, Published by Elsevier Ltd, vol. 10, nos. 3/4, pp. 175–186.

[4]Sangeetha M, Purushothaman B M, Babu S S (2014). Estimating Cellphone Signal Inbtensity & Identifying Radiation Hotspot Area for Tirunelveli Taluk Using RS and GIS, International Journal of Research in Engineering and Technology, Vol. 03, Issue 02, pp 412–418.

[5]Persai P, Katiyar S K (2015). Telecommunication Utility Analysis Using GIS and GPS Technology. Journal of Geomatics, Vol 9 No.2, pp. 203–212.

last years, GIS has become an effective tool for businesses, because of its unique ability to tie attribute data to specific locations and usage, to plan, build, and operate telecommunication networks and associated services as well as many other use cases. The expansion of services depends on many factors, for instance choosing the appropriate location of the tower, which depends on other factors, including the distance between the towers and the number of participants expected, to revenues in the coverage area and so forth. There are a number of problems in the field of telecommunications. Areas with mountainous or low terrains are lacking signals and have weak access to them. This affects outgoing and incoming phone calls and internet services, which is the biggest problem for many people.

The main aims of this article are to apply GIS techniques in the field of telecommunications, especially in those areas of low terrain to identify the most important factors related to this field. It is shown, after a thorough analysis, that the signal coverage of towers is related to the impact of topographical range factors, and how it affects disability of broadcasting and receiving signals. The powerful capability of handling geospatial data through Remote Sensing and GIS techniques is the best solution among many of the solutions used to analyze the towers signal length and strength.[6] Recently, researchers have utilized several GIS techniques in improving communication services,[7] demonstrating effective and accurate results.[8] Some researchers have proved, that using GIS tools reduces manual fieldwork and overtime costs that may be a result of inaccurate information in the legacy paper maps and facilitated maintenance operations through information provided by topographic maps.[9] Moreover, it helps in the distribution of signals of towers, checking of the frequency and strength of signal passes by a device attached on the antenna, defining the appropriate angle, at which the maximum signal is provided to the customers.[10] The difference of signal strength depends on many factors, so GIS tools help to facilitate bidders to expand their services, especially in remote and rural areas.[11] Using GIS analysis tools to determine the visibility of cell towers, the results obtained showed a varying degree of signal strengths based on different terrain[2]. Areas with a high population density and different terrain structure are exposed for poor signal reception or not having the support of good visibility.[12]

[6]Naveenchandra B, Lokesh K N, Usha, Bhat H G (2011). Signal Strength Measurements and Coverage Estimation of Mobile Communication Network Using IRS-IC Multispectral and CARTOSAT-1 Stereo Images, Hyderabad India, Geospatial World Forum.

[7]Goodchild M C (2010). Twenty years of progress: GIScience in 2010. Journal of Spatial Information Science, No. 1, pp. 3–20.

[8]Isong E, Uduak A U (2015). A GIS Performance Analysis of a 3G wireless Cellular Network. Information and Knowledge Management, Vol.5, No.7, pp. 124–130.

[9]Merghani A A, Mukhtar F (2012). Selection of the Most Suitable Locations for Telecommunication Services in Khartoum. King Fahd University of Petroleum and Minerals.

[10]Aslam M F (2011). Cellular Signal Distribution through Spatial Visualization, National GIS Symposium, Saudi Arabia.

[11]Khan S, Farooq U B (2011). GIS as a planning tool for the USF Co rural telecom and E-service project in Pakistan. Procedia Social and Behavioral Sciences, Vol. 19, No 5, pp. 11–20.

[12]Lan L, Gou X, Xie Y, Wu M (2011). Intelligent GSM Cell Coverage Analysis System Based on GIS. Journal of Computers, Vol. 6, No. 5, pp. 897–904.

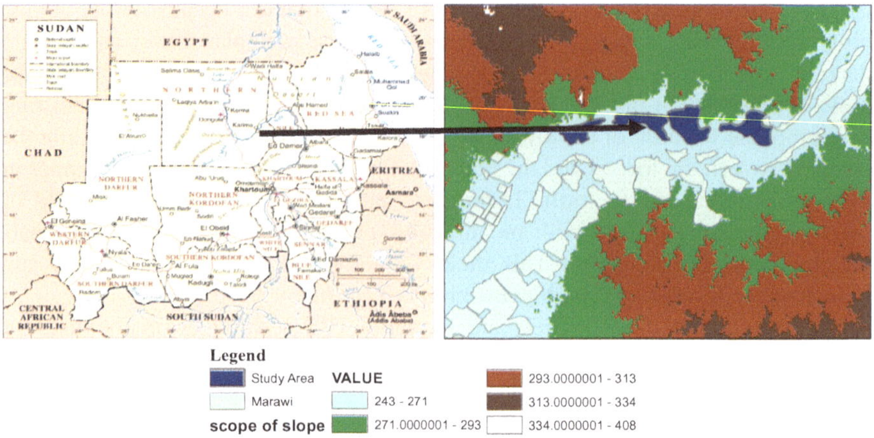

Fig. 1 Color map classification of the study area elevations

2 Materials and Methods

2.1 Study Area

First of all, there is a railroad station with its facilities, comprising the ticket office, platforms, etc. for loading and unloading train passengers and freight near the cities of Karima and Merawi,[13] (shown in Fig. 1).

The area extends to about 8–10 km in length and 2–2.5 km in width, in addition to (Al-swyqat and Al-kooa) areas,[14] the total population of the study area is estimated to be about 4350 inhabitants. It is divided into (Kassingar Gable) with a population of about 1640 persons and 1505 persons in the second part (Kassingar

[13]Getamap 2006–2018, search for map, getamap, viewed 20 March 2016.7:00 pm http://www.getamap.net/maps/sudan/%28su30%29/_kassingar.

[14]Sudan Central Bureau of Statistics (CBS): provides updated, accurate, timely, reliable, comprehensive and value-added statistical products and services by adopting the best criteria and the best practices at the global level in support of the decision-making process, http://cbs.gov.sd/index.php/en/home/index.

Table 1 Details of towers in the study area

Company	T.Name	Longitude	Latitude	Technology	Power (dB)	Sectors no	Distance/Km
A	P–A1	31.960669°	18.544511°	2G	43	2	7.82
A	P–A2	31.972911°	18.614461°	2G	43	3	6.18
A	P–A3	31.842667°	18.552389°	2G	43	3	8.95
A	P–A4	32.043665°	18.660158°	2G	46	3	15.7
A	P–A5	32.039222°	18.605333°	2G	43	2	12.8
B	P–B1	31.957090°	18.562593°	2G	46	3	5.71
B	P–B2	32.042975°	18.659467°	2G	46	3	15.2
C	P–C	31.974350°	18.560890°	2G	49	6	7.28

bahary). The neighboring Al-kooa area has a population of about 560 persons, in addition to the 640 persons in the Al-swyqat areas.

2.2 Data Set

The data of key interests pertaining to the study area is firstly the Digital Elevation Model (DEM) for Kassingar from obtained from the STRM mission,[15] (shown in Fig. 1). In Sudan, there are three companies working in the field of telecommunications only (Zain, MTN, and Sudani). In order to not mentioning names, we used the aliases to symbolize each company and each tower and details of the towers in and near the region of the study area, (shown in Table 1).[16] In order to determine the nearest tower, it is supposed to cover the affected area within the space of coverage, and knowledge of the factors that impede the signal transduction process (height, rise, fall ...etc.).[17] Table 2 shows the height of the towers with regard to mean sea level and a comparison with the elevation of the study area that was collected by Google Earth Pro (shown in Table 2).

[15]The CGIAR-CSI GeoPortal (CGIAR-CSI): is able to provide SRTM 90 m Digital Elevation Data for the entire world. These are available in both ArcInfo ASCII and GeoTiff format to facilitate their ease of use in a variety of image processing and GIS applications, viewed 25 March 2016. 12:55 am http://srtm.csi.cgiar.org/SELECTION/listImages.asp.

[16]Telecommunications and Post Regulatory Authority: Organizing and developing communications and postal services to keep abreast of developments in the field of telecommunications in Sudan, and encourage and protect the investment in telecommunication and postal services and applications, regulate free competition (https://tpra.gov.sd/).

[17]Boye C B, Peprah M S, Kodie N K (2018). Geographic Assessment of Telecommunication Signals in a Mining Community: A Case Study of Tarkwa and its Environs, Journal of Technology, Vol. 2, No. 2, pp. 41–49.

Table 2 Comparison of elevations between towers and the study area

Tower Name	Site height (S.h) Feet's	Tower height (T.h) (m)	Total Height (S.h + T.h) (F)	Study area height (F)
A1	891F	30	989	838
A2	941F	60	1137	838
A3	863F	30	961	838
A4	879F	30	977	838
A5	869F	30	967	838
B1	858F	30	956	838
B2	881F	30	979	838
C	856F	30	954	838

Table 3 Signal coverage levels (Standard)

Signal level	Best signal (dB)	Signal Quality	Coverage (km)	Color
Level 1	>= –75	V-Good signal	6	Green
Level 2	>= –85	Good signal	12	Yellow
Level 3	>= –95	Weak signal	18	Orange
Level 4	>= –105	Faded signal	24	Pink

3 Data Analysis

3.1 Profile Graph Analysis

A profile graph of the study area was created in Esri's ArcMap 10.1, as a result of a steepest path surface analysis.[18] This profile passes through the study area with a max elevation of about 370 m and min. elevation of about 240 m. The study area has its lowest heights at about 245 m (shown in Fig. 2). This indicates, that the study area is located at a lower elevation compared with neighboring areas, which fall in the same line and thus is supporting the hypothesis of this study. The next steps are calculating the elevation profile graph of each tower and analyze the affected areas, delivering the height of each point in the line at sea level (shown in Fig. 3).

[18]Esri 2018, ArcGIS for Desktop, Esri, viewed 24 July 2018 http://www.esri.com/ArcGISforDesktop/.

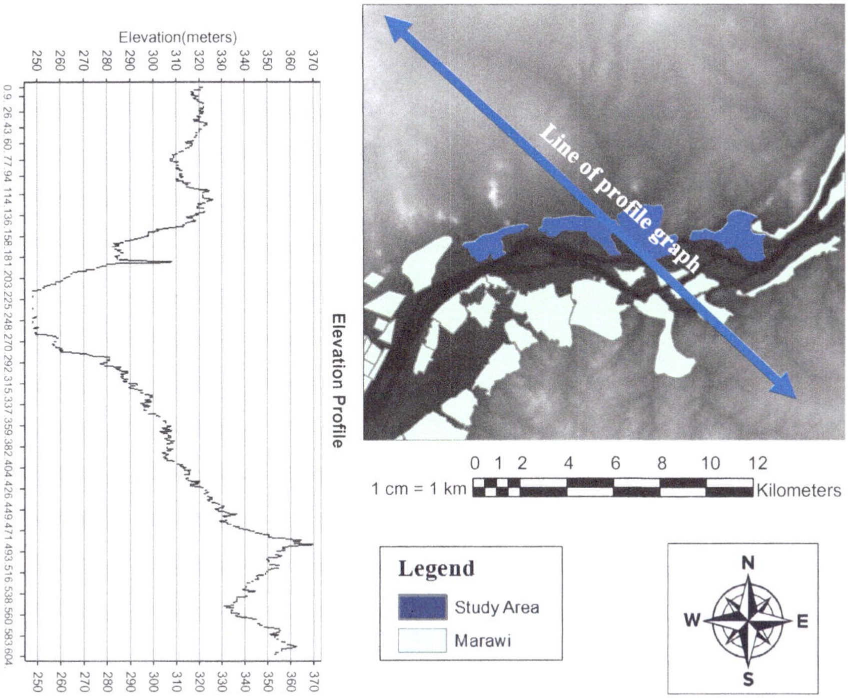

Fig. 2 Elevation profile graph of the study area

3.2 Signal Coverage Analysis

Distance is a very influential factor in the strength of the telecommunication signal. Physically, the more we move away from the tower, the more probable is the existence of other natural barriers. Including four distance levels covering up to 24 km, we know from experience, that the third and fourth level is most critical. This is because we find quite often a possible existence of barriers avoiding or disturbing the arrival of the signal up to a potential loss. For this reason, we introduce a classification for the signal level (shown in Table 3).[19] Thus, we can rely confidently on using a buffer zone in Esri's ArcMap to present the coverage of each tower up to a distance of 12 km (first and second level), where we receive a strong or very good signal (shown in Fig. 4a).

[19]Telecommunications and Post Regulatory Authority: Organizing and developing communications and postal services to keep abreast of developments in the field of telecommunications in Sudan, and encourage and protect the investment in telecommunication and postal services and applications, regulate free competition (https://tpra.gov.sd/).

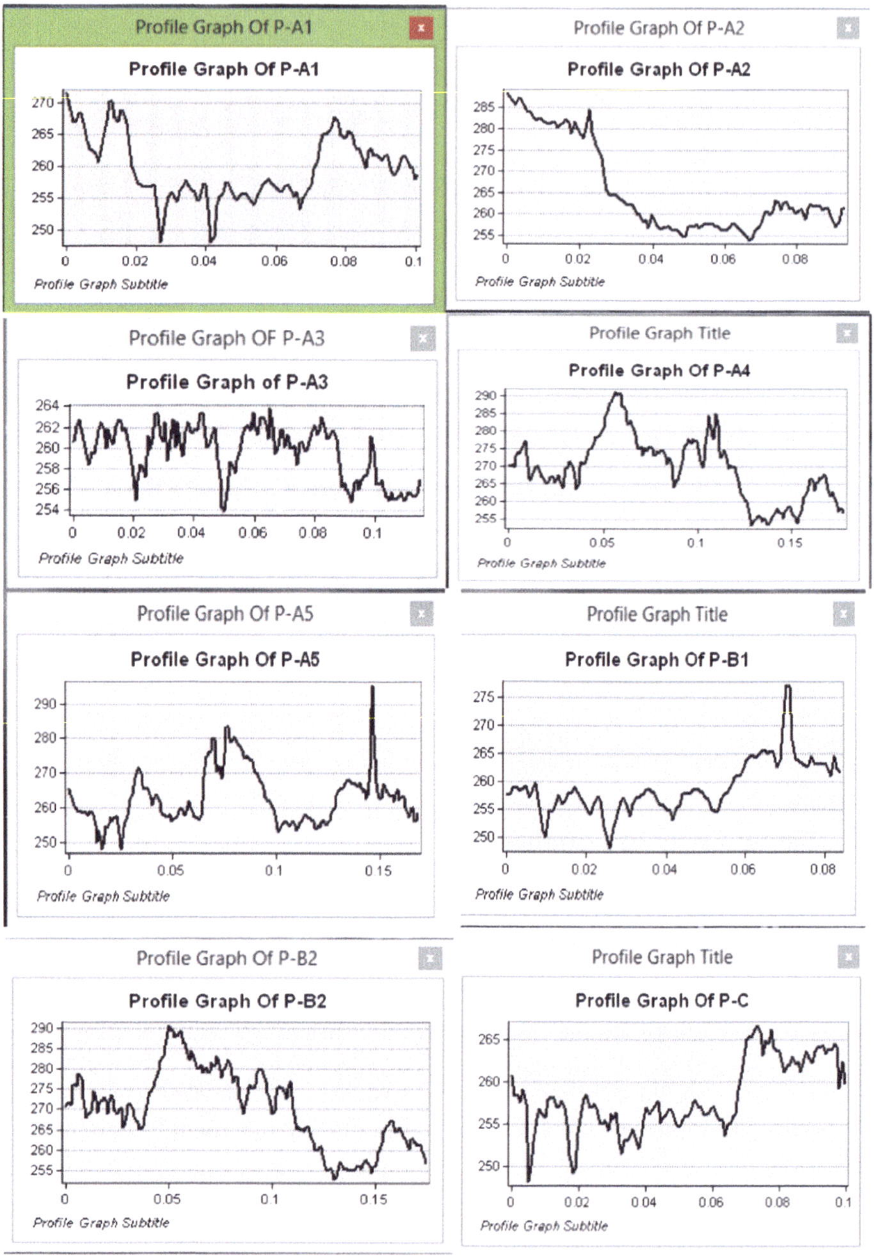

Fig. 3 Elevation profile graph of towers

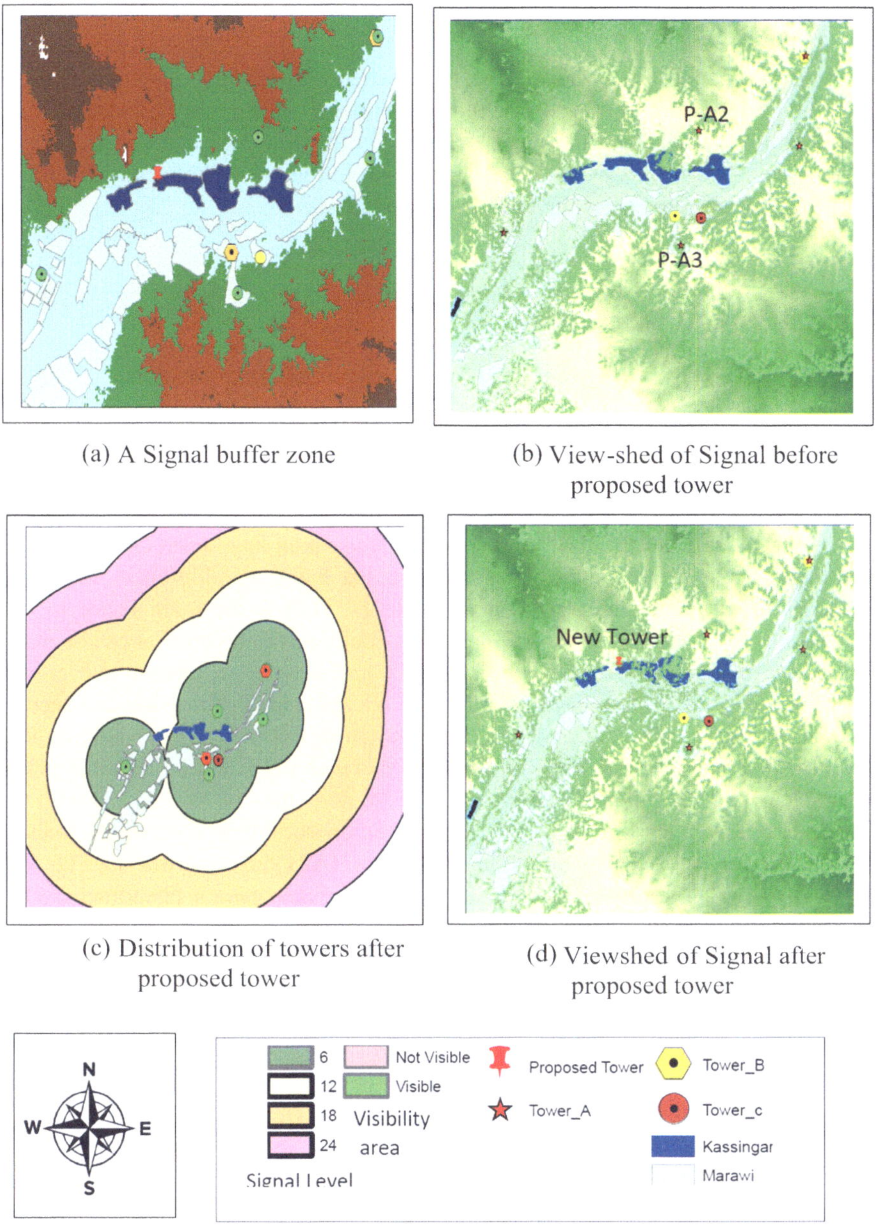

(a) A Signal buffer zone

(b) View-shed of Signal before proposed tower

(c) Distribution of towers after proposed tower

(d) Viewshed of Signal after proposed tower

Fig. 4 Signal coverage analysis of the study area

3.3 View-Shed Analysis

The View-shed tool in ArcMap is very useful when we want to know which objects might be visible. Here it was used to find out the points of the study area that will be visible or not visible from the tower location. On the other hand, we would like to see the view-shed results using all towers (shown in Fig. 4b). The visible points are given by yellow color from the elevations model.

4 Results and Discussion

4.1 Results

The elevation analysis showed, that the study area is located at the lowest level between "300–320" km for the profile graph line, (shown in Fig. 2). We found, that the number of towers surrounding the area is eight towers, (shown in Fig. 4b).

The distances between each tower and the study area have been calculated and compared to the height of each tower above sea level, in addition to its height relation with the study area (shown in Table 2). That is the main factor in our analysis using the view-shed tool, which finally delivers the points that can be made visible by green color, and the buffer zones of signals coverage, (shown in Fig. 4b).

We found that the study area is within the coverage area of the first and second level of the two towers (A2 and A3) only (shown in Fig. 4a), after excluding that towers which are most-distant in terms of the distance ratio and are existing on the other side of the River Nile. Here, we should also note, that the tower (A3) is on the other side of the River Nile, which means that the probability of receiving signals in the study area is currently based on the tower (A2) only (shown in Fig. 4b). Obviously, the land north of Sudan is hilly and there are natural phenomena, such as rocks, mountains, and slopes that hinder the arrival of telecommunication signals and to make use of them.

4.2 Discussion

This study aimed at providing signal coverage for the study area (Kassingar) using GIS techniques. The motivation for this study is that the population in the study area is suffering from the lack of telecommunication services, which is now one of the most important issues in modern life. Our results confirmed the relationship between the regression or elevation of the study area, and the weakness or lack of signal coverage. This means the broadcast signal is starting from a height point of about 346.82 m, compared with the height of the study area that was estimated to be about 255.42 m above mean sea level. Therefore, we propose to add new tower in the study area that can be located at the coordinates [longitude 31.912523°, latitude 18.596873°] and height of 876 feet, which is estimated to be about 262.8 m (shown

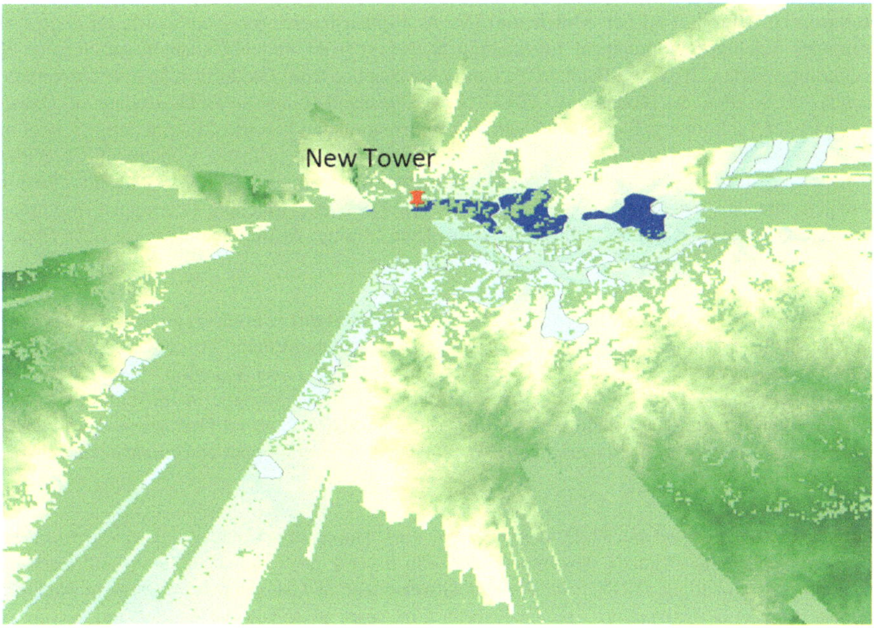

Fig. 5 View-shed of the proposed tower after editing the heights

in Fig. 4c). With that proposal, we can provide high-quality signal coverage in the study area and its surroundings, which has been proven by reusing the view-shed tools integrating this new tower overseeing the height points of the study area of about 10–15 m (shown in Fig. 5).

The results obtained for the proposed tower deliver for the study area a first level signal quality (shown in Figs. 4d and 5). In addition, the view-shed analysis has shown an increase in the set of points that can be visible without any obstacle (shown in Fig. 5).

5 Conclusions

After accomplishing a thorough analysis we can say, that there is a very strong inverse relationship between the distances to the tower and a decline in influencing the quality of the signal. Therefore the decision was made to propose the construction of a new tower bearing in mind the special nature of the area under study. We can also say, that using GIS techniques in the analysis and planning of cellphone networks to build or provide better telecommunications services is an economic factor in terms of cost, time and accuracy of the results and outputs.

Mohammed Alfateh TajElsir Abdelraoof (M. A. Tajelsir-Raoof) received his Master degree of Computer Science (Geographical Information Systems) from Sudan University of Science & Technology-Sudan, 2016 and Bachelor of Computer Science from Omdurman Islamic University, Faculty of Science & Technology, 2011. Now a researcher and a Ph.D. student at Dalian University of Technology, Faculty of Electronic Information and Electrical Engineering, School of Computer Science and Technology. Lecturer at University of Alfashir College of Computer Science and Information Technology, Department of Computer Science, Alfashir, Sudan. Research interests include Neural Networks (NNs), Deep and Machine Learning, Geographic Information Systems (GIS), Distributed Systems, Data Mining, Artificial intelligence (AI), Expert systems, Internet of Things (IoT).

Dieter Fritsch is Research Professor at the University of Stuttgart (Germany) and was for 25 years Director of the Institute for Photogrammetry. He published more than 400 publications, dealing with Computer Vision, Geographic Information Systems (GIS), Laser Scanning, Photogrammetry, Signal Processing and Statistical Inference. Furthermore, he is interested in Intercultural Affairs and Life-Long-Learning technologies making contributions to Cultural Heritage and university developments. Based on his experience he is requested as international consultant for academia and the private sector.

Rifaat Abdalla received his Ph.D. degree from York University, Canada and MASc from the University of Regina, Canada. His experience is in diverse professional, research and teaching environments; including DRDC, the Federal Government of Canada research arm in National Defence and York University, where he spent four years as a Tutorial Leader and served as Adjunct Faculty afterward. He was also employed by the Geological Lab of the Provincial Government of Saskatchewan. Prior joining Sultan Qaboos University (SQU) in 2017, he served for five years with King Abdulaziz University in Saudi Arabia for two terms between 2006–2008 and 2013–2017. Dr. Abdalla has worked for the industry in various capacities from entry-level to senior consultant. Dr. Abdalla is an active professor, interested researcher and eager professional combining multiple career paths.

A System of Enquiry for the Establishment of a Developmental Agenda for Space in Africa that Could Ensure Positive Economic Contributions for African Societies

Anton de Waal Alberts

Abstract

The African continent, as many other regions on the planet, is hamstrung by poverty and many developmental issues. It may seem counter-intuitive to investigate the space capabilities of the continent and other less developed regions as space activity is usually associated with developed states and regions. However, given the fact that the global space industry is set to grow in future and generate huge dividends due to various forms of New Space activities, it is a valid question to investigate how the African continent can insert itself into the existing and future unfolding space economy. This study endeavours to establish a generic system of enquiry whereby any area or entity—in this case the African continent and its societies—can identify the dominant industries that can be used as a point of leverage for insertion into existing space economic value-chains and to identify any low barriers of access for insertion into especially New Space economic value-chains. This system of enquiry may light the way to economic development through space where no poor state will be left behind.

A. de Waal Alberts (✉)
Member of the Gauteng Provincial Legislature (MPL),
Ph.D.-student at the University of Cape Town, Rondebosch, South Africa
e-mail: anton.alberts@prevoyance.co.za

A. Froehlich (ed.), *Space Fostering African Societies*, Southern Space Studies,
https://doi.org/10.1007/978-3-030-32930-3_16

1 Introduction

The world has made tremendous strides in reducing poverty on a global scale. Despite these achievements, the African continent remains one of the poorest places in the world, especially sub-Saharan Africa.[1] Many types of developmental interventions have been made since the advent of decolonisation with varying degrees of success. However, Africa is still dangerously behind in technological development in relation to the developed world and can become permanently mired in an orbit of dependence and remain a mere consumer instead of a creator of value that can compete in the global market. To a great degree this has become the case in various economic sectors of which mining is an example where raw materials are exported with little to no value added and consumer goods made from these materials imported again at great cost.

Many of these economic positions came about due to path dependent factors that unfolded over time. Mining, for instance, is a part of the heritage from the colonial era where certain power positions were established that were detrimental to African societies and the continent. These patterns still play themselves out today, although the complexity has increased due to new ownership patterns within African societies in addition to the unequal power relations with developed states. Much of this is due to internal and external politics with self-interest as the driving force. Thus, African societies remain poor.

Perhaps positioning Africa and its states and societies to focus on new technologies on a continental level can break some of the shackles of the past that bears down on it? Communications technologies for example, like cell phone networks, have proven to be hugely successful in Africa. It assisted the continent to leapfrog landline communications technology to a great degree and network operators are some of the highest employment creation drivers on the continent. It has also enabled other value-added services on the communications network as a technological platform, like online banking, trading and even education via the internet.

Space-based technologies, though not completely new, have also played a significant role in African societies over time. Satellites have enabled telecommunications and broadcasting thus opening the world to isolated African societies. Remote sensing data has also become a source for better economic planning and disaster management. However, most African states have no space assets themselves and have been reduced to mere consumers in this domain. The advent of New Space and the explosion of more cost-effective technologies in the form of pico-, nano- and micro-satellites (including cube-satellites) that can perform some functions of the larger legacy satellites, have opened new possibilities for African states and societies. Some African states have engaged in developing their own satellites to build human capacity which later morphed into the design and building of proprietary functional satellites. Other sectors of the space industry have also

[1]United Nations Department of Economic and Social Affairs, *United Nations Sustainable Development Goals*, undated, https://sustainabledevelopment.un.org/sdg1 (accessed on 20 May 2019).

developed, such as the use of more cost-effective launch vehicles and the development of the Internet-of-Things via space communication swarm satellites. Space-faring states like the United States and private companies are planning a (permanent) return to the Moon, trips to Mars and even working on space mining projects.

Many of these developments are still new and the question arises whether Africa as a whole or at least some of its constituent states and/or societies, can strategically insert itself into this unfolding new economic eco-system to ensure economic growth. This question is important as the space economy is set to become huge with reports that economic power in the 21st century will emanate from space activity.[2]

This question sets the stage to ascertain whether a developmental framework can be created that will assist Africa and its societies to set out an agenda and plan strategically with a view to be a legitimate player in the New Space economy. This time around, can we say no African state and/or society will be left behind?

Given this question this study sets out to create a system of enquiry by making use of a succession of tables that can, if populated with correct data, indicate linkages between existing African economic capacity and possible space capacity. It should also provide an indication for opportunities with low barriers to entry.

2 The System of Enquiry

The system of enquiry proposed herein is a first version that should develop incrementally over time as the need arises and as it becomes aligned with realities. It is generic in nature and can be applied globally, per continent or regionally, or per country or society/community.

2.1 Components

The system of enquiry consists of an array of tables that build on each until domains or opportunities are discovered that may allow for the entry into space industry fields that could ensure economic development. It provides insight into those opportunities that can be referred to as 'low hanging fruit'. These opportunities should allow for ease of access to space and concomitant economic development opportunities on a cost-effective basis without a disproportionate large risk. The framework is thus designed to search for the following two opportunities:

- Access to space opportunities via fields of existing capacity that can mitigate costs and overinvestment in a new and uncertain field;
- Access to space opportunities via fields with low barriers to entry.

[2]Oxford Analytica. *Understanding the Space Economy*. 2008. http://isulibrary.isunet.edu/opac/doc_num.php?explnum_id=290 (accessed on 25 June 2019).

Table 1 The entity's economic industries

Sectors	Industries	Dominant industries (2019)	Emerging industries (2019)
Primary sector (Extractive industry)	Mining, Oil, Agriculture etc	1. Agriculture 2. Mining	1. Solar Power
Secondary sector (Manufacturing industry)	Food, beverages, chemicals, auto, cement, petroleum, apparel etc	1. Food 2. Petroleum 3. Auto	1. Software[a] 2. Auto Components 3. Industrial & Business Machinery 4. Chemicals 5. Agri-Processing 6. Cement Production 7. Clothing & Footwear
Tertiary sector (Service industry)	Tourism, financial services, education, health, transport, telecommunication, television & film, retail, professional services, informal trade	1. Tourism 2. Retail, 3. Telecommunication 4. Financial services 5. Informal services	1. Advanced telecommunication-based services via cell phones
Quaternary sector	Internet, data, telemedicine, online learning, satellites	1. Internet	1. Advanced telecommunication-based services via cell phones

[a]Landry S. 2018. The potential of manufacturing and industrialization in Africa. *Africa Growth Initiative*. Brookings Institute

The system of enquiry operates in accordance with the following six steps:

(i) **Entity Identification**: Identifying the entity to be studied, in this case the continent of Africa. It can be applied on behalf of a global entity, such as a multinational corporation or political organisation such as UNCOPUOS, a continental entity such as the African Union, a regional entity such as the South African Development Community (SADC), or countries or smaller societies/communities/companies within countries;

(ii) **The Entity's Economic Industries**: Identifying the dominant and emerging industries of the entity (Table 1);

(iii) **Global Space Capacity**: Identifying the current and planned and probable future global space capacity (Table 2);

(iv) **Space Economic Linkages**: Comparing Table 2 with Table 2 in the form of Table 3 and identifying linkages between the entity industries and global space capacity, current, planned and probable. This includes identifying the dominant connections that are indicative of space opportunities via existing capacity or fields with low barriers to entry;

(v) **Entity Space Capacity**: Identifying the current space capacity of the entity for use in step 6 in assistance to establishing a basis of space opportunities (Table 4);

Table 2 Global space capacity

Sectors	Current global space capacity	Planned & probable future global space capacity
Satellite sector	1. Legacy satellites from the Cold War-era 2. Increasing numbers of satellites owned and operated by government and private enterprise 3. Increasing numbers of small satellite owners and operators with swarm satellites a developing reality[a] 4. Increasing number of experimental satellites by government, private enterprise and universities[b]	1. Low-earth-orbit (LEO) swarm satellite constellations consisting of small satellites are set to increase with the first set launched by SpaceX.[c] These lower-cost satellites can increasingly perform global communications and remote sensing services at a fraction of the cost of legacy satellites. Many of these satellites are built from less costly off-the-shelf technology
Ground-segment sector	1. Increasing ground segment sector for communications as space-based assets increase 2. Increasing number of astronomical research assets as witnessed by the Square Kilometer Array (SKA)	1. The ground segment sector is set to further expand as space-activities increase 2. Over time this sector will also be established on the Moon, Mars and asteroids as developments to set up permanent settlements gain momentum
Launch sector	1. Legacy launch systems continue with the US government developing new versions for manned space flight to the Moon and Mars 2. New Space launch systems are developing apace with SpaceX already operational, and Blue Origin and Virgin Galactic in testing phases 3. Micro-launchers are in development for the launch of smaller payloads, like small satellites	1. Private enterprise will in partnership with government or on their own devices develop new reusable launch systems that are more energy efficient, cost-effective, and powerful for greater carrying capacity, reuse and economies of scale 2. An increase of improved micro-launcher activity will take place 3. Alternative launch systems will be tested and used, such as single stage rockets and space planes
Space-segment sector (Non-satellite)	1. Most space activities outside of satellite projects consist of those related to the International Space Station (ISS), planetary exploration, and other scientific missions 2. Legacy space vehicles are used for payload and manned flight. Since the mothballing of the US space shuttle systems, the Russian Soyuz systems have been used to carry cargo and astronauts to the ISS	A whole range of projects are planned and in certain cases the first few steps have been initiated:[d] 1. **Moon and Mars manned missions and settlement**: NASA, SpaceX, the European Space Agency, Blue Origin etc 2. **Space Mining:** this is a burgeoning new field of space activity where research and preparation is growing in intensity

(continued)

Table 2 (continued)

Sectors	Current global space capacity	Planned & probable future global space capacity
	3. New Space private companies, like SpaceX, are now also being used to ferry cargo to the ISS	3. **Space Energy Generation**: generation in space for the various planned activities is also at the forefront of the space agenda Without it very little advances can take place in the other planned space activities 4. **Construction**: in-space and celestial body construction ability is of utmost importance as not all infrastructure can be launched from earth. This includes additive building processes like 3-D printing 5. **Tourism**: this activity is one of the first dreamed up for civilian space participation and is set to grow once Virgin Galactic is fully operational with other niches being filled over time, such a space hotel or Moon hotel

[a]Kulu E. Nanosats Database. 2019. https://www.nanosats.eu/#figures (accessed on 25 June 2019)
[b]Ibid 10
[c]SpaceX. 2019. Starlink Mission. https://www.spacex.com/news/2019/05/24/starlink-mission (accessed on 25 June 2019)
[d]Landon J. Schneider E. 2019. These 5 industries will be first to do business in space. World Economic Forum. https://www.weforum.org/agenda/2017/11/industries-will-make-money-in-space/ (accessed on 25 June 2019)

(vi) **Space Economic Opportunities**: Synthesising the results of the linkage exercise in Table 3 and the current space capacity as set out in Table 4 within a concluding table (Table 5).

The remainder of this work is illustrative of the functioning of the system of enquiry.

2.1.1 Step 1: Entity to be Studied

The continent of Africa.

2.1.2 Step 2: Table 1—Target Entity's Existing Economic Capacity—the African Continent

This table focuses on the target entity itself and provides an overview of its own current economic capacity. This is the logical starting point of the system of enquiry to ascertain whether it can access space opportunities for economic development.

Table 3 Space economic linkages (Across Tables 1 and 2)

SECTORS	Dominant Industries (2019)	Emerging Industries (2019)
PRIMARY SECTOR (Extractive Industry)	1. Agriculture 2. Mining	1. Solar Power
SECONDARY SECTOR (Manufacturing Industry)	1. Food 2. Petroleum 3. Auto	1. Software 2. Auto Components 3. Industrial & Business Machinery 4. Chemicals 5. Agro-Processing 6. Cement Production 7. Clothing & Footwear
TERTIARY SECTOR (Service Industry)	1. Tourism 2. Retail 3. Telecommunication 4. Financial services 5. Informal services	1. Advanced telecommunication-based services via cell phones
QUATERNARY SECTOR	1. Internet	1. Advanced telecommunication-based services via cell phones

SECTORS	CURRENT GLOBAL SPACE CAPACITY	PLANNED & PROBABLE FUTURE GLOBAL SPACE CAPACITY
SATELLITE SECTOR	1. Legacy satellites. 2. New Satellites. 3. Small and Swarm Satellites. 4. Experimental satellites.	1. Low-earth-orbit (LEO) swarm satellite constellations.
GROUND-SEGMENT SECTOR	1. Increasing ground segment sector for communications. 2. Increasing number of astronomical research assets.	1. Expanding ground segment sector. 2. Also, on the Moon, Mars and asteroids as developments to set up permanent settlements gain momentum.
LAUNCH SECTOR	1. Legacy launch systems. 2. New Space launch systems. 3. Micro-launchers.	1. New reusable launch systems. 2. Improved micro-launcher activity. 3. Alternative launch systems.
SPACE-SEGMENT SECTOR (Non-Satellite)	1. ISS, planetary exploration, and other scientific missions. 2. Legacy space vehicles. 3. Private space vehicles.	1. **Moon & Mars missions and settlement.** 2. **Space Mining.** 3. **Space Energy Generation.** 4. **Construction.** 5. **Tourism.**

Table 1 is based on the accepted four main economic sectors used in economic analysis, namely the primary sector consisting of the extractive industries such as mining and agriculture, the secondary sector consisting of the manufacturing industry where the extracted materials are turned into products like vehicles (heavy industry) and food (light industry), the tertiary sector consisting of the services industry with financial services, health care and retail as examples, and lastly the quaternary sector—a fairly new sector spun off from the tertiary sector—that

Table 4 The entity's space capacity

Sectors	Africa space capacity
Satellite sector	1. Various African states own and operate satellites. South Africa, Nigeria, Egypt, Algeria and Kenya own satellites of various sizes and capabilities. However, most African states are users of satellite capability owned by states and companies outside Africa 2. Most satellites are large satellites procured from outside Africa 3. South Africa, Algeria and Kenia have developed local capability to design, build and operate small satellites (including pico, nano and micro-satellites)
Ground-segment sector	1. Most African states have general satellite signal reception capability for television (even among the public: direct to home television) and telecommunication purposes 2. South Africa has advanced ground-segment capability emanating from the years supporting NASA's space flight operations 3. South Africa has also developed unique astronomical capability by the implementation of the MeerKat and SKA projects designed to create the world's largest radio telescope[a]
Launch sector	1. No African state currently has space launch capability 2. South Africa does have legacy launch platforms that is not in use anymore but can be commissioned again
Space-segment sector (Non-satellite)	Not active

[a]Square Kilometre Array Organisation, Africa, 2019, https://www.skatelescope.org/africa/ (accessed on 20 June 2019)

represents the elements of the knowledge economy, namely telecommunications, information technology, research and development and education, amongst others.[3]

Table 1 was compiled by tracking the economic activity of the whole African continent as an aggregate of all constituent states. Obviously, every state's economic structure differs, but overall broad trends are discernible across the continent. However, research can also be limited depending on the available data on the target entity. In this case finding research that can describe the scope of all services in the tertiary sector is severely lacking forcing one to use estimates based on existing data.[4] In fact the informal sector is still the greatest employer in Africa.[5] Therefore, the data used in Table 1 is broadly illustrative only.

Accordingly, an analysis of the African continent's existing economic capacity can take on the following form.

The next step is to ascertain the status of global space capacity by way of Table 2.

[3]Kennessey Z. 1987. The Primary, Secondary, Tertiary and Quaternary Sectors of the Economy. *The Review of Income and Wealth.* Wiley Online Library.

[4]Dihel N. Goswami AG. 2016. The unexplored potential of trade in services in Africa. *World Bank Group.*

[5]Ibid 4.

Table 5 Space economic opportunities

Sectors	Proposed space capacity for africa
Satellite sector (Growing sector)	While the existing activities can be expanded, it will make sense to explore the increased development and manufacturing of small satellites. Swarm constellations should be consideration by stronger space-states or in partnership between various states. These constellations can enable financial services and internet connectivity across the continent, amongst others. The manufacturing expertise in the auto industry can be utilised to set up the constellation manufacturing and to produce parts that can be sold globally. The financial services and internet industry can assist to finance the constellation. The financial benefits could be huge as international clients will ensure an influx of foreign exchange and the internet can be expanded across the continent with its concomitant benefits of trade
Ground-segment sector (Strong developing sector)	The ground sector will witness development as new satellites, including the swarm constellations, grow in number and provide direct services to the African continent. New astronomical developments could also see this sector grow, but the most potential will be satellite driven
Launch sector (Weak sector)	The launch sector can also experience growth in two manners: 1. Africa can market itself as a launch venue of choice based on its geographic positioning; 2. The exponential growth in satellite development, especially easily launched small satellites, and the need to launch regularly can lead to the development of the continent's own launch capability. This can be developed by the stronger space-faring states or in partnership with each other, even cross-continentally under the leadership of the African Union or another special purpose organisation. The development of proprietary re-usable small rockets will be hugely beneficial as international clients will ensure much needed foreign exchange influx into the continent. The petroleum and automotive manufacturing sectors can play a huge role in this domain
Space-segment sector (Non-satellite) (Non-existent sector)	Despite Africa's absence in this sector, it can use its existing industrial base as leverage for points of insertion into future space value-chains: 1. The agri-processing sector should immerse themselves in any activities regarding food research to be used in future space activities like long-term or permanent settlements on the Moon and Mars 2. The mining sector should likewise search for points of entry into the unfolding plans and value-chains for mining on celestial bodies 3. The possible developments in the launch sector alluded to above in this Table 4, can also find new niches within this sector. For instance, the launch capability in both rockets and launch facilities or newly developed spaceports can witness the proliferation of manned space launches and even the full-scale development of Africa's own proprietary space vehicles for manned and unmanned purposes

2.1.3 Step 3: Table 2—Current, Planned and Probable Future Global Space Capacity

The space industry is a growing one with much potential. By 2013 the value of global space activities amounted to $314.17 billion.[6] Most activity today is commercial in nature and this will increase with the acceleration of the New Space era. This table will set out this information as the background against which the target entity in this case, the African continent, will be measured. Importantly it will also later assist to indicate any linkages that might exist or arise for the African continent to insert itself into the space economy.

This table will make use of the manufacturing-base analysis. However, this form of analysis is limited given the range of space activities currently taking place. As a result, the table has been adapted to include additional activities under the heading "Space-Segment Sector" that includes all in-space-related activities outside of the satellite sector.

The above table, if populated with the correct data, should establish a summary of the current and planned and probable space activities on a global basis. Once this process has been completed, the last steps must now be implemented, namely the comparison between the African continent's economic industries (Table 1) and the global space capacity (Table 2) to establish linkages that can provide a basis from which space activities can be identified as points of insertion into the industry due to existing capacity or low barriers of entry.

2.1.4 Step 4: Comparison of Tables 1 and 2 and Linkage Discovery

In order to perform this comparative analysis within the confines of this study, Table 2 will be presented in an abridged form to allow for ease of comparison. This can be performed for any table that is laboriously large. Table 1 will thus from this point on not use the column titled "Industries" as the linkages will only be based on those industries that are dominant—as possible industry points of insertion and possible low barriers of entry—and emergent—as possible low barriers of entry. The dominant industries present the so-called 'real low hanging fruit' that allows for insertion and ease of access based on experience, and the emergent industries present the opportunity to identify access to new space activities based on possible low barriers to entry, the so-called 'possible low hanging fruit'.

The linkages are performed by drawing connecting lines between the entity's dominant and emerging industries in Table 1 and the various forms of global space capacity in Table 2.

The identification of the linkages is based on logic. For instance, mining as a dominant industry (Table 1) connects seamlessly with prospective space mining (Table 2). More complexly, the dominant industry of agriculture (Table 1) connects to food and agri-processing within Table 1 as parts of a value-chain across to the planned Moon and Mars settlements (Table 2). The Table 1 industries can therefore

---

[6]Space Foundation, Space Report, 2018, http://www.spacefoundation.org/programs/researchand-analysis/space-report (accessed on 20 June 2019).

connect directly with a form of space activity or cross from one piece in the value-chain to the others within Table 1 and then connect across to a single form of space activity or more in Table 2.

These examples as applied logic is illustrated below.

The linkages identified show the following trends:

- The same industry can link with more than one global space activity/sector;
- Dominant and emerging industries can combine to link with a global space activity;
- Some industries can link up with one or more of a global sector's space activities, thus including current and planned and probable space activities;
- Interestingly, almost all industries were able to link with a space activity, although as will be shown *infra* not all linkages are equally strong;
- The linkages are not necessarily complete as another analyst might come to different conclusions based on other available data.

The following are the strongest linkages identified in terms of dominant industrial base and low barriers of entry with explanations of choice:

Agriculture → Food → Agri-Processing = Moon & Mars Settlement: Africa has a long history of agriculture and still predominantly relies on agriculture as major economic activity. It is thus a strong base to find a point of insertion into any future food production value-chain necessitated by Moon and Mars settlements. As this linkage is future-oriented a well-chosen point of entry in the future value-chain can offer a low barrier of entry.

Mining = Space Mining: Africa is well-known for its natural resources and mining. The same arguments as those of agriculture applies to mining's value-chain.

Petroleum → Chemicals = Launch Sector & Space Segment Sector (Non-Satellite): Not all African states are involved in the petroleum industry, but those that are have a strong legacy. Likewise, this industry mirrors those of agriculture and mining.

Auto → Auto Components = Satellite Sector & Launch Sector: The auto industry mirrors the petroleum industry.

Financial Services = Low-earth-orbit (LEO) swarm satellite constellations: This industry is not evenly advanced across the African continent but presents some of the most state-of-the-art services where available. As much of these services take place via cell phones, connectivity is of paramount importance and LEO swarm satellites might be a low barrier of entry.

Internet = Satellite Sector: This industry mirrors that of the financial services.

The lesser linkages can also be analysed for future development, but for sake of brevity will be left out of this discussion.

2.1.5 Step 5: Table 4—Target Entity's Existing Space Capacity—the African Continent

This table provides a second focus on the target entity by establishing the space capacity thereof. The space industry can be divided and analysed in various ways for instance by manufacturing-base (satellite; ground-segment; launch vehicles), by social function (scientific, military and commercial), and even a simplistic binary division into the satellite and non-satellite sectors. Given the growth of New Space, traditional earth-bound sectors have invaded the space industry in the form of space tourism, commercial space transport, and space mining.

For the purposes of this study the space industry will be analysed by way of the manufacturing base-method.

This analysis reveals that while Africa is active within the space industry it is on a very limited scale. The question is how far Africa is lagging the rest of the space industry?

Africa's strength currently vests in the ground-segment sector and to an increasing degree in the satellite sector. It has latent ability in the launch sector and no presence in the space-segment sector.

The last step of the system of enquiry can now be applied.

2.1.6 Step 6: Table 5—Synthesis and Conclusion

During this step the linkages identified during step 4 is combined with the results of step 5 in the form of the last concluding table, namely Table 5, as follows.

The conclusions reached in Table 5 by the synthesis of information in the previous tables has realised interesting insights that can serve as a point of departure for further research and exploration. This system of enquiry can thus open new ways of thinking about possibilities and provides a sense of future, even within a poorer continent like Africa. If used correctly, it can also unlock much inspiration and a sense of hope for a future filled with technological advances and economic development that eradicates endemic poverty.

3 Conclusion

The system of enquiry devised herein is a first version set of tables that can bring about insights through its methodology of linkages between local industries and the global space environment. The enquiry's starting point of analysing the entity of choice's—in this case the African continent—economic environment is of utmost importance to ensure the aggregation of correct data to ensure proper global space industry-linkages are identified in the next step. Thereafter the identification of the current status of the entity's space capabilities provides further context to ensure that not only new linkages are identified, but that existing space capability is taken into account in the last step where opportunities are identified to access space sectors based on current strengths and low barriers of entry. This version of the system of enquiry will see further adaptation as it used in practice and should develop into a practical tool for identifying real opportunities that did not seem apparent at first.

Anton Alberts is admitted as an advocate/barrister of the High Court of South Africa specialising in the legal fields of media law, ICT and space law. He was a Member of Parliament in South Africa (2009–2019) and served as a full member on the Parliamentary Portfolio Committee on Trade and Industry where he, amongst others, promoted the development of the country's space industry. Currently he serves as a Member of the Provincial Legislature (MPL) in the Gauteng province, South Africa's economic hub, where he promotes the airspace industry. He received his legal education at the University of Johannesburg where he obtained the degrees, BA (Law), LL.B, and LL.M (International Law (Cum Laude)), as well as an M.Phil in Futures Studies from the University of Stellenbosch. He is currently studying towards a Ph.D. in Space Law at the University of Cape Town. He is a prolific researcher and has published several legal works. Anton's focus is now increasingly on Space Law and its development for a new era of cooperation between government and private industry.

African Traditional Concepts for Property Rights in Outer Space

Annette Froehlich

Abstract
On-going discussions about property rights and exploitation of natural resources in outer space have an international dimension. Even though this is a topic driven by space faring nations, it is time that emerging space faring countries also become involved and advance their way of thinking to develop new approaches to property rights. In this regard, African traditional concepts related to property rights may be a good source of inspiration insofar as they have always striven to achieve a balance between natural resource exploitation and preservation of the environment, since this balance and harmony within the community have been vital for the survival of the whole group.

1 Introduction

Since technical developments seem to be making possible the exploitation of space resources in the future, discussions have begun on the legal status of these space resources especially as some private missions to Mars are being scheduled. The question of how to settle and acquire property on other celestial bodies has emerged in order to avoid a run on space resources. Unfortunately, it seems that humans are fixed in their ways of thinking, and are mostly advancing the same arguments (similar to those defended on Earth in view of other propriety related issues) that have historically caused conflicts, mistrust and wars on Earth. It seems no lessons have been learned from the past mistakes of colonization and conquest of foreign territories, since always the same concepts are put forward when it comes to the

A. Froehlich (✉)
University of Cape Town (SA), Rondebosch, South Africa
e-mail: Annette.Froehlich@uct.ac.za

A. Froehlich (ed.), *Space Fostering African Societies*, Southern Space Studies,
https://doi.org/10.1007/978-3-030-32930-3_17

acquisition of property. However, on Earth other concepts of property rights have emerged (and thus do not need to be invented/created). Regrettably, these more recent improved concepts of property right are not prominent in current discussions, despite being very efficient in several communities. Therefore, attention must be drawn to the long-standing concepts of property rights in some indigenous populations in Africa. These could guide or at least be a source of inspiration for on-going discussions on property rights in outer space.

Therefore, this first part will describe the points of contention in regards to the use of outer space resources and property rights in the field of international space law in order to analyse, in a second part, some African indigenous concepts, especially of the San population, in view of possible outer space resource mining.

2 The Freedom Rights for Space Activities and Property Rights in Outer Space

Art. I Outer Space Treaty (OST)[1] grants freedom rights for space activities, especially the freedom to explore and to use outer space. Therefore, Art. I-2 OST stipulates that *"outer space, including the Moon and other celestial bodies, shall be free for exploration and use by all States without discrimination of any kind, on a basis of equality and in accordance with international law, and there shall be free access to all areas of celestial bodies."*[2] However, space endeavours *"shall be carried out for the benefit and in the interests of all countries,"* (Art. I-1 OST).[3] Moreover, Art. II OST sets boundaries to this right by stating that *"Outer space, including the Moon and other celestial bodies, is not subject to national appropriation by claim of sovereignty, by means of use or occupation, or by any other means"*.[4]

The subsequent Moon Agreement[5] of 1979[6] took notice of the possibility of resource mining in space by foreseeing a special regime for the exploitation of space resources. Article 11-2 MOON reiterates that "*the Moon is not subject to national appropriation by any claim of sovereignty, by means of use or occupation, or by any other means*".[7] Article 11-3 MOON is even more precise: *"Neither the surface nor the subsurface of the Moon, nor any part thereof or natural resources in place, shall become property of any State, international intergovernmental or non-governmental organization, national organization or non-governmental entity*

[1]Treaty on Principles Governing the Activities of States in the Exploration and Use of Outer Space, including the Moon and Other Celestial Bodies of 27 January 1967 (Status of ratification as of 1.1.2019: 109 states).
[2]Art. I-2 OST.
[3]Art. I-1 OST.
[4]Art. II OST.
[5]Agreement Governing the Activities of States on the Moon and Other Celestial Bodies of 18 December 1979.
[6]Twelve years after the OST.
[7]Art. 11-2 MOON.

or of any natural person. (...) The foregoing provisions are without prejudice to the international regime referred to in paragraph 5 of this article."[8] Indeed the State Parties to the Moon Agreement decided under Art. 11-5 MOON to establish an international regime to "govern the exploitation" of resources on the Moon "as such exploitation is about to become feasible".[9] However, it must be underlined that this clause conferring rights and obligations for space resource exploitation is only binding upon the sixteen states that have so far ratified the Moon Agreement. In this context, it should also be mentioned that while the Outer Space Treaty is the most widely adopted of the five UN space related treaties, the latest, the Moon Agreement, has received far fewer signatures and ratifications[10] and may eventually even be considered as failed treaty due to its extreme limited acceptance. The United States, like most space faring nations e.g. the Russian Federation, China, Germany, United Kingdom, has not signed it. Among the countries of the African continent, only Morocco has signed and ratified this treaty. Therefore, due to the small number of ratifications, the regulations of the Moon Agreement and especially its Art. 11 have not gained the status of international customary law.[11]

Nevertheless, questions around the possibility of space resources mining remain and have been raised in recent years due to technological developments. And, despite these international treaties, in 2015 the U.S. passed its Space Act[12] that was welcomed by several companies as a supportive framework for the sustainable development of outer space and the beginning of humans as a multi-planetary species. However, critics have described this act as the beginning of a gold rush in space, and have recalled the negative impact of previous colonial imperialisms on Earth stating that this 'First come, first served' approach brings back a kind of 'Wild West mentality'.

Indeed under § 51303 entitled "Asteroid resource and space resource rights," the US Space Act stipulates *"A United States citizen engaged in commercial recovery of an asteroid resource or a space resource under this chapter shall be entitled to any asteroid resource or space resource obtained, including to possess, own, transport, use, and sell the asteroid resource or space resource obtained in accordance with applicable law, including the international obligations of the United States".* Sec. 403: 'Disclaimer of extraterritorial sovereignty' seeks to

[8]Art. 11-3 MOON.
[9]Art. 11-5 MOON.
[10]Status of ratification as of 1.1.2019: 18 states, i.e. Armenia, Australia, Austria, Belgium, Chile, Kazakhstan, Kuwait, Lebanon, Mexico, Morocco, Netherlands, Pakistan, Peru, Philippines, Saudi-Arabia, Turkey, Uruguay and Venezuela. Four further states have signed, but not yet ratified: France, Guatemala, India, and Romania.
[11]Even though the MOON was adopted by consensus at the UN General Assembly, in their statements on UNCOPUOS some delegations such as the U.S. refer to four core treaties, indicating that the Moon treaty does not belongs to the "core" of UN space treaties.
[12]U.S. Congress, 2015, H.R.2262 – U.S. Commercial Space Launch Competitiveness Act, https://www.congress.gov/bill/114th-congress/house-bill/2262/text (accessed 14.05.2019).

further justify this approach: *"It is the sense of Congress that by the enactment of this Act, the United States does not thereby assert sovereignty or sovereign or exclusive rights or jurisdiction over, or the ownership of, any celestial body."*

These new US regulations have provoked international discussions since the Outer Space Treaty stipulates the non-appropriation principle clearly in the aforementioned article II OST. However, article II OST only addresses "national appropriation", private actors are not expressly mentioned. This vague wording has led to uncertainty, and since space activities are very investment and cost intensive, companies want to get certainty before financing space resource mining.

In fact, these US regulations grant property rights in outer space to U.S. companies. But the question that arises is: How can a state grant rights to its private actors that, from an international point of view, the state does not itself have? Unfortunately, at the international level the only common understanding is that the non-appropriation clause of art. II OST refers to "territorial" claims, thus creating uncertainty as to whether the taking and removing of resources is also covered by this non-appropriation principle. Art. I-2 OST is relevant insofar as it grants the right of free exploration and use of outer space and its celestial bodies but it requires that this right be exercised *"without discrimination of any kind, on the basis of equality and in accordance within international law"*.[13] Regrettably this right of free use and its restrictions are not further defined leading to wide disputes at the international level. From the point of view of equality, it is unclear if the right of free use includes the right to use, take away, and consume non-renewable natural resources. This question is unresolved not only concerning resources in outer space. Given the immense waste and exploitation of natural resources on Earth, it seems that human beings' unreasonable behaviour will be transferred or exported even beyond Earth to further celestial bodies. However, studies also claim that the use of space resources could ameliorate various problems on Earth, such as environmental degradation.[14]

Therefore, concerning the US Space Act, it may be argued that "in view of the absence of a clear prohibition of the taking of resources in the Outer Space Treaty one can conclude that the use of space resources is permitted. Viewed from this perspective, the new United States Act is a possible interpretation of the Outer Space Treaty. Whether and to what extent this interpretation is shared by other States remains to be seen".[15]

---

[13]Art. I-2 OST.

[14]The harvesting and exploitation may have less adverse effects on Earth environment, less pollution etc.

[15]International Institute of Space Law (IISL): Position Paper On Space Resource Mining Adopted by Consensus by Board of Directors on 20 December 2015.

3 Concept of Resource Governance of Indigenous Populations in Africa

Since several countries[16] have now declared their intention to adopt similar national space regulations to grant rights to space resource exploitation to their own private entities, reflections on past experiences of conquering foreign territories on Earth must be urgently undertaken to avoid the same mistakes. Moreover, past experiences that led to tremendous conflicts, horrific wars and massive injustices must be critically examined. And it is also time to look at other concepts of property that already exist among communities, especially in Africa, to try to find a more just and equal model of natural resources exploitation for the benefit of all.

In this context, it is worthy to cast a glance on the first nations of South Africa, the San (hunter gatherers whose ancestors lived in the southern parts of Africa since the Early Stone Age). They have a different concept of property that may differ from currently known concepts but could be relevant to the space arena, especially since their decision-making process is similar to the adoption process within the UN Committee on the Peaceful Uses of Outer Space (UNCOPUOS) sessions. Indeed, UNCOPUOS relies on the principle of consensus, and it was this committee, in charge of the new topic of the peaceful use of outer space, that was the first entity within the UN system to use this method of decision-making. This principle is also the basis of the long-standing San community rules. "The San have no formal authority figure or chief, but govern themselves by group consensus. Disputes are resolved through lengthy discussions where all involved have a chance to make their thoughts heard until some agreement is reached."[17]

Therefore, if these two communities, the San and UNCOPUOS, already share the principle of community rules in the decision making process, it is worth investigating if further rules within that African community could inspire or lead other areas of the UNCOPUOS work on the peaceful use of outer space, such as the question of property issues.

Indeed, "San are largely egalitarian, sharing such things as meat and tobacco. Land is usually owned by a group (…). Membership in a group is determined by residency. As long as a person lives on the land of his group he maintains his membership."[18] The concept that an individual person or entity cannot acquire property is very common in societies based on long traditional communities where good co-existence and harmony among all members of the group are vital for the survival of the whole community.

[16]See United Emirates or Luxembourg, the latter has actually adopted similar national regulations 'Loi du 20 juillet 2017 sur l'exploration et l'utilisation des ressources de l'espace', http://legilux.public.lu/eli/etat/leg/loi/2017/07/20/a674/jo (accessed 14.05.2019).
[17]Siyabona Africa, San, http://www.krugerpark.co.za/africa_bushmen.html (accessed 30.5.2019).
[18]Siyabona Africa, San, http://www.krugerpark.co.za/africa_bushmen.html (accessed 30.5.2019).

Those local or indigenous rules and the concepts behind them reveal an approach that may lead to another way of resources management. Since local communities have depended over thousands of years on those resources, they have had and still have an extraordinary interest in striving for a balanced system among its members to guaranty stability, equality, and peace within this local community.

In this regard, the rules on Madagascar on the use of maritime resources should be mentioned, as they try to allow the use of these natural resources while preserving the environment. Indeed the Fokonolona, the customary community structure, has the exclusive right to this resource exploitation. However, a contract between this local community and the public authority determines the conditions of access and use of those resources. In this system, the right of exclusivity can be for specified resources (for example the octopus, which is a very precious resource in this region) so that fishing generally is allowed but with restrictions on octopi, or an exclusive right to fishing areas (without species distinction) can be given but with restrictions in the license for fishing on the number of trawlers (which may rotate).[19] However, in both cases, it is group management and not individual decision.

The same could be applied to outer space resources so that the resources belong to the whole community, humankind, but with dedicated regimes for special resources. With reference to the above mentioned collective group management, at the international level no state could by national act decide upon the resources of the whole group (of the whole of humanity) since there is also a need for harmony and coexistence, which is not only vital for the survival of the beings in outer space but also for those on Earth i.e. among the state community. Also the African philosophy of *ubuntu* may be inspiring since it reflects a philosophy of life which is the concept of humanity, charity, a sense of community as well as the experience and consciousness of being part of the whole, which is still today an important part of inter-state relations in Africa.[20]

4 Conclusion and Way Forward

In conclusion, it is time that Africa advances its established community rules at the global level to discuss them in the frame of current topics such as the use of outer space resources. In this context the Workshop on SocioEconomic Benefits of Space

[19]Alasdair Harris, "To live with the Sea", Development of the Velondriake Community—Managed Protected Area Network Southwest Madagascar, Madagascar Conservation & Development, Vol. 2, Issue 1, 2007, p. 43, available from: https://www.researchgate.net/publication/26589494_To_live_with_the_Sea_Development_of_the_Velondriake_Community_-_Managed_Protected_Area_Network_Southwest_Madagascar.

[20]See further: Annette Froehlich/André Siebrits, Space Supporting Africa, Volume 1: A Primary Needs Approach and Africa's Emerging Space Middle Powers, Springer 2019.

Resources Utilization—African Perspectives (Pretoria, South Africa, 24–25 May 2019)[21] was a good initiative to start and foster this process in view of formulating not only the needs of Africa but also strengthening what it can contribute. As one of the results of this workshop "it was also noted that emerging spacefaring countries need to determine proactively the contributions that they might make to space resources activities, rather than solely focusing on attempting to stop or delay activities in the established spacefaring countries."[22] Therefore, raising awareness and capacity building among emerging space countries is of utmost importance at the national and international levels. In this regard the establishment of a working group to discuss this matter more at the global level was proposed within UNCOPUOS.[23] However, a consensus among UNCOPUOS members could so far only be reached on continuing the discussions in this regard and to have next year a single item for discussion on "General exchange of views on potential legal models for activities in the exploration, exploitation and utilization of space resources".[24]

Dr. Annette Froehlich is a scientific expert seconded from the German Aerospace Center (DLR) to the European Space Policy Institute (Vienna), and Honorary Adjunct Senior Lecturer at the University of Cape Town (SA) at SpaceLab. She graduated in European and International Law at the University of Strasbourg (France), followed by business oriented postgraduate studies and her PhD at the University of Vienna (Austria). Responsible for DLR and German representation to the United Nations and International Organizations, Dr. Froehlich was also a member/alternate head of delegation of the German delegation to UNCOPUOS. Moreover, Dr. Annette Froehlich is author of a multitude of specialist publications and serves as a lecturer at various universities worldwide in space policy, law and society aspects. Her main areas of scientific interest are European Space Policy, International and Regional Space Law, Emerging Space Countries, Space Security and Space & Culture. She has also launched as editor the new scientific series "Southern Space Studies" (Springer publishing house) dedicated to Latin America and Africa.

[21]This workshop was organized by Secure World Foundation partnering with the South African National Space Agency (SANSA), https://swfound.org/events/2019/workshop-on-socioeconomic-benefits-of-space-resources-utilization-african-perspectives?mc_cid=73d16ec2a2&mc_eid=cf325dc2f2.

[22]Space in Africa, Insight – African Perspectives on the Space Resources Dialogue (Press Release from SWF), 6 June 2019, https://africanews.space/insight-african-perspectives-on-the-space-resources-dialogue/.

[23]During the 58th Legal Subcommittee (LSC) session of UNCOPUOS, Vienna, 1-12 April 2019, a working paper by Belgium and Greece was presented on "Proposal for Working Methods and Work Plan of the Working Group on Legal Aspects of the Exploration, the Utilization and the Exploitation of Space Resources", A/AC.105/C.2/2019/CRP.26.

[24]Committee on the Peaceful Uses of Outer Space, 58th session, Vienna, 1-12 April 2019, Report of the Legal Subcommittee, A/AC.105/1203.

African Woman Competition

Temidayo Oniosun, Ndéye Marie Aida Ndieguene, Mwenya Mwamba, Sharon Kendi Amugongo, Oluwafunmilayo Oluwayomi Olateju, Charlette N'Guessan Désirée, Gift Jedida Ndede, Botho Modukanele, Liepollo Arcilia Letooane, Simpliste Grâce Ninahazimana, Anita Antwiwaa and Gracious Ernest

Abstract
In celebration of the 2019 International Day of Women and Girls in Science, Space in Africa asked women and girls in Africa between the ages of 18–35 to write an essay on enhancing women and girls participation in space science and technology. 106 entries were received from 20 African countries, this chapter highlight recommendation submitted by various participants on how to enhance women and girls participation in space, precisely in Africa.

The African space industry had a history of relatively quiet, sporadic progress, but it is now blossoming into enormous growth. The aggregate GDP of the continent has doubled in the last 10 years, to over US$2.2 trillion. The African space market is now worth US$7 billion annually, and we project that is likely to grow over 40% in the next five years to exceed US$10 billion by 2024.

Space in Africa is the authority on news, data and market analysis for the African Space Industry. Space in Africa produces authoritative business and market analysis reports for the different segments of the African Space Industry. The company covers the business, technology, discoveries, events and political news about the African space and satellite industry. Space in Africa's mission is to provide an amazing journey celebrating the African Space Industry.

T. Oniosun (✉)
Space in Africa, Lagos, Nigeria
e-mail: temidayo.isaiah@africanews.space

A. Froehlich (ed.), *Space Fostering African Societies*, Southern Space Studies,
https://doi.org/10.1007/978-3-030-32930-3_18

The multibillion dollar industry is dominated by the male gender and it is no mystery that women are underrepresented in various science, technology, engineering and mathematics (STEM) sectors around the world. A UNESCO report reveals that women enrollment in STEM studies is low, ranging from ICT (3%); natural sciences, mathematics and statistics (5%); as well as engineering, manufacturing and construction (8%). But what about when it comes to space career? Globally, the gender gap in space exploration is far below the 35% average for all STEM fields. According to a Big Think infographic on Women in Space, since 1961 almost 550 people have been to space, but less than 11% have been women. The situation is not different in Africa.

Africa is witnessing a growing interest in the space sector, however, there is no indication of growth in the number of women who are joining their male counterparts in the space industry. The gender gap is a concern for the few women who are active players in the industry. In commemoration of the International Day of Women and Girls in Science, Space in Africa called on female students and young professionals in Africa to submit an essay to address the gender gap in the African space industry.

In 500–700 words, over 100 girls and women between age 18–35 from 20 African countries submitted essays on how we can get more women in Africa to pursue a career in space science and technology; the title of the essay was "What will it take to enhance the participation of women and girls in space in Africa?". The essay was evaluated based on originality, innovation and creativity, feasibility and relevance, scalability, clarity in writing and presentation and impact on development and potential impact on jobs.

Recommendations from this competition have been shared with policy makers, concerned government institutions including space agencies and organizations, the African Union Commission department of science and technology, United Nation Office for Outer Space Affairs and other relevant institutions strategically positioned to implement policies and undertake initiatives to get more women involved in space science and technology across Africa.

Majority of the institutions are already driving initiatives to achieve this; for example, the South African space agency is about to formally launch the Women in Space Chapter and with dedicated programmes for female learners while a good number of the young people accepted by the African Union Commission department of science and technology for the newly formed Africa youth advisory board on disaster risk reduction are women.

In the following some of the entries from the competition.

1 Women Among the Stars

Ndéye Marie Aida Ndieguene
Pan-African Polytechnic Institute of Dakar, Sénégal
e-mail: hitomikgm@gmail.com

She was sitting in the darkness,

Little girl,

The stars in the eyes,

She was there,

In the dark night,

Fixing the sky,

Wanting to catch them with his hands,

It was decided,

One day,

She would discover them!

The stars,

It's decided,

One day,

She would be a woman among the stars …

"Science is not made for women!" Does this sentence schock you? It reflects the thinking of many people in Africa … But not only.

The notion of gender has always marked our social relations. But it also marked our connection with some professions. So be a woman and an engineer, mathematician, or astronomer is an exceptional fact.

By definition, science isn't male!

In French, as in many languages, words like humans have a gender.

And in French, the word "science" is by definition female.

Science is a woman!

Science is made for women!

All the more reason for there to be even more women in science!

Okay … This argument sounds weird. But is it not so much to put female genius at the service of powerful science?

How?

By fighting the social, institutional and absurd barriers that exist between women and science.

We must destroy the popular beliefs that keep girls out of science. By promoting female examples!

On TV, in the media, everywhere, the example of the woman who is given is often that of the housewife, taking care of the children and her husband.

Little girls grow up with a very simplistic image of the woman. Is not it time to put ads with women in laboratories, on construction sites or scientific meetings. The influence of the environment and images is no longer to be proven. So let's play on this influence and cultivate the love of science by presenting female scientific models! The same observation is made at school! We do not talk enough in class about women in science. Pythagor, Thales, Pavlov etc …

Fantastic!

And women?

Where were they during the great scientific discoveries? They were there but we do not talk enough!

Mae C. Jemison, first female astronaut, do you know? Marie Currie who was just as brilliant as her husband! Jennifer Doudna, Claudia Megan Urry, first woman head of the Yale Department of Astronomy, Tiera Guinn! You know?

Never you will these women be mentioned in our schools. Why?

Well, because they are women!

How do you want girls to become scientists if we give them only male models. It must stop!

If you want women among the stars … you have to give them examples to identify with.

What do you think of the creation of a space hub dedicated to women?

Where our astronomy enthusiasts can set up collaborative projects. Women must be able to exchange, collaborate, create around their common passions. So they will encourage other women to do the same thing.

The creation of an elite African academy for women astronomers: Why not?

There are many schools of excellence for men. Why not create an academy of excellence dedicated to women.

Positive discrimination that can create generations and generations of women astronomers. And that will push girls to choose science.

Centuries ago, Aglaonice of Thessaly was the first woman astronomer. Without technology, she was able to describe phenomena related to stars. Her end was tragic as she was treated as a witch and executed. Even today, in 2019, little girls have to fight for science. Stop by the sexist barriers, they have to fight twice more to reach a profession as an astronomer. Why? Is not humanity supposed to be evolved?

Oh but I forgot, in French, like many other languages, words have a gender, just like humans. The word star is female! All the more reason to have … Women among the stars.

2 My Journey to Space

Mwenya Mwamba
University of Lusaka, Lusaka, Zambia
e-mail: mwambamwenya22@gmail.com

My passion of becoming an Aerospace Engineer started since I was young. In my third grade, my father usually picked me up from school to watch a NASA documentary or a space shuttle leaving for its voyage in space. He told me stories of his tour of The Kennedy Space Centre in Florida, USA back in the late 70's. The detailed explanation of how a space shuttle looked from within made me think he was a NASA astronaut himself. The thrill that the three Zambian employees working for the space center got upon seeing his name, and realizing that he was Zambian, how a 29 year old man was interested in such a diverse field of science and technology that was beginning to take center stage in the 20th Century. And there, I fell in love with aeronautics; science and technology. He made me dream, not to be content with so little, but to reach out for something big.

Talking to women and girls about space related activities from a young age opens up their minds to the possibility of a career in the space industry. This transcends the social perception that women are just meant to pursue "easy" fields.

In my tenth grade during senior high orientation, the guidance teacher came to our class to tell us what was expected of us in high school and life after high school. What subjects we were required to concentrate on, so that we are able to get into the highest tertiary institutions in the country. During his sermon, he decided to find out what our career prospects were, so that, he could give an example of what that career per say needed in terms of subjects. I was asked to state my career choice and during that encounter, I informed him that I wanted to become an aerospace engineer. Suddenly the room was filled with unison laughter from him and my classmates. He then told me to wake up from the fantasy dream I had, as it was nearly impossible for a woman or girl to get into such a field in Zambia. He "advised" me to critically think about my life, and it was best that I changed my career path. These comments later had a profound effect on what I wanted to pursue as a career.

There is lack of proper career guidance in schools. Guidance teachers to my understanding have one of the most important roles in shaping one's career choices as there is need to; study each pupil, learn their interests, provide support (encourage them to; follow their dreams, think outside the box, learn to explore opportunities) and creating platforms for pupils to grow in their crafts and develop new ones. Guidance teachers should not only be available when pupils perform poorly; preaching about studying harder, and spending too much time trying to make students focus on being someone important rather than doing something important.

On the 25th of June, 2015, I had the amazing opportunity to meet and interact with Astronaut Dr. Michael Barratt (Astronaut, Aerospace Engineer, and Medical Doctor) through the US Embassy's Student Exchange Program. As opposed to when I had earlier introduced my aspirations of being an Aerospace engineer, I was received with a different air on this occasion and introduced as someone who had been accepted to well renowned universities in this field. Everyone looked at me in awe and they clapped. The interest Astronaut Barratt showed in me throughout the exchange program, wanting to know more about me and my passion was humbling, to have inspired such a great man and to have him believe in my story and requesting to mentor me was inspiring.

I have discovered throughout my journey that in as much as we say, 'in this age of information, ignorance is a choice', it is not always the case as access to information with regards to opportunities in the space industry, is still inaccessible in countries that are not taking part in activities surrounding groundbreaking efforts being made in this industry. Student exchange programs are one way in spreading this news, but there is need for it to spread at grass-root level where people least expect to find someone with such a dream. The space industry has a lot of importance in the achievements taking place in science and technology and seeing women take part in this making of history is a story worth telling.

3 Take It Up, Embrace Your Journey

Sharon Kendi Amugongo
Samasource Kenya, Nairobi, Kenya
e-mail: kendi.sharon92@gmail.com

The pre-existing disparities regarding gender, ethnicity, class, and post-colonial underdevelopment has largely contributed to less participation of women in space. Most women in Africa face a common disease called FEAR in life. They do not believe in themselves. This can be largely attributed to the fact that, in our continent, there is pre-existing educational disparity between males and females, hence, equity and equality between the sexes is lagging.

Consequently, women do not tend to be confident and they fear taking risks to fight for their rightful places; we are each other's enemies; jealousy towards our own gender, making us tear each other down instead of building ourselves. To enhance the participation, we need to strategize on how to fight the norm in our African culture.

This has led to a big gender gap for many years. Its upon me, as a woman to stand strong and say NO to this attitude I have in the world. The world is me, it all starts with me. I need to grow a thick skin and have a bold face to fight for similar positions as men. I believe the change starts with me as a "WOMAN".

The right attitude is an important step in this journey to participate in Space in Africa. We need to stand as one, "A WOMAN". Let us not allow our weakness to determine our fight in the gender gap in Space exploration. We need to cheer each other while climbing the ladder. Team spirit is raised by a great cheering squad. Women have great ability, that's why it's said; "behind every successful man, is a great woman". Therefore, as a woman I have a great mind, but due to fear of implementing and working on my goals, I let the men take charge and claim victory. As a woman, I need to have the right attitude, support my fellow woman and appreciate her creative mind. Champion her in the world of science and technology. It is not all about man power, it is about concept and technique on how to grow in Space.

While growing up in a small village called Jegereni, in western Kenya, I learned a woman is a very hard-working person. I saw a lady ploughing land while pregnant, under the scorching sun, to provide a meal for her 3 daughters, as is expected of her. I later learned that she was due when she did not show up for work the next day. She delivered that night. This means, it is our nature, as African women, to work hard for our families. However, due to small wages, the reward is little. We need to build the spirit that, "if I can plough while pregnant, my daughter can be an Aeronautical engineer". We need to mentor our upcoming generation to think big and know we have the same opportunities as men. We need to encourage young girls to be bold and courageous enough to go for challenging opportunities in life. They need to see life beyond the poverty and alienation they face.

Exposure to life beyond the village will be key in supporting our young women. We cannot depend on the government to do this but understand education is the key to opening many doors. As we excel, we will be forced to look above and beyond, for what the world has to offer. Our Government tends to recognize this and helps by providing education loans. Bringing awareness in primary schools and encouraging our girls to go for it at a lower age might be an effective way to do it. I believe, if the habit is introduced while one is still young, they learn and adapt. We need to introduce grants and scholarships to support education for low income families.

Finally, having female role models to interact with, as mentors, is a powerful motivator. For most part, successful women do not spend much time in the villages, which makes it difficult to harness the talent. Having a quarterly forum where role models are invited to give talks in schools would be one way to let girls know they can do anything they set their mind to.

4 Fostering Women and Girls Participation in African Space Industry

Oluwafunmilayo Oluwayomi Olateju
National Space Research and Development Agency, Abuja, Nigeria
e-mail: duduyemmy82@gmail.com

An often-heard quote says, when you educate a woman, you educate a nation. To increase the participation of women and girls in space science in Africa, more girls need to be educated.

It has become clear that space science remains a fundamental key and solution to a lot of societal problems and a veritable tool for development in the global community. Recently, many African countries are catching up in the space race by continuous investment in space science for sustainable development of the economy.

Women remain less represented in Science, Technology, Engineering, and mathematics (STEM) even though they constitute about half of the world population. African countries are no exception with a fewer number of women participating in space science compared to men.

However, on the population side, if women and girl's representation in space sciences increases, it would foster national productivity in many nations and there will be collective gains in income, health, quality of life and other benefits that scientific advancement has to offer.

To increase the participation of girls and women in space science and technology in Africa, the following must be considered.

Firstly, African countries need to address the cultural barriers that play a huge impact on our societies. For instance, some are still of the belief that a girl should not be educated because she will end up being a man's property. Also, there are beliefs that men are suitable for certain jobs than women, and that women are to take part-time jobs to care for family and many more.

These cultural beliefs have limited women from following their passion and reaching the peak of their careers. African societies need to be egalitarian for women to be well represented in space sciences and technology to enjoy the development of space technologies which have far more-reaching socio-economic benefits.

Secondly, there is a need for investment in human capital development, more girls and young women should be assisted to acquire knowledge in the areas of telemedicine, rocketry, remote sensing, and similar space programmes through scholarship and funding of research including innovations. This would provide opportunities for many young women to earn a degree, kick start a career in space technology and it would stimulate technical and commercial innovation in our nations.

In addition, African countries need more specialised institutions offering space science courses, this would reduce the number of young women that separate from their families to study overseas and might increase education tourism between Africa countries.

Thirdly, there is a need for women working in the African space industry to mentor girls and young women to develop and transfer knowledge of space science into applied industries such as Agriculture, Telecommunication, Geology and Healthcare. This could be realised if the female space scientists in African countries come together to form an association with aims to promote diversity, inclusion, mentoring and coaching on possible career path available to young women in space science.

Also, having more women occupy leadership positions in space-related fields would stand as a reference point to younger women and spurred them with the hope to break the glass ceilings in the space science field.

Furthermore, the Governments and Stakeholders in space industries need to recruit more STEM female graduates into related job positions. Availability of job positions after graduation for women will increase their representation in space industries. Employers in space industries need to make workplace friendly to women. Challenges like sexual harassment, pervasive stereotype, bullying, and similar inappropriateness that could deter women from pursuing a career in space sciences and similar fields should be avoided in workplaces for women to contribute to the national development.

Lastly, the media have a crucial role to play in the representation of more women in space sciences. On screens, radios and dallies, we see and hear more of male characters portrayed in Science, Technology, Engineering, and Mathematics (STEM) professions than female characters portrayed in similar professions. African female space scientists need adequate coverage by the media to project their research works to inspire more young women and girls in STEM.

In conclusion, to foster the participation of women in space science, all hands including young women, parents, government, media, society, stakeholders and educators must be on deck to inspire and support girls to follow their passion to succeed in pursuing a career in space science for the benefit of our nations.

5 Is Technology Attractive Enough for Women?

Charlette N'Guessan Désirée
Tech Blogger-AdventInTIC, Ivory Coast, Accra/Ghana
e-mail: charlette1693@gmail.com; charlette@bacegroup.com

In the past, coding was considered work for women. As surprising as it may seem, before 1971 coding was entrusted to women. Indeed, at that time coding or programming was a low status office task. For Vogel, coding was a gender issue since from 1971 to 1985, the number of women in the computer sector tripled and accounted for 38% of the workforce.

The influence of women in computer science is so obvious that we owe them certain inventions. In 1843 Ada Lovelace invented the algorithm that made it possible to manufacture the first calculating machine. This was the birth of the code. After the Second World War, six women conceived the code for the very first entirely electronic computer: the ENIAC. They were KathleenMcNulty, Jean Bartik, Marlyn Wescoff Meltzer, Frances Bilas, BettySnyder and Ruth Teitelbaum.

In 1984, a sharp drop in the proportion of women in computer science. Some experts in the field explain by the arrival of microcomputers on the market. According to Isabelle Collet, at that time the advertisements targeted the men and one heard phrases of the same kind as: "if you want your son to get a good job, make him study computer science!"

More than 30 years later, what happened? What approach do we have to this gender disparity in the technological field?

Based on my contributions and experience in the field, everything starts with the education of the girl child. School, family and society make us believe that to succeed in the technological field, you need to have the male mindset. Over the years, the woman in tech is described as physically unattractive and not feminine. The image of a geek leads to this mis-perception of the tech woman. The geek, this unfriendly young man has been so much presented in movies, novels and series that when a woman does not present this image, she tends to be rejected.

These prejudices poison the growth of the girl's mind. She says she is less able to embrace a tech career because she fears her life will end in front of a screen and she would miss her life as a woman. Barriers are thus created in the girls mind. Feeding and promoting the idea that you have to think like a man to succeed in a technological career stifles the female spirit.

In some countries and regions, girls do not have access to school because they are resolutely destined for the noble career of housewives. In Africa, this career is called duty. For me, it is about inequalities in access to education. And when a girl gets to school, she has to deal with the system. This infernal system of education what misleads us in our underdeveloped countries. Once the baccalaureate is in the bag, students are misinformed and worse, misguided. Educational counselors do not play their role. You will agree with me that if we have a high number of boys than girls in science classes, it is obvious that we will have more men and fewer women in technology careers.

Young girls are looking for models of women to be inspired, but where are they? How many young women are able to share stories of inspiring women in the field. It is very common to hear as models, Steve Job, Bill Gates, Mark Zuckerberg etc...., great men who have marked the history of technology. But what about these women who have and continue to contribute to the evolution of the computer field? How can we connect the spirit of these women to these young beginners? How can we promote success stories? Questions that each of us must ask ourselves.

Let's admit that technology trades require enough time, sometimes extra hours of work and more physical and mental resources.

Nowadays, it is no longer a problem of gender but of qualification because the market is very competitive. Technology companies are looking for competence first and foremost. Some women are not as invested in the job as some men. It is clear that if you do not have the skill, you will not be able to respond to the offer.

Among these low rate of women who have had the guts to start a technological career, many face discrimination, sexist messages, and harassment within technology companies.

It is time to bring down the curtains and reveal the darkness of tech companies. I share with you some results of a survey made about women in technology within silicon valley companies: The elephant in the valley.

We asked more than 200 women… Our respondents occupy positions of power and influence: 25% are CXO, 11% of founders, 11% of companies at risk… we also have employees of large companies, including Apple, Google and VMWare.

- 47% were asked to perform lower level tasks that male colleagues are not asked to do (e.g. taking notes, ordering food, etc.)
- 90% witnessed sexist behavior at off-site meetings and/or industry conferences
- 65% of women who report unwanted sexual advances received advances from a superior, half receiving multiple
- etc …

These facts are real and are happening around us. In a voluntary or involuntary way, we are all part of those career killers of women in technology. It's time to say STOP. You're a woman, passionate about technology so remember Marianne Williamson's words: 'Nothing will trap you except your thoughts. Nothing limits you except your fears. And nothing controls you except your beliefs. You want to see change, so get out of your comfort zone and affirm yourself as a woman in technology.'

6 Advancing Africa's Space Exploration: Conquering Partriachy

Gift Jedida Ndede
Moi University, Kenya
e-mail: giftjedida@gmail.com

"Sometimes people have already decided who you are without your story shining through; do not let that define you." Mae Jemison, astronaut and first African American woman to go to space. As the African space race intensifies, the absence of women in the field is conspicuous. Africa, in concert with Mae Jemison's words, has defined a story for its women which is far flung from astronomy and space exploration. Its people seem to believe that because men were the first to explore space and succeed in it, it should be reserved for men. It is chauvinistic thoughts like these have led to an extremely male-dominated presence in the science, technology, engineering, and mathematics (STEM) field. As a young woman interested in this field, I refuse to be held down by these ideologies. I believe that our role as women is to ensure we get seats at the table and create tables where there are none.

While there has been some advancement in the inclusion of women in the field with African women like Abimola Alale making great strides in the space field; this advancement is slow paced and non-universal. Gender stereotyping is still a pertinent discussion. It is no secret that women in space and science at large still face gender inequalities. Because a "masculine culture" pervades the work environment in the space industry, it is an uphill task for women to fit in. The pursuit of gender parity should not be a battle women fight alone. This is a communal issue and therefore should be fought for by the entire society men included. When women are the only advocates of equality, no progress is made. African men in the space field need to step up to encourage support and mentor their female colleagues. If men do not encourage women to come up then, improvement in this area will be unfeasible. African women have not shown a shortage in passion for space exploration, it is just the intimidation they face that bars them from joining the spatial field.

Secondly, to enhance participation of African girls and women in the field, professionals and senior women need to mentor their juniors. Having the guidance, encouragement and support of an experienced mentor provides a mentee with a broad range of personal and professional benefits. It should be the duty of all women to support each other professionally as together we make great strides. By exposing young girls to female role models for mentorship, they gain the courage and confidence they need to pursue a career in the spatial field. Mentorship does not only imply holding lectures but also the call to female professionals to ensure they provide opportunities for apprenticeships and internships. Spatial science should be introduced early in the elementary level of study so as to ensure that students going up levels are aware of the market needs and the opportunities available for them to ensure that they make sound career choices and explore the possibility of pursuing a career in space.

African governments have a pivotal role to play in making spatial studies more attractive to girls in school. The gender disparities faced by women in space should be adequately addressed by governments. National administrations should also support initiatives that aim to close the gender gap. Considerations should be made to encourage the private sector both internationally and locally who wish to empower women to enter the field. Countries should make commitments to change the status quo through policy making, programs, and projects. Adopting legislative policies that motivate more women to enroll in space courses is critical. Government policies should be put in place to ensure girls have access to education, hiring and career development in the space industry.

In conclusion, while space exploration is still a novel idea in Africa, there is need to put in measures to ensure women do not find themselves in the classical disadvantage. This therefore means a call to action for African people to be educated to eradicate the societal precepts that promote gender disparities. All efforts including affirmative action and mentorship must be made to redress this situation for the benefit of the African continent. When African women thrive, Africa will thrive. This message should be made loud and clear; it is time for women to be given the SPACE they need to excel in this field.

7 Cultivate Her Mind for Space

Botho Modukanele
Botswana International University of Science and Technology (BIUST), Botswana
e-mail: botho.moduks@gmail.com

Women have for a lifetime lead in the house hold relentlessly, multitasked on a daily and were able to raise a very fierce and strong generation that became pioneers in all the industries and more pleasingly in the science, engineering and technology fields. I remember when I was about five years old and becoming more aware of my surroundings, all I ever wanted to do was go touch the sky, I wanted to collect all the shiny little dots I saw in the sky at night. I once asked my mother if I could go up into the sky and collect all the stars I wanted, see what the moon was made of and know where it went when it was not visible on some days. My mother told me that I would go when I am older, well, I did grow older and with each year, I learnt more about what was going on around me, how great minds have created what could only be a crazy man's idea back in the day and how stars were actually just a little piece of the puzzle of what is actually out there in the outer space and beyond.

The space industry in African countries does exist but, develops very slowly compared to countries such as United States of America, China or Russia. And because this is often an industry that is filled with men because well, engineering and science fields that make up this industry are considered to be a "man's" profession so girls from a young age often shy away from their sciences and mathematics, we tend to see less women in this industry. A child's enthusiastic mind and the harsh reality imposed on ones dreams as they grow up is strong enough to let someone believe they will never amount to anything especially in a certain field, all because of two simple things, either race or gender. In all my years of schooling, I realized that being able to mold a child and show them that they too could be one of the first few women to explore space, that they too are smart enough to take part in projects such as the square kilometer array as data scientists and programmers is could go a long way. Yes, there are a lot of projects that reach out to students to encourage them to take part in lower levels of education as well as higher levels but, the one mistake made is they deal more with teaching theory more that carrying out the actual practical work.

Introduction of a programme that could encourage participation of women both in lower levels and higher levels of education to be more interested in space could be key to seeing more women in this industry. This programme would not only be run by women who are already in this industry but, would have an online magazine or blog that will be run by women as well as female student volunteers who will also contribute to its content creation. This programme would not only encourage females to take STEM courses but would have frequent practical seminars that would invite people from well-established space agencies to come teach basic skills such as coding, data analysis and all about Artificial intelligence as well as give advice to women who want to excel in the space industry. An official mascot would be ideal for this programme to help with familiarizing and advertising.

If this is executed well on time, it will portray a united front of women that will inspire even more women of all ages to participate more in the space industry and encourage young girls to take up STEM courses that will pave way to a career in the space industry and ultimately reduce the huge difference between numbers of men and women in this industry.

8 African Women and Girls in Space Technology

Liepollo Arcilia Letooane
National University of Lesotho & UNISA, Lesotho
e-mail: liezal.letone@gmail.com

In actual fact there has been a gender gap when it comes to science and technology when it comes to women and girls, because it had been the field that was dominated by men for a very long time and women still coming behind because of insecurity; of not believing in ourselves and the fear of knowing that I am a woman and I am not strong like men; men have always been considered to be the toughest gender, also nature has also contributed the likes of pregnancy and sensitivity of our bodies when it comes to go through Space.

One of the major barriers for women and girls is inaccessibility to information and education about what exactly it is when we talk of Space not only particularly in Africa but what it encompass and its benefits or impact that we could make a living out of it. We do not have access to tertiary institutions even if an individual could have passion in Space Science and Technology. Lack of information about institutions that offer studies in Space Technology is a concern. Apart from that, space technology remains unrealized due to limitations in expertise, infrastructure, equipment and education. Regional competitions should be held annually so as to shed some light about space technology in Africa and access to scholarships specifically for African women and girls would enhance their participation.

Media is an ideal tool that could be used in our times to enhance the participation of Women and Girls in Space Africa, if there could be an application that could be programmed specifically as "Space in Africa for Women and Girls" then it could reach as many people as possible; where role models and mentors could share information and be followed. We are living where media technology has replaced many things such as e-libraries and e-learning. Many young people are able to access smart phones and information is easier to be accessed.

The curriculum in high schools should have (geophysics) subject so that students could have an idea from their high schools because it is from high schools that students are able to make their choices of courses to purse at tertiary levels. Our curriculums should be relevant to address both the current and future challenges on the continent. Most students at African schools are not exposed to the basic knowledge in astronomy and space science. The African continent students should have the opportunity to study space exploration techniques, become familiar with sensors that are used to analyze the surface and interior of planets; and learn the effects of space environment on satellites. The application of space science in real life situations for the benefit of communities, range from climate change and weather to natural resources, agriculture, health, water, risk assessment and disaster response had been recognized. It is clear that space science and technology is an important tool ensuring the sustainable use of natural resources and the creation of high technology in industrial sectors. Lack of technology in our continent contributes largely, to the participation of space technology in Africa and once in individual pursue such a course it becomes difficult of where she would apply such skills in the home country after studies.

9 Vital that Both Genders Understand the Need of Their Respective Input

Simpliste Grâce Ninahazimana
Youth Building in Synergy to end Poverty (YBSP), Tujenge Africa Foundation, Burundi
e-mail: sninahazimana@gmail.com

When we talk about underrepresentation of women and girls, be it in economy, technology, politics, or education, there is only one thing responsible for this state of affairs: society. Even though the male gender could be targeted to be the cause, the female gender also plays a significant role in it. Therefore, we are all responsible for the underrepresentation of women and girls in those areas, especially in scientific fields. So, men and women will have to deepen their understanding of the main root causes of this gender gap in Space programs in Africa.

The first question to ask is: Do men have interest in the participation of women and girls in space programs in Africa? Ideologies and affirmations tend to become the reality we base ourselves on to measure our abilities. Men have been for so long the center of attention. They have been hearing all of their life that nothing is impossible for a man. Then, those affirmations became their reality. So, how about their women counterparts? Probably, they have been hearing the contrary to what men have been told.

But still, the main question does not lie in spotting the existence of such an imbalanced privilege between the two genders. It lies in contemporary men's recognition of the importance of women's participation in all those fields in which men have traditionally held a monopoly. In other words, what are men doing to welcome women's participation in space programs in Africa?

However, saying that only men have the responsibility to advocate for women and girls to participate in space programs would be a way of affirming that, in fact, men have power over women and girls to the point of being the ones to hold whatever future women can aspire to. All this is to say that women also should—and have to—actively take part in their own battle.

Africa is for the most part cultural, and traditional practices are essential for its identity. Moreover, even though Africa has been evolving to conform itself to some modernity, it is still harshly conservative in whatever concerns women and girls. That is one of the many reasons why there is a perpetual restriction to the promotion of women and girls to participate in what has been taboo to them in the past.

It will take women and girls to take control of their own promotion and overcome society's judgments. They cannot rely on their oppressors—men—to forever maintain interest in the fight, because the women and girls' case would be lost. It is important that women and girls take ownership of their own struggles in a movement they themselves lead.

Even if considered as having no interest at all, men are nowadays having an increasing understanding of the importance of women's participation in science and related areas such as space programs, a field in rapid expansion in Africa. Nevertheless, if one day we want to see a considerable part of women and girls in space programs in Africa, women and girls will have to learn to go for their cause however difficult the journey will be.

Women and girls will have to be aggressive and firm on their opinions so as to shape their future; women and girls have been given the door's key. It is now up to them to open it, once and for all, while being totally sure of themselves.

I think that for women and girls to participate in space programs in Africa, men will have to give opportunities to women and girls to hear that they are valuable people at the same level as men. Additionally, it will take women and girls to actively defend their beliefs and hold on to their ideologies. With men and women working together, Africa will see successful science and space programs that flourish because all talented individuals are welcome to participate.

10 Harnessing the Potentials of Girls and Women in Space Technology and Its Related Activities in Africa

Dr. Anita Antwiwaa
All Nations University-Space Systems Technology Laboratory (ANU-SSTL), Ghana
e-mail: draantwiwaa@anu.edu.gh

Women and Girls who participate in the African Space Science and Technology which is one of the key discipline which employs all other technological areas in implementing projects that see through the lens of all human needs and wants and brings up solutions which addresses them and advances the sustainable development goals for societal benefit has been on the lower side over the years. This area has proven to be an essential contributor to achieving the SDGs and African Union Agenda 2063 and women cannot be left out. The world population of women is almost exceeding that of men meaning that workforce and contribution of women is needed in all technological areas to achieve sustainable development. There is therefore the need to educate African women in the area of practical oriented Space Science, Technology, Engineering and Innovation [SPACE-STEMI] to attain a sustainable wide scope of addressing human needs through Space Science and Technology.

Over the years, fewer African girls remain in the area of STEMI education after primary school. Less percentage of these girls get the opportunity to remain in STEMI after secondary school leaving very few in the career of STEMI which thereby leaves few women in the area of space Science. African women are not able to joyfully embrace space science due to the African culture perspective of a woman which perceive the woman as a weaker gender and are therefore restricted from getting involved in adventurous activities that can broaden their spatial analysis which will lead them to choose a career in Space Science. Most African parents consider education in space science as a "tough" and a daring area therefore they do not encourage the women whom they consider to be a weaker gender to pursue such courses.

Secondly, most young girls losses their self-confidence because they see their male counterparts as having intelligence and boldness in this area. If these girls are given the right mentorship at that stage, it will boost their confidence thereby choosing a career in the area thereby erasing the primitive ideology that Space technology is not a venture for women.

Moreover, there are fewer women who have achieved successful careers in space in Africa for the younger female generation to model after. These young girls find themselves in an environment with women successful in other career lines other than Space Science.

Lastly, the space industry in Africa is still in its infant stage and has lesser job prospects on the African continent. Most African countries are not actively into this area therefore there is less motivation of choosing a career which have uncertainty of job security afterwards. All these hinders the participation of more women in Space.

It is very difficult to boast the interest of a person to take a career in Space Technology when the subject matter is studied theoretically in most African countries. Solving this problem requires the use of miniaturized satellites like CANSAT and HEPTASAT as a model to teach Space technology in all levels because it is easier to building and it is not costly. Girls in lower grades can be empowered to choose a career in Space by exposing them to CANSAT activities at a younger age will boost their interest in choosing a career in Space Technology.

Secondly, there is the need for the women who are already in Space to get involve with the young girls by mentoring them so as to generate their interest in taking a career in Space Science and Technology in the future.

Furthermore, the space institutions can get involved by organizing space competitions, conferences and seminars for girls in primary and high schools, so as to create awareness and to sprout their interest in choosing a career in Space and its related applications. Scholarships should also be made available for girls who wants to choose a career in this field.

Finally, the space institutions and industries should come up with policies that will give talented young women who completes the university with space related degrees the advantage over their male counterparts in securing jobs in this field. This will help to increase the participation of more women in Space and its related activities in Africa.

11 Getting African Women and Girls to Reach for the Stars

Gracious Ernest
Africa & For Creative Girls, Nigerian Law School, Abuja, Nigeria
e-mail: ernestgraciousanita@yahoo.com

"*But girls don't go to space*" replied my six-year-old niece, when I asked her if she would love to walk on the moon too someday, as she delightfully sang a song she had learnt from school about Neil Armstrong.

Generally, little is ever mentioned of Valentina Tereshkova and Mae Jemison as much of we hear of Neil Armstrong and Yuri Gagarin; and if we must encourage more girls to explore space, we must talk more about the many outstanding women that have made important contributions to human spaceflight.

When Valentina Tereshkova went to space in 1963, she made history as the first lady of space. And when Mae Jemison went to space in 1992, it was a feat of double achievement not just for women, but for black women the world over.

However, the challenges facing women in space globally range from negative stereotypes about the biological effects on women, to the entrenched patriarchy which can be noted in Randolph Lovelace's theory in the 1960s; he believed that having a large space installation would necessitate women going to space in order to occupy the pink-collared positions like secretaries, lab assistants, telephone operators, nurses, et al. All gender-stereotyped roles.

The space sector in Africa is not as large as its counter-parts in first world countries, but in recent years, it has begun to develop, and the trend is spreading across Africa.

Women in the science, technology, engineering, and mathematics (STEM) fields in Africa, is also becoming part of the larger conversations about gender equality, and women and girls are beginning to take interest in these fields; however, gender equality in the space sector globally and in Africa still lags behind.

If we want to enhance the participation of African girls and women in space exploration, we need to get them to know they can, through; mentorship, space related programs, opportunities, and conferences for girls, jobs diversity and inclusion, and by redesigning our educational curriculum to include more space-related courses.

According to Oprah Winfrey, a mentor is someone who allows you see the hope inside yourself. Mentorship is thus crucial to inspire girls, boost their confidence, and challenge them to take the right direction in life. To encourage more girls in the STEM fields and specifically in the space-sector, we must inspire and show them women like themselves, who are a reflection of what they can be. There is need for the few women in the space sector in Africa and globally to mentor other girls and women. Mentorship programs can be set up, and made accessible in schools and online platforms. Through inspiration, we can spark a significant interest of women and girls in the space sector.

Another solution is in organizing more local space related programs, opportunities, and conferences for girls across Africa. Workshop groups and trainings in schools are also a great way to bring the space sector closer to women and girls in Africa. Through these programs, more girls are enlightened and taught about the different advantages and opportunities in the space sector and this goes a long way to pique their interest in space.

Collaboration between the government and private sector is also necessary in creating space related opportunities for girls, in order to encourage their participation in the space sector. There is need to partner with space agencies in Africa to further enhance space-related programs for girls and women.

A certain issue to address is the educational curriculum in Africa, which should be redesigned in order to include more courses related to science and space exploration, and make them more practical. Furthermore, scholarships and financial aids for girls to study these courses either locally or internationally would serve as great incentives.

It is indeed essential to break all stereotypes facing women in space exploration. These stereotypes which are baseless, non-factual and entrenched in patriarchy need to be dismantled. Women and girls need to be taught that women can go to space, without any adverse effect on their health.

The space sector is developing in Africa, and females must constitute an integral part of it. We must break the narratives and factors that make the STEM fields male dominated in Africa.

In a few years from today, when the first African goes to space, it can be a woman because- women go to space too.

Dry, the Beloved Country: Space and Water: The Cape Town Water Crisis

Nicolas Ringas, James Wilson, Asim Raza, Bafowethu Setheli, Barbara Amelia King, Jahanzaib Hussain, Luke Colvin, Mirza Waqas Baig, Maureen Tanner, Mohammad Naveed, Muhammad Ebtisam Ahmed, Muhammad Mubeen Anwar, Nasir Mehmood, Nauman Majid, Okeletsang Mookeletsi, Saeed Ur Rehman and Saqib Kabeer

Abstract

The Western Cape Province in South Africa is facing one of its worst recorded droughts. This study explores how satellite-based technologies can be used to address the water crisis using three specific approaches, namely: (i) alien tree mapping and removal at catchment areas, (ii) rainfall pattern monitoring and (iii) rainwater harvesting. The study proposes the creation of alien tree mapping software that uses remote sensing data to automatically identify, classify and

N. Ringas (✉) · J. Wilson · A. Raza · B. Setheli · B. A. King · J. Hussain · L. Colvin · M. W. Baig · M. Tanner · M. Naveed · M. E. Ahmed · M. M. Anwar · N. Mehmood · N. Majid · O. Mookeletsi · S. U. Rehman · S. Kabeer
University of Cape Town, Rondebosch, South Africa
e-mail: rngnic002@myuct.ac.za

J. Wilson
e-mail: wlsjam015@myuct.ac.za

A. Raza
e-mail: rzxasi001@myuct.ac.za

B. Setheli
e-mail: sthbaf001@myuct.ac.za

B. A. King
e-mail: kngbar003@myuct.ac.za

J. Hussain
e-mail: hssjah001@myuct.ac.za

L. Colvin
e-mail: clvkyl001@myuct.ac.za

M. W. Baig
e-mail: bgxmir003@myuct.ac.za

A. Froehlich (ed.), *Space Fostering African Societies*, Southern Space Studies,
https://doi.org/10.1007/978-3-030-32930-3_19

rank alien trees located in catchment areas to allow for planned tree removal programmes which will increase the amount of water collected at each catchment. Rainfall pattern monitoring and investigation in to rainwater harvesting techniques showed that almost 30 million litres of water can be collected during the rainy season using appropriate rainwater harvesting systems. It is proposed that the government investigate this further, as well as perform an in-depth rainfall pattern analysis to identify potential areas for new catchments. Additional remote sensing applications which can assist with drought mitigation strategies are discussed within the chapter and include: subsidence monitoring at groundwater extraction plants, leak identification in water pipelines, monitoring the environmental impacts of the planned desalination plants and planning future water infrastructure projects based on urbanization trends.

1 Introduction

The Western Cape Province in South Africa is facing one of its worst recorded droughts in history. Water restrictions have been implemented by the City of Cape Town (CCT) to reduce the City's water consumption and avoid "Day Zero". This study explores how satellite-based technologies can be used to address the water crisis in CCT and the province as a whole. The study first describes the current

M. Tanner
e-mail: mc.tanner@uct.ac.za

M. Naveed
e-mail: nvdmoh001@myuct.ac.za

M. E. Ahmed
e-mail: ahmebt001@myuct.ac.za

M. M. Anwar
e-mail: anwmuh002@myuct.ac.za

N. Mehmood
e-mail: mhmnas017@myuct.ac.za

N. Majid
e-mail: mjdnau002@myuct.ac.za

O. Mookeletsi
e-mail: mkloke001@myuct.ac.za

S. U. Rehman
e-mail: rhmsae001@myuct.ac.za

S. Kabeer
e-mail: kbrsaq001@myuct.ac.za

N. Ringas · J. Wilson
University of the Witwatersrand, Johannesburg, South Africa

water crisis in detail and discusses the City's plans to combat the drought. The general principles of drought management schemes are investigated through a case study of the River Murray-Darling River Basin in Australia.

The publication then discusses how space-based technologies are currently being used within South Africa to aid in water management. This information was obtained through correspondence and interviews with key players in both private and public entities. The existing legal framework governing water resource management is explored on a national, provincial and local governmental level.

Three space-based applications that can be employed by the Western Cape Province and CCT in battling the water crisis are investigated and analysed. The first application involves using satellite imagery to identify alien tree species (specifically pines, eucalypts and wattles) situated in catchment areas around the six major dams in Cape Town, with the aim of quantifying the amount of water these trees are preventing from reaching the dams.

The second application involves using satellite-based rainfall prediction technologies to analyse the rainfall patterns around the Cape area to determine the efficiency of existing catchment areas and possibly propose the installation of new catchments in high rainfall areas. This is linked to the third application which uses satellite imagery to quantify how much water CCT could collect using rainfall harvesting systems. This application estimates the surface area of suitable roofs within the City for a small target area and then uses the information to look at the whole City.

The next section looks at how other satellite applications can be used to supplement and aid other water management techniques currently in use by the City. The analysis contains a discussion and recommendations section which lists the authors' recommendations on which satellite applications the Western Cape should consider implementing and offers suggestions on how they can be achieved.

2 Background

Cape Town is the second most populous city in South Africa (SA) after Johannesburg; reaching 4 million inhabitants in 2018, from 3.74 million in 2011[1]—an annual population growth of 0.98% over 7 years. With beautiful beaches, endemic flora and fauna and the famous Table Mountain it's the most seasonally visited city in SA. These aspects have a direct effect on the demand for resources required by the city and its citizens, particularly potable water.

The Western Cape is currently experiencing[2] the worst drought in approximately 311 years, with a 90% confidence that it is the worst drought between 105 and 1280 years. The Western Cape and CCT have implemented Level 6B water restrictions, requiring savings of 45% by residential users.

[1]Western Cape Government—Regional Development Profile (2013) https://wazimap.co.za/profiles/municipality-CPT-city-of-cape-town/ [accessed: 22/06/2018].

[2]CCT, 'Water Outlook 2018 Report'. Department of Water and Sanitation, Rev. 24-17 April 2018.

There are currently augmentation management plans being carried out on a provincial (CCT and Western Cape Government) and national (Department of Water and Sanitation—DWS) level to alleviate the drought effects, however, as shown in this report the plans neglect the use of space-based technology and particularly Earth Observation (EO) satellite data.

This publication looks at the current drought situation, how it has been dealt with in comparison to droughts experienced previously in Australia and how relief schemes and water management can be assisted by satellite data. It aims to add valuable complementary information and tools which can form a part of the water management plan for CCT and the Western Cape.

3 Contextualization

Although the drought took many Capetonians by surprise, there were early warning signs as far back as the 1990s by numerous scientists and experts. For example, the drought was predicted by the scientist and UCT graduate, Gordon Maclear in his 1995 master's thesis. Mr. Maclear spoke in response to the National Water Response Campaign, advising CCT that tapping into the Cape Flats Aquifer Unit (CFAU) will take pressure off the mostly runoff dependent Western Cape Water Supply System (WCWSS).[3] This was the first of many warnings that were seemingly inadequately responded to by authorities. Another researcher, Mr. David W. Olivier suggested in a more recent article,[4] the issue of delayed response and denial was partly a political issue.

How the drought situation has developed recently: On 22 January 2018, models predicted Day Zero (13.5%) to occur on the 12 April 2018, with weekly dam level drawdowns at 1.4% and agriculture exceeding CCT's daily demand. By 22nd March, the weekly drawdown had reduced to 0.4% with agriculture using only 4% of water from the system, resulting in 13.5% dam levels being projected into August. The Day Zero calculations were based on conservative assumptions of consumption beyond the City's control, including releases to agriculture, urban demand, evaporation and low rainfall. The projected Day Zero was updated weekly using the values from the previous week and not adjusting for the reduced demand or potential rainfall. DWS also stopped water release to irrigation boards, dramatically reducing drawdown from the system. Furthermore, a transfer of 7–10 million cubic meters (Mm^3) was made by private farmers in the Groenland Water

[3]L.G.A. Maclear, 'Cape Town Needs Groundwater'. *Geohydrology Directorate*, Department of Water Affairs and Forestry, Cape Town. Technical report No. Gh3868, August 1995.

[4]David W. Olivier, Biz News (Dec 2017) 'Cape Town water crisis: Blame rests with ANC government, not the DA—Expert'. https://www.biznews.com/undictated/2017/12/15/cape-town-water-crisis-anc-da/ [18 June 2018].

Users Association (GWUA) based in an adjacent catchment.[5] All aspects mentioned pushing Day Zero beyond the anticipated start of the rainy season, hence beyond the foreseeable future.

3.1 Cape Town Water Infrastructure

The demand of water for Cape Town is fulfilled by the Berg River Basin, which is located at Western Cape Region of South Africa. This system fulfills the demands for households, industrial and commercial use and provides water for almost 15,000 ha of cultivated crops (Fruits, Vegetables, Wine Farms) used in Cape Town metropolitan region.[6] The WCWSS also supplies industries producing livestock, meat and dairy and fields of crops like wheat, barley and canola for which plenty of water is required. The Berg River catchment is running out of water due to high demand for water in this region and reduction in the rainfall over the previous few years.

Therefore this demand for Cape Town is relying more heavily on using all 6 major dams, Bergriver, Steenbras Lower, Steenbras Upper, Theewaterskloof, Voëlvlei and Wemmershoek through WCWSS.

3.2 Water Crisis in Cape Town

There is a lot of discussion about the nature of the water crisis in Cape Town, but there is a general consensus on certain factors such as the fact that 2015, 2016 and 2017 has been the driest three year period[7] with regards to the rainfall in the dam catchment areas in recorded history.

Also, the water supply system is based on 98% assurance (or 49 out of 50 year preparedness) of unrestricted supply, obviously being insufficient for this 1-in-311 year drought.[8] The combination of these facts has resulted in the hydrological drought that is being faced by Cape Town. Figure 1 shows the amount of water stored in the 6 large dams in the Cape area and how they have been declining since 2014.

Climate change alone cannot be blamed for the drought as shown in the historical measurements taken by one station in the catchment of Theewaterskloof

[5]Jenni Evans, 5 Feb 2018. 'Fruit Farmers opening Sluice for Cape Town'. News 24, https://www.news24.com/SouthAfrica/News/fruit-farmers-opening-sluices-for-cape-town-20180205 [18 June 2018].

[6]Jabavu C Nikomo, Bermard Gomez, "Assessments of Impacts and Adaptations to Climate Change (AIACC)" 2006, International START Secretariat, USA.

[7]Piotr Wolski (22 Jan 2018). Facts are Few, Opinions are Plenty… on drought severity again. http://www.csag.uct.ac.za/2018/01/22/facts-are-few-opinions-plenty-on-drought-severity-again/ [accessed: 23 June 2018].

[8]P Wolski, What Cape Town learned from its drought, Bulletin of the Atomic Scientists, 16 April 2018, https://thebulletin.org/what-cape-town-learned-its-drought11698 [last accessed 18 June 2019].

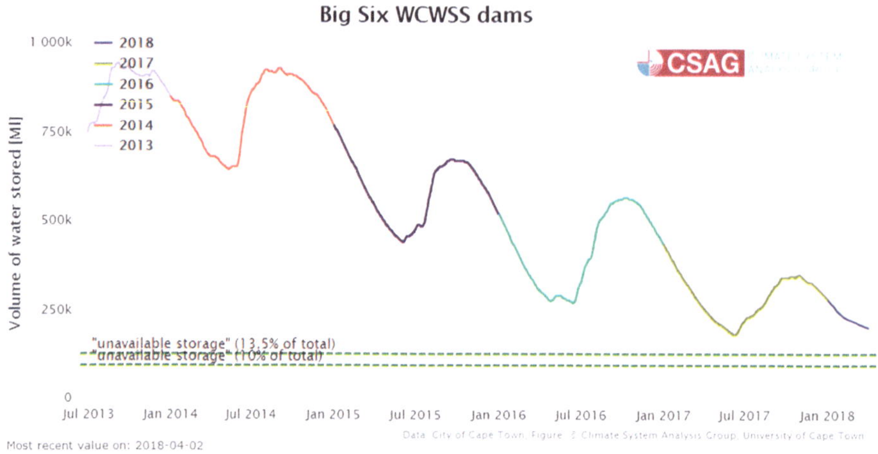

Fig. 1 Documented dam levels (2013–2018)

Dam. There have been many dry years in the past, with the most recent being the 2003–04 period. Moreover, there have been many direct warnings from the experts from 2002[9] onwards for the need to utilize new water sources. This resulted in the construction of the Berg River Dam in 2009. However, the population of CCT is fast growing and subsequently the water demand has increased, as such the construction of the new dam did not prove to be sufficient for the City's water supply needs. So apart from climate change and variability, there are other factors like lack of preparation, lack of planning, monitoring, and forecasting that need to be considered as contributors to this drought.

3.3 General Drought Management Strategies

There are many strategies that can be applied for drought management. Drought management strategies around the world are different based on the local climate, geography and needs. But the overarching aim of drought management remains essentially the same for all the strategies. Drought management strategies are selected based on the type of the drought (see Fig. 2). A meteorological drought is the term used when the actual rainfall is below the average rainfall over some period of time. The time period varies from monthly to annual. The effects of prolonged shortfall of rainfall can result in a hydrological drought. It is also important to consider what constitutes the beginning or the end of the particular drought. Another important clarification is that the term drought refers to a shortage

[9]W S Conradie, Why is Cape Town running out of water?, Climate System Analysis group (CSAG) at the University of Cape Town, 21/02/2018, http://www.csag.uct.ac.za/2018/02/21/cape-town-running-out-of-water/ [last accessed 18 June 2019].

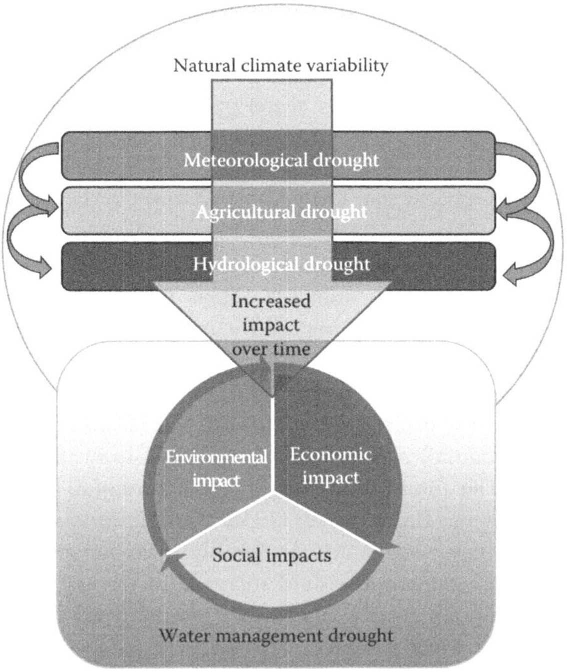

Fig. 2 Different types of droughts (S Eslamian et al., Principles of drought and water scarcity, CRC Press, Isfahan University of Technology, 2017)

of water for the purpose of irrigation or agriculture whereas the term water scarcity refers to a shortage of clean drinking water. Drought management refers to all the activities relating to the management of droughts from early warning, planning, and eventually the mitigation strategies.

Case Study: Drought Management in South Australia

Australia was affected by an extended period of drought (termed the "Millennium Drought")[10] which started from low rainfall in the 1996 season but continued almost until 2010. The region that was worst effected was the region supplied by the Murray-Darling Basin. The drought mostly affected environment, irrigation and urban water supply in the basin. Australia implemented numerous different strategies and technologies that provided resilience in such a severe drought. There are many lessons that can be learned from the drought management strategies implemented in South Australia especially in the Murray-darling Basin. One of the key factors is the full implementation of the Integrated Water Resource Management

[10]Australian Government Bureau of Meteorology, Recent rainfall, drought and southern Australia's long-term rainfall decline, April 2015, http://www.bom.gov.au/climate/updates/articles/a010-southern-rainfall-decline.html [accessed June 2019].

(IWRM) plan. However the main reasons for the successful implementation of the IWRM lies in the level of education, devolution of authority down to catchment level, and stake holder participation. Apart from IWRM, there are other steps such as water sharing, water use restrictions, water treatment (in the form of desalination and waste water management) and urban rain water harvesting, which have been proven successful in the face of severe drought.

Also, in order to assess water availability of surface water and groundwater resources of the Murray-Darling Basin, the Australian National Water Commission contracted CSRIO to perform the world's largest basin-scale investigation in the form of Murray-Darling Basin Sustainable Yields Project.[11] The project assessed the impact of climate change and variability on the water resources. Also, the impact of ground water extraction and catchment development were considered. CSRIO utilized many models including rainfall-runoff models, NDVI patterns, irrigation areas as well as evapotranspiration of wet and dry lands. The analysis revealed the average rainfall and run-off has declined from the historical average from 1997 to 2006. The future estimates of average rainfall and average run-off over the Murray Darling Basin up to 2030 were uncertain. The project also determined a decline in surface water availability over the span of the study and predicted ground water extraction would increase in the region. However some parts of the basin would not sustain any further ground water extraction due to the reduction in the ground water resource.

3.4 Summary of the Water Outlook 2018 Report by CCT

CCT is part of the WCWSS, a system of dams that supply Cape Town (64%), Western Cape agriculture (29%) and other urban areas (7%).[12] The catchment is made up of six major dams (Fig. 3) that are primarily filled with rain water runoff (88%) and are supplemented by the current augmentation scheme making use of groundwater and aquifers.

The Department of Water and Sanitation (DWS) manages the largest three of the six dams in the system and is responsible for planning and implementing water resource schemes to meet water demand for cities, industries, mining and agriculture on a national level. DWS plans to accelerate the planned 2022/3 augmentation schemes for WCWSS, which will transfer water from the Bergriver Dam to the Voelvlei Dam, providing about 60 million litres of water per day.

"DWS plans on a 1 in 50 level assurance", which means that any drought experienced that is more severe than another experienced in the previous 50 years

[11]Commonwealth Scientific and Industrial Research Organization (CSIRO), A basin-scale investigation into water availability in the Murray-Darling Basin and the impacts od development, water extraction and climate, https://www.csiro.au/en/Research/LWF/Areas/Water-resources/Assessing-water-resources/Sustainable-yields/MurrayDarlingBasin/Overview [last accessed 18 June 2019].

[12]CCT, 'Water Outlook 2018 Report'. Department of Water and Sanitation, Rev. 24-17 April 2018.

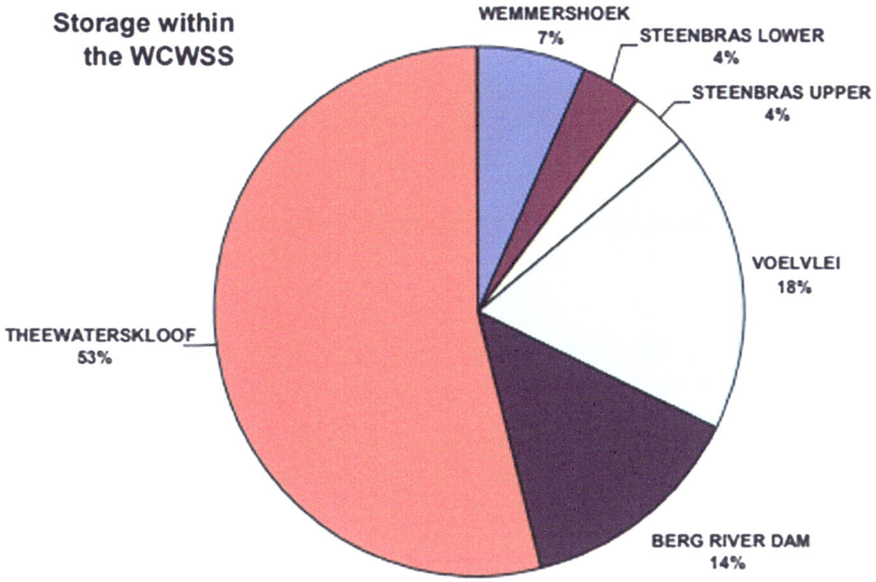

Fig. 3 Six major dams storage contributions within the WCWSS (Department of Water Affairs and Forestry: Western Cape Water Reconciliation Strategy, Newsletter. March 2009 [accessed: 22/06/18])

requires restrictions to be implemented to reduce the demand. To cope with the current drought situation, the Department introduced restrictions to maintain the dam levels above 15%. CCT must then meet DWS restrictions by implementing restrictions on consumers, which it has now done. Level 6B restrictions are currently in place across the city, which aim to reduce resident's consumption by 45%, individuals are restricted to 50 L per day and agricultural 60% reductions—although some irrigation boards have been cut off after reaching quotas. These efforts are part of Phase 1 of CCT's Disaster Management Plan, established in response to the drought. Phase 2, would be triggered if the dams became critically low (i.e. less than 13.5% capacity) causing major disruptions, would halve residents current supply of 50 L per day and require individuals to collect the other 25 L per person from designated distribution points. This would provide just three months' worth of water at a reduced volume supply of 350 ML/day (MLD).[13]

CCT developed a drought management programme to try avoid Day Zero. The programme focusses on three main categories: (A) better management of existing water supplies, (B) reducing water usage as much as reasonably possible, and

[13]City of Cape Town Safety and Security Department, Critical Water Shortages Disaster Plan—Public Summary, October 2017.

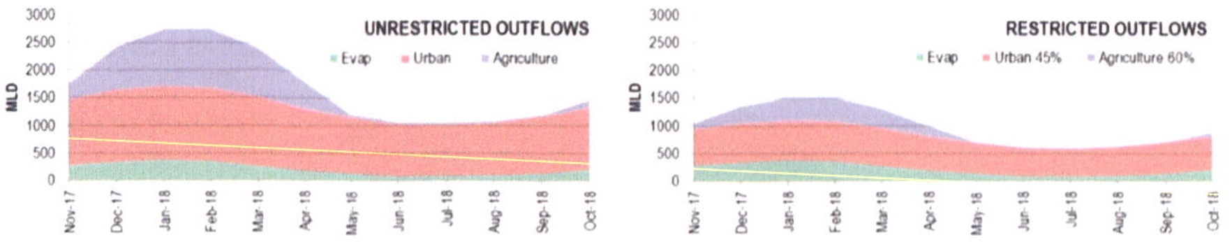

Fig. 4 Diagrams of unrestricted and restricted water outflows (taken from CCT online water dashboard)

(C) identifying and developing new water sources. These are discussed briefly in the following sections.

3.4.1 Managing the Remaining Water in Dams

Dam behaviour is modelled conservatively on 2017 runoff data. The three main water users resulting in dam level decrease are the agricultural sector, urban uses (i.e. CCT and other municipalities in the Western Cape), and evaporation (as shown in Fig. 4). So far curtailing urban use has been successful, however the WCWSS is still under strain due to low rainfall levels experienced in 2017.

Dam level changes are published weekly and are tracked on the CCT's water dashboard along with urban and agricultural demand and can be compared to allocations. Decisions are made on dam level tracking and anticipated behaviour. The situation becomes more critical each day when the 450 MLD target is exceeded.

Additional dam management aspects:

- WCWSS covers West Coast district municipality and local municipalities of Drakenstein, Stellenbosch and Witzenberg, all urban use has been restricted to 45%. The dams operate as a system, connected by pipelines, canals and tunnels.
- CCT controls three of the six WCWSS dams: Steenbras Upper, Steenbras lower and Wemmershoek. Steenbras Upper can distribute over the widest area of the three and is therefore kept the fullest, and is also required to maintain efficient use of the Steenbras hydroelectric power station.
- CCT manages some smaller dams, such as those on top of Table Mountain, with a storage capacity of around 4.4 Mm^3.
- Domestic use makes up 70% of the water consumption within CCT.
- Informal settlements in Cape Town use 4% of water.

3.4.2 Managing Demand to Reduce Water Usage

Individual use must be curtailed and limited to specific volumes regardless of the individual's whereabouts. CCT is implementing measures to reduce water usage by:

- Restriction Level 6B: Level 6 was enforced on the 1st of January 2018 and intensified to Level 6B by the 1st of February 2018. The target of these restrictions being 450 MLD, which is in line with the DWS Level 6 tariffs, assuming 4 million citizens use 50 L per day (200 MLD) and approximately 150 MLD is used by government, commerce and industry. This leaves a surplus of 100 MLD, which is required as restrictions have not been adhered to thus far.
- Communications campaign: ensure each individual understands that the situation is a crisis. 'Household leak detection' and 'How to use 50 litres' are some of the campaigns launched by CCT. Further radio, print and social media campaigns have been launched in an attempt to reach the 450 MLD goal.
- Household flow regulators: installing water regulators for debt management has been ramped up, especially in areas where households have not adhered to restrictions. The expected usage of a household of four people is 6 kL per month, houses above 10.5 kL per month will be targeted and particularly households using more than 20 kL per month. Many cases of over use were found to be related to undetected leaks. CCT has installed approximately 250,000 water management devices in the last decade. Due to low price of water smart metering has been unviable.
- Punitive tariff: tariffs are linked closely to restrictions as services must still be provided while lower volumes are being distributed. Stepped tariffs result in progressively increased cost for higher volume use.
- Adaptation: CCT to engage with large and small businesses to incentivise reduced usage. Private borehole usage and rainwater harvesting schemes to be included in the augmentation system.
- Information campaigns and dissemination to drive behavioural change: the star rating in buildings and the City Water Map help visually display user's usage and encourage reduction in consumption.

3.4.3 Water Source Development

CCT aims to diversify its supply sources to increase the resilience of its water supply network. Furthermore, new sources of water which will help mitigate the drought effects are identified. In this regard, CCT have done the following work:

- Prioritise groundwater extraction: extraction of water from Cape Flats Aquifer (CFA) and Table Mountain Group (TMG) Aquifer. These aquifers have significant water storage volumes. Although they are subject to drought conditions, the effects have a significant delay.
- Pursue permanent desalination at optimal scale: Plan and execute permanent desalination at an optimal scale, modules of 120–150 MLD and not bigger than 200 MLD. Explore alternative procurement, competitive bid turnkey approach, using private sector and water purchase agreements. This will provide lower costs per unit of water and be quick to implement provided regulatory processes are fast tracked.

– Implement water re-use: the high cost of groundwater extraction and desalination, justifies the re-use of water to maximise benefits. Suitable locations have already been identified.

Further surface water augmentation: DWS is implementing the Bergriver Voelvlei Augmentation Scheme (BRVAS) which is expected to add 23 Mm^3 (60 MLD) into the WCWSS in 2021.

3.5 Project Investigation

As a prelude to determine which space-based technologies could be employed to assist with the drought mitigation, an investigation was performed to determine what applications regarding the drought were already in use by various public and private-sector entities. Therefore, the team engaged with pivotal actors in both governmental and commercial sectors involved in ameliorating the water crisis.

3.5.1 Department of Science and Technology

The Department of Science and Technology (DST) implements the National Space Strategy and the South African Earth Observation Strategy, and is governed by the guidelines of the National Space Policy, an instrument of the Department of Trade and Industry (DTI).

The DST oversees and funds research and development in strategic and emerging focus areas, such as space science. Under its aegis is the Academy of Science of South Africa, the National Research Foundation, the Council of Scientific and Industrial Research (CSIR) and the South African National Space Agency (SANSA).[14]

In terms of the evolving water shortage, many entities were called upon in one way or another to contribute knowledge and expertise in resolving the crisis. But the prime mover to apply space based information to monitor the problem at hand is SANSA.

[14]Department of Science and Technology, South Africa, website: www.dst.gov.za [last accessed 18 June 2019].

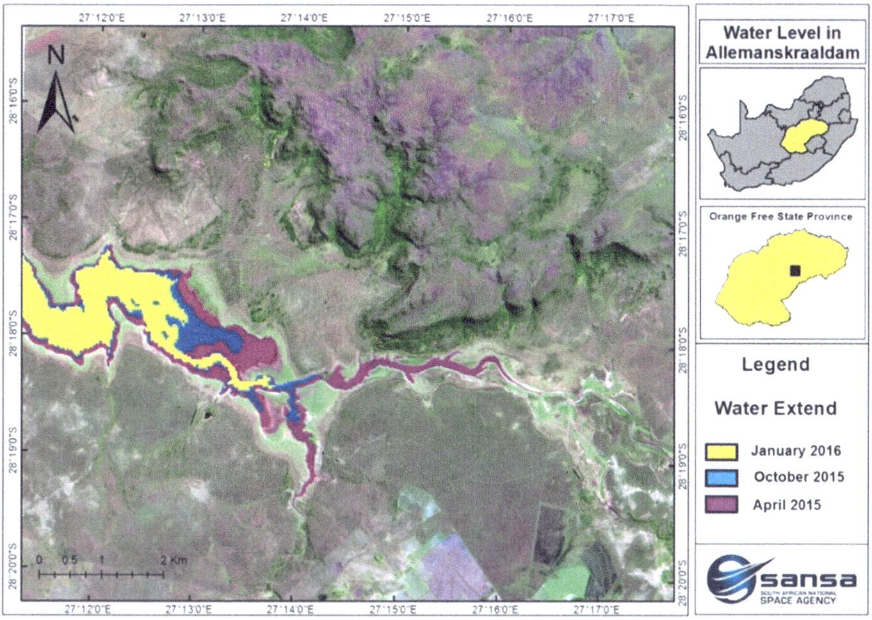

Fig. 5 Example for water level mapping in the Allamanskraal dam, *Image Credit* SANSA

3.5.2 South African National Space Agency (SANSA)

SANSA was established in 2008 through the South African National Space Agency Act to promote space science research and cooperation in space-related activities country wide.

The expertise of SANSA's Earth Observation Programme focusing on utilisation of space to address societal needs, (including resource and environmental management, disaster management, food security, global change monitoring, health, safety and security), came to the fore.[15]

SANSA provides state of the art ground station services and data sets to the government and civil entities, as well as globally recognised space missions. To address the drought crisis, SANSA uses SPOT 6/7 and LANDSAT 8 data to perform alien species mapping, water body mapping (Fig. 5) and irrigation monitoring.

Digital elevation models (DEMs) along with population density information are employed to perform flood mapping. The DEM data is from the Shuttle Radar

[15]South African National Space Agency, South Africa, website: www.sansa.org.za [last accessed 18 June 2019].

Topography Mission or SRTM, with 30 m spatial resolution. This is in partnership with TIGER-NET which is an ESA initiative that aims to help African countries use remote sensing data to better manage their water resources. Flood mapping is part of the disaster management which is a central focus of the National Disaster Management Center, to plan, execute, manage and advise initiatives aimed at alleviating impacts of natural disasters such as drought and floods.

SANSA currently does not perform any groundwater prospecting using remote sensing. However, there has been one local project in the Ramotswa Aquifers' area that used remote sensing data to identify areas that potentially had underground water aquifers along the Crocodile River in Limpopo and Botswana.

Lastly, at the request of Rand Water (the water utility in Gauteng, and the largest in Africa), SANSA developed a commercial product that employs RS data to monitor, track and predict algal blooms in their open-surface dams and reservoirs. This service is now carried on in the commercial sector by CyanoLakes.[16]

3.5.3 CyanoLakes

One of the space-based applications used by the DWS, CSIR and the Water Research Commission (WRC) is the measurement of the quality of public water sources. CyanoLakes in Pretoria is a monitoring facility that compares the levels of biopollution and health risks in public water bodies against an index developed by the World Health Organization.

CyanoLakes utilizes ESA's Sentinel 3 imagery and interprets it through an algorithm that distinguishes between potential cyanobacteria and other algae blooms. Dr. Mathews developed this specialized application to measure the levels of cyanobacteria and nutrient pollution within water bodies.[17] As a downstream data provider to the DWS and the WRC, CyanoLakes provides empirical data to underpin any user's decision-making strategies. CyanoLakes uploads data on its website daily for continuous monitoring by the agencies that need it, to be combined with statistics and GIS layers for multi-use graphic material.[18]

3.5.4 GeoTerra Image

GeoTerra Image is a water resource monitoring service that provides a monthly wall-to-wall countrywide inventory of all surface water features across South Africa, utilized by the DWS, CSIR and the WRC. It primarily relies on ESA's 20 m resolution Sentinel2 satellite imagery, which allows all surface water bodies typically >0.25 ha to be identified and mapped (see Fig. 6). The Sentinel2 satellite system provides five-day image observations which are composited into an optimal, minimal cloud obscured dataset from which the monthly total surface water areas are determined.[19]

[16]SANSA, Earth Observation, Oupa Malahlela, interview held at SANSA Pretoria, May 2018.
[17]Cyanolakes, Monitoring cyanobacteria blooms in Earth's waters using satellite remote sensing, http://www.cyanolakes.com/ [last accessed 18 June 2019].
[18]Ibid..
[19]GEOTERRA IMAGE, Innovative Geospatial Business Products, www.geoterraimage.com, [last accessed 18 June 2019].

Fig. 6 Example for GeoTerra image showing mapped water body

Commercial subscribers have access to actual digital monthly water maps (in GIS compatible raster format, based on 20 m cells), and comprehensive water data containing monthly and long-term surface water area statistics per quaternary catchment, from October 2015 to present.[20]

3.5.5 A Provincial Crisis and a National Disaster

In the mid twentieth century, the current DWS's predecessor, the Department of Water Affairs, deemed water conservation to be "of national importance to generate new knowledge and to promote the country's water research purposefully, owing to the view held that water would be one of South Africa's most limiting factors in the 21st century."[21]

What predicated that announcement was a brief period of water shortage in the late 1960s. In response to it, the then Department of Water Affairs also established the WRC in terms of the Water Research Act (Act No. 34 of 1971).[22]

And today, nearly five decades on, the crisis they were hoping to avoid in the 1970s has finally reached a crescendo, despite the WRC's best efforts. The DWS has been grappling with this extreme water shortage in three of the country's provinces which had previously been declared Provincial Disasters.

The Western Cape's Provincial Disaster status was brought to an apex when Minister Zweli Mkhize of The Department of Cooperative Governance and Traditional Affairs declared the drought in the Western Cape a National Disaster on

[20]GEOTERRA IMAGE, Remote Sensing Services, http://www.geoterraimage.com/solutions-remote.php, [last accessed 18 June 2019].

[21]Water Research Commission, Mandate, http://www.wrc.org.za/about-us/mandate/, [last accessed 18 June 2019].

[22]Ibid.

February 13, 2018.[23] The National Disaster reclassification enables the province to claim from the six-billion-rand allocation set aside for drought relief, and also shifts the responsibility for coordination and management to the National Executive.

The reclassification extends over a three-month time period.[24] However, Minister Mkhize did not renew the National Disaster classification when the time period ended, as all the funding and planned interventions had been implemented and others will be continuing.[25] So Cape Town's water shortage problem is back to being termed a Provincial Disaster, and is again, under the auspices of the DWS.

4 Earth Observation for Water Resource Management

Satellite-based Earth Observation (EO, including both optical and radar imaging) provides essential data to effectively evaluate, monitor, assess, and forecast terrestrial water resources and is used extensively around the globe to assist in water resource management. The EO data is indispensable in the current situation of extreme water scarcity in the Western Cape region and CCT in particular. Different ways in which EO data can be used to compliment water resource management are discussed below.

4.1 Hydrological Network Mapping

Knowledge of hydrological networks (i.e. rivers, lakes and coast lines) is used to determine and distinguish land and water bodies such as the area of the land, the length of rivers and the perimeter of lakes and coastlines. Information about hydrological networks is essential for water resource inventory and management. With EO data, changes in hydrological networks can be identified and investigated using temporal data spanning up to 45 years back in time. Both Synthetic Aperture Radar (SAR) and optical sensors can be utilized to identify the changes in the boundaries of water resources over time. In this case, SPOT series data, historic Landsat data, and Sentinel data with high temporal frequency can be used. Figure 7 shows an example of using Landsat data to monitor changes in the Poyang Lake in China.

[23] N Odendaal, South Africa's drought declared a state of emergency, Engineering News, 13/03/2018, http://www.engineeringnews.co.za/article/south-africas-drought-declared-a-state-of-disaster-2018-03-13/rep_id:4136, accessed June 18, 2019.

[24] Ibid.

[25] News 24 Online, Ngqakamba S, Government Lifts National Drought State of Disaster June 13, 2018 news24,/desktop/government%20lifts%20national%20drought%20state%20of%20disaster%20_%20news24.htm accessed June 18, 2018.

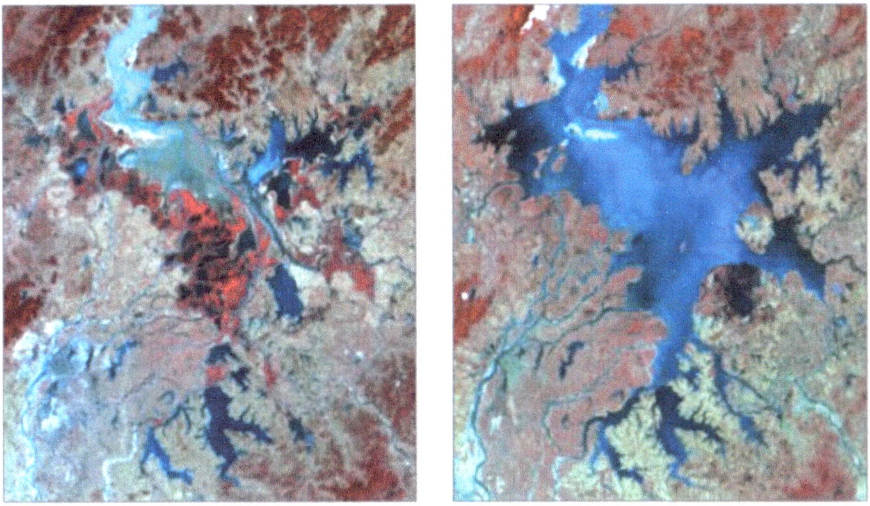

Fig. 7 Landsat false color composite infrared images of the Poyang Lake (Wu, Guofeng & De Leeuw, Jan & Best, Elly P.H. & Barzen, Jeb & Venus, Valentijn & Burnham, James & Liu, Yuelian & Ji, Weitao. (2008). A conceptual framework to integrate a simulation model of plant biomass for Vallisneria spiralis L. with remote sensing and a geographical information system. 187–199)

4.2 Surface Water Monitoring

Satellite-derived information supports the determination and monitoring of water resources from catchments, watersheds and rivers. Surface water monitoring services focus on the identification of water bodies and wetlands as well as their seasonal changes. EO based services can be utilized to estimate the discharge and recharge of water. They can also be used to determine the water levels of open reservoirs, as well as long-term and short-term changes in water levels. Such information supports water authorities and basin commissions in managing resources such as dams and lakes.

In order to implement mapping of water bodies, both passive and active remote sensing can be utilized. Also, in order to estimate the water body levels, space altimeters can be used. Water has a unique property in that it absorbs most of the energy in the visible and infrared (VIS-IR) band. The most suited band for observing water bodies is the shortwave infrared (SWIR) band, as in this case the water quality does not affect the water characterization. However, the one negative aspect with this band is that it is affected by cloud cover and atmospheric water vapor content. Spatial resolution is an important factor in measuring the extent of the bodies. For reservoir management, high to very high resolution imagery is required. So in order to estimate historic trends of water bodies, Landsat-5 can be utilized. Whereas for current estimation of changes, Tera, Landsat-7, Landsat-8 and Sentinel 2A/2B/2C can be used.

In order to overcome the sensitivity of VIS-IR to cloud cover and water vapor, one could rather use SAR based sensors. Another advantage of SAR sensors is that they can also be used to penetrate areas with high canopy cover. Again, high to very high resolution SAR is required for effective reservoir management. For SAR in the case of historic trends Envisat data can be used. For current water estimation the Sentinel 1A/1B can be used.

In order to measure the water levels in the reservoirs, satellite-based radar altimeters are used. For this case, radar altimeters data from missions like JASON 1 and TOPEX can be used for historic trends. For current estimation, data from missions like the JASON 2/3 can be utilized. In the future, the data from the SWOT mission will be very useful for water level estimations. In this case the Global Reservoirs/Lakes (G-REALM)[26] and Hydroweb[27] can be used for water levels.

4.3 Settlement Characterization and Change Assessment

Urbanization represents a major challenge for water, sanitation and risk management. A key component when planning infrastructure upgrades is the identification of high risk areas regarding water access. EO can serve as an active tool in the decision-making process for urban water demand and supply. The EO products can serve as reference information for short-term evaluation of water needs, and long-term decision-making processes regarding water demand and water supply as well as predicting urbanization rates and patterns.

In this case, high resolution satellite data is required; hence IKONOS and QUICKBIRD can be utilized. There are many limitation factors in the estimation, like the different number of sensors and different types of bands that these operate on. Also SAR sensors (e.g. TanDEM-X) can be used to estimate urban settlement patterns.

4.4 Vegetation and Land Cover Characterization

Vegetation and Land Cover (VLC) are intrinsically coupled with the water cycle as the distribution and productivity of the terrestrial surface is controlled by the local water balance. In turn, the composition and distribution of VLC is of fundamental importance for evapotranspiration and runoff generation. Identifying, delineating and mapping VLC is therefore of paramount importance to the successful and

[26]Dr. C Reynolds, Global Reservois/Lakes (G-REALM)—10-day near real time products, US Department of Agriculture, https://ipad.fas.usda.gov/cropexplorer/global_reservoir/, [last accessed 18 June 2019].

[27]Theia Data and Services Centre for Continental Surfaces, Time series of water levels in the rivers and lakes around the world, CNES available http://hydroweb.theia-land.fr/, [last accessed 18 June 2019].

sustainable management of water resources as it establishes the baseline from which monitoring can be performed.

For VLC, very high spatial resolution imagery is not required. So data derived from the MODIS, Landsat-7/8, and Sentinel 1/2/3 can be used. Vegetation characterization supports the monitoring of both meteorological and agricultural droughts.

4.5 Ground Water Mapping and Characterization

Groundwater plays an essential role for water supply, and is often used for irrigation, industrial and domestic uses. As groundwater lies below the land surface, monitoring of groundwater from space is challenging, and that's why indirect techniques or proxies are used to derive information about aquifer depletion and recharge.

The main technique used for estimating ground water levels of known underground water supplies is through satellite-based gravity field mapping. This technique is used to measure the changes in the gravitational field over geographic areas and its relationship with the ground water storage. In this case, the GRACE mission of NASA provides great details and insight.

4.6 Hydrological Modelling and Monitoring

Hydrological Modelling and Monitoring (HMM) is a key element of scientific decision support to water resources management, as it provides estimates of future water availability in the basin. HMM depends on many types of climatologies including precipitation, snow cover, evapotranspiration and soil moisture content. EO can be used to quantify the amount of precipitation over a geographic range. Both GEO and LEO-based satellite sensors can be utilized. Satellite-based precipitation estimation works best when used in combination with ground-based gauging areas, especially where there are fewer than four geo-reference squares per one degree by one degree.[28]

Other climatologies like the snow cover, evapotranspiration and the soil moisture are also critical. Estimates of these factors define the hydrological condition of the basin and the rainfall-runoff modelling of the catchment area. These factors are also critical for the hydrodynamic modelling of river flows.

[28]García, Luis; Rodríguez, Juan Diego; Wijnen, Marcus; Pakulski, Inge. 2016. Earth Observation for Water Resources Management: Current Use and Future Opportunities for the Water Sector. Washington, DC: World Bank.

4.7 Water Quality Monitoring

Satellite-based remote sensing is recognized as a reliable and cost-effective technique for water quality monitoring of inland and coastal waters. EO can be used to monitor chlorophyll content (indicative of the phytoplankton), cyanophycocyanin (indicative of toxic algal blooms), Total Suspended Matter (indicative of the water turbidity), and water temperature (which is required to assess the metabolic rates of life forms and solubility of water).[29]

5 Space Products Related to Water Resource Management

There are a number of EO-based information products that are being used in South Africa for water resource management. Some of the critical EO products related to the drought situation in the Western Cape and especially CCT are mentioned here.

5.1 TIGER-NET Initiative

It is evident from the previous discussion that satellite data is of paramount importance in water resource management. Considering this importance, ESA initiated the TIGER project with the aim of assisting and empowering African water authorities in their water resource management activities through EO related products. TIGER-NET is part of the overall TIGER project. The basic aim of TIGER-NET is to provide open source Water Observation and Information System (WOIS) software that can be utilized by African authorities to derive EO products for water resource management. Through WOIS, African authorities have access to historic and future Sentinel satellite data. WOIS software is being used locally by DWS for the following applications[30]:

1. Water discharge forecasting
2. Land cover based water demand mapping
3. Water management related land cover change mapping.

In order to address the drought and water scarcity situation in CCT, the following applications need to be developed and performed on a periodic basis by DWS:

[29]UNESCO (in partnership with the International Hydrological Programme and the International Initiative on Water Quality), IIWQ World Water Quality Information and Capacity Building Portal, http://www.worldwaterquality.org/.

[30]TIGER-NET, Department of Water Affairs of South Africa, http://www.tiger-net.org/index.php?id=34, [last accessed 18 June 2019].

1. High and medium resolution basin characterization
2. Urban sanitation planning support
3. Hydrological monitoring (precipitation, evapotranspiration, soil moisture, water level)
4. Long term seasonal variation of the wetlands.

5.2 Assessment of Irrigated Areas

Assessment of water required by the irrigation agricultural sector is critical for the overall estimation of water resources in countries where there is a low amount of average annual rainfall. South Africa has an average amount of 495 mm/year rainfall, so the overall irrigation sector requires 62% of the overall water resources.[31] The existing methods for determining the overall water usage for irrigation are time consuming and are not cost effective. In order to address this issue, a research project was performed by the University of Stellenbosch, under the funding of the Water Research Commission (WRC). The aim of the project was to determine the existing state of affairs related to water usage in the irrigation sector. The project utilized the evapotranspiration dataset (ET) and Geographical Information System (GIS) to map the irrigable land and the subsequent water usage. The project utilized EO products based on the MODIS, Landsat 8, and SPOT 5 and 6 imagery. The outcome of the project can be used by DWS to help aid their water resource management activities. The overall conclusion of the project is that the EO products are critical in automated mapping of the irrigable land and its water usage. Also, EO products can play a vital role in reducing the overall latency of this estimation. The report also suggests water accounting for water allocation and that the irrigation water estimation needs to be performed on a continuous or monthly basis and EO products can play a key role in this.

6 Legal and Regulatory Framework

This section looks at the legal framework governing water resource management within South Africa. First, the Department of Water and Sanitation (DWS) is discussed, which has national jurisdiction. After that, the provincial and local regulatory framework is analysed along with the relevant laws and regulations.

[31] van Niekerk A; Jarmain C; Goudriaan R; Muller SJ; Ferreira F; Münch Z; Pauw T; Stephenson G; Gibson L (2018). An earth observation approach towards mapping irrigated area and quantifying water use by irrigated crops in South Africa. Research Report No.TT 745/17.

6.1 National Department of Water and Sanitation

The National Department of Water and Sanitation (DWS) is the custodian for all water resources in South Africa. Its primary focus and responsibility is to formulate and implement policies in this sector. It strives to make sure that all South Africans have access to clean water and dignified sanitation.

The vision of DWS is to be "a dynamic, people centered department, leading the effective management of the nation's water resources, to meet the needs of current and future generations," and its mission is to serve the people of South Africa by[32]:

- "making a positive impact on our country and its people as custodians of our water and sanitation resources, and as innovative and committed partners in the drive for sustainable development;
- being service and delivery oriented. We strive to get it right the first time, every time, on time—ensuring that our citizens are provided with water and sanitation services they deserve;
- leading our sector and enable partners with knowledge and capacity to ensure that all water services are delivered;
- being committed to innovation and use cutting edge technology as a catalyst of positive change, connecting our people and enabling them to work anywhere anytime;
- having a heart that values our investment in our people. We provide them with a caring and trusting environment that encourages personal development and is a breeding ground for talent."

6.1.1 Reporting Structure

There are several water boards in South Africa, the three biggest ones include Umgeni Water, Rand Water and Overberg Water. These water boards are all government-owned and they operate bulk water supply infrastructure, dams and some wastewater systems. The Water Service Authority (WSA) is the authority whose responsibility it is to make sure that there is access to water service. The Water Service Provider's (WSP) primary duty is to provide water services in terms of the Constitution and the Water Services Act. Water board members are chosen by the Minister, and they may not be on the board longer than four years.[33] The Minister of Water and Sanitation controls the budget. The National Treasury distributes budgets of every government department. Table 1 shows the budget summary until 2020.

[32]Department of Water and Sanitation, About Us, South Africa, http://www.dwa.gov.za/about.aspx#vision, [last accessed 18 June 2019].

[33]Department of Water and Sanitation, Water Service Institutions, South Africa, http://www.dwa.gov.za/IO/wsi.aspx, [last accessed 18 June 2019].

Table 1 South African national budget summary relating to water (South African National Treasury, Budget 2017 Estimates of national expenditure—Vote 36 Water and Sanitation, Republic of South Africa, 22 February 2017, available at: http://www.treasury.gov.za/documents/national%20budget/2017/enebooklets/Vote%2036%20Water%20and%20Sanitation.pdf)

R million	2017/2018				2018/2019	2019/2020
	Total	Current payments	Transfers and subsidies	Payments capital assets	Total	Total
Administration	1,628.4	1,542.3	22.0	64.1	1,658.0	1,755.3
Water planning and information management	816.5	739.6	1.2	75.6	884.0	949.8
Water infrastructure development	12,251.7	623.1	8,090.1	3,538.5	13,499.3	14,318.3
Water sector regulation	410.8	396.5	1.1	13.2	519.5	491.4
Total expenditure estimates	**15,107.4**	**3,301.5**	**8,114.5**	**3,691.5**	**16 560.8**	**17,514.8**

6.1.2 National Water and Sanitation Master Plan

The National Water and Sanitation Master Plan is currently being developed for purposes of investment planning, development of water resources and providing water and sanitation services up until 2030 and beyond. The Master Plan includes a specific aspect of water security which has its own initiative called the National Integrated Water Security Framework driven by the National Planning Commission. The water security initiative deals with identifying future water sources and additional development options of water resources to meet the increasing demand due to population growth and increased agricultural and industrial applications.[34]

Drought Interventions

The DWS has approached the drought situation in South Africa in four phases including groundwater optimization, desalination, water conservation and demand management, and water re-use optimization. Where Cape Town is concerned, the DWS has recorded that its efforts include putting in place Level 6 water restrictions as of January 2018. The City of Cape Town has further intensified its water consumption program by putting in place Level 6B restrictions since February 2018. This level 6B restriction drops the daily use per person to 50 L.[35]

The South African Water Research Commission (an entity of DWS) has created a drought management water pack which includes documents detailing, among other things: water saving manuals, water efficiency case studies, manuals and

[34]Department of Water and Sanitation, National Water and Sanitation Master Plan, South Africa, http://www.dwa.gov.za/National%20Water%20and%20Sanitation%20Master%20Plan/default.aspx, [last accessed 18 June 2019].

[35]Department of Water and Sanitation, Drought Interventions, South Africa, http://www.dwa.gov.za/drought/default.aspx, [last accessed 18 June 2019].

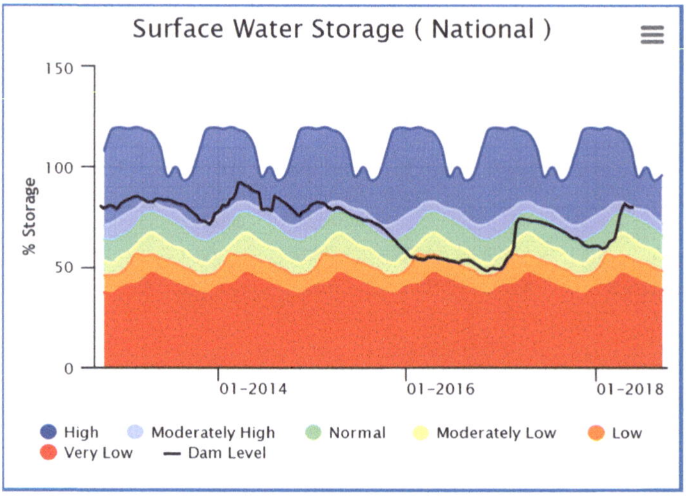

Fig. 8 Surface water storage dashboard spatially shows the dam levels and a time-series table of dam storage trends (as of 1 June 2018)

guidelines for best practices in water management and documents on adapting to climate change.[36]

Managing Risks in Strategic Projects—TCTA

The Trans-Caledon Tunnel Authority (TCTA) is accountable to the DWS, it was established in 1986 to bring to life the Treaty on the Lesotho Highlands Water Project. The regulatory framework of the TCTA is found in the National Water Act of 1998 and the Public Management Act of 1999. The TCTA is mandated to implement "large scale bulk water schemes on behalf of DWS."[37]

The DWS provides on its website weekly data of water capacity and water storage levels in dams for all nine provinces. Figure 8 provides the dam level trends and rainfall trends. One is able to download charts showing annual rainfall levels including national monthly cumulative annual rainfall trends. One is also able to access current dam level data from DWS's website.

In light of the above, one can see that the DWS focuses on the country as a whole to provide the most current data about water. It then stands to reason that

[36] A. J. Jordaan, Vulnerability, adaptation to and coping with drought: the case of commercial and subsistence rain fed farming in the Eastern Cape, Volume I, Water Research Commission and the Department of Agriculture, Forestry and Fisheries, South Africa, April 2017 (ISBN 978-1-4312-0884-5).

[37] Andrew McDonald and Jessica Fell, Water sector risk governance—a compendium of South Africa and International case studies, p. 49 http://www.wrc.org.za/Pages/Drought/4.2/Emergency%20response/water%20sector%20risk%20governance%20case%20studies_TT%20668-16.pdf.

each province should take that data and analyse it to cater for its own needs. The WSA and WSP are the boards which report to the Minister as they are appointed by the Minister.

6.2 Local Government

The constitution states that "everyone has the right to have access to sufficient food and water", it goes further to say, "the state must take reasonable legislative and other measures, within its available resources, to achieve the progressive realisation of each of these rights".[38] This makes access to water a basic human right. This right is currently under threat because of the drought in the Western Cape Province. The much dreaded 'Day Zero', the day when all taps run dry—was continuously being monitored and displayed on CCT's online water dashboard. It was last expected placed on the 9th of July 2018 but has now been pushed further into 2019 with a possibility of it never happening because of consumers, the agricultural and business sector successfully making efforts to save water.[39] At the time of publication, CCT's water dashboard no longer displays a countdown to "Day Zero", and the water restrictions have been reduced to Level 3, which increases the daily personal water allowance to 105 L.[40]

All three spheres of government, namely the national, provincial and local government have a vital role to play. This requires the national, provincial and local government, which are the three spheres of government, to be separate and independent, while ensuring cooperation with each other to perform their functions. These governments must adhere to the notions of unity, decentralisation and cooperation. Cooperative governance depends on the acceptance of the following principles: no single actor can effect change, complementary and competing interests must be recognised, new structures should be established to promote cooperative behaviour amongst various stakeholders, and the importance of clarifying the responsibilities of all stakeholders.

It is the role of government to establish structures and institutions to promote and facilitate cooperative governance as well as define mechanisms and procedures to promote and facilitate intergovernmental relations.[41] A good example of cooperative governance at work is the establishment of Catchment Management Authorities to decentralise water resource governance between themselves, local government

[38]Republic of South Africa, Constitution of the Republic of South Africa No. 108 of 1996, Bill of Rights, Chapter 2.
[39]L. Chutel, How Cape Town delayed its water-shortage disaster—at least until 2019, Quartz Africa, 9 May 2019, https://qz.com/africa/1272589/how-cape-town-delayed-its-water-disaster-at-least-until-2019/.
[40]City of Cape Town, Water Dashboard, https://coct.co/water-dashboard/; Accessed on 18 June 2019.
[41]Watson, D., & Goga, S. (2016). Natural Resource Governance Systems in South Africa, Water Research Commission, WRC Report No 2161/1/16, South Africa, July 2016.

and other actors of water resource management at a catchment level. Water governance in South Africa falls under the DWS under Minister Nomvula Mokonyane. This department obtains its mandate from the 1998 National Water Act. The Act also allows the Minister to "for as long as it is necessary deal with an urgent situation or emergency".[42] The Western Cape Government Drought Management and Water Resilience Committee was formed that consists of various stakeholders.

In August 2017 the national government granted 75 million Rand to the Western Cape Province to deal with the water crisis. According to the Western Cape provincial government this occurrence should have been declared a national disaster earlier as the Western Cape government asked for a disaster declaration already in 2015.[43] Despite the idea of corporative governance being in the constitution, as well as the National Water Act, it has not been easily implemented in the Western Cape area, where the National Government is led by the African National Congress (ANC) while the local and provincial spheres are under the opposition party, the Democratic Alliance (DA). The DA's provincial government blames the National government for taking too long to declare the drought as a national disaster and thereby allow funding and human aid to better deal with the crisis,[44] while on the other end, the National government says the mandate rests on the municipal government to provide water and sanitation to its residents and that it should have properly planned and foreseen the problem developing. The drought was declared a national disaster on the 9th of February 2018 by an inter-ministerial task team. This declaration made it possible for financial and humanitarian aid to be granted by the national government. Under Premier Helen Zille of the Democratic Alliance (DA) the provincial governments' role is to provide oversight, monitor and support the national and local government in their task of supplying water. The disaster declaration enables the provincial government to instruct municipalities to impose water restrictions.

The South African Constitution states, "local government oversees municipal water services".[45] The National Water Act concentrates on the duties of the national government while the Water Service Act of 1997 concentrates on the tasks of the local governmental sphere.[46] Although she survived a motion of no confidence, DA

---

[42]Republic of South Africa, National Water Act (Act 36 of 1998), Chapter 6, Part 3, August 1998.
[43]Water Crisis in Cape Town: Lessons to be Learnt PART 2: The Responsibilities' Of The Three Spheres Of Government 3 Konrad-Adenauer-Stiftung e.V. SOUTH AFRICA DOROTHEA GIBSON February 2018 www.kas.de/suedafrika/ relevant provision to intervene.
[44]Ibid.
[45]Republic of South Africa, Constitution of the Republic of South Africa No. 108 of 1996, Bill of Rights, Chapter 2.
[46]Water Crisis in Cape Town: Lessons to be Learnt PART 2: The Responsibilities' Of The Three Spheres Of Government 3 Konrad-Adenauer-Stiftung e.V. SOUTH AFRICA DOROTHEA GIBSON February 2018 www.kas.de/suedafrika/ relevant provision to intervene.

Mayor for Cape Town Patricia de Lille was replaced by Deputy Mayor Ian Neilson as head of the water crisis response team in January 2018.[47,48,49]

Legislation which impacts on the role of local government in water services are the South African Constitution the National Water Act, the Water Services Act, the Municipal Structures Act, the Municipal Systems Act, the Municipal Finance Management Act and the Division of Revenue Act. The Constitution vests the executive authority for water services in local government. One of the objectives of local government is to ensure the provision of services to communities in a sustainable manner.

The Constitution of the Republic of South Africa

The Constitution of the Republic of South Africa vests the executive authority for water services in local government whose objectives include ensuring the provision of services to communities in a sustainable manner. These mandates can be found in the Constitution[50] in the following sections:

Section 152 (1) states the constitutional mandates of local government as:

- The provision of democratic and accountable government for local communities;
- Ensuring the provision of services to communities in a sustainable manner;
- The promotion of social and economic development;
- Promoting a safe and healthy environment;
- Encouraging the involvement of communities and community organizations in local government matters.

Section 153 requires a municipality to:

- Structure and manage its administration and budgeting and planning processes to give priority to the basic needs of the community, and to promote the social and economic development of the community;
- To participate in national and provincial development programmes.

Section 156 (1) gives the municipality executive authority as well as the right to administer:

- Matters listed under Part B of Schedule 4 and Part B of Schedule 5 of the Constitution including municipal planning, storm water management systems in built-up areas, water and sanitation services limited to potable water supply systems and domestic waste-water and sewage disposal systems, local sport facilities, refuse removal, refuse dumps and solid waste disposal;

[47]Ibid.
[48]Ibid.
[49]Ibid.
[50]Republic of South Africa, Constitution of the Republic of South Africa No. 108 of 1996, Bill of Rights.

- Any other matter assigned to it by national or provincial legislation;
- The making and administering of by-laws for the effective administration of matters that it has the right to administer;
- Matters in Part A of Schedule 4 and 5 which relate to local government may be assigned by national government and provincial government to the municipality either by agreement and subject to any conditions if the municipality has the capacity to administer it or the matter would most effectively be administered locally.

Therefore Section 156 assigns executive authority on the municipalities for matters that are contained in Schedule 4 and 5 Part B which relate to local government including municipal planning, storm water management systems in built-up areas, water and sanitation services limited to potable water supply systems and domestic waste-water and sewage disposal systems, local sport facilities, refuse removal, refuse dumps and solid waste disposal as well as any other functions that have been assigned to the municipality by national government where they will be most appropriately performed at local level and if the municipality has the capacity to perform these functions.[51]

Water Services Act 108 of 1997

Water Services Act 108 of 1997 is enacted in terms of Section 27 of the Bill of Rights in the Constitution which provides, amongst other rights, that everyone has the right to have access to sufficient food and water and that the State must take reasonable legislative and other measures within its available resources, to achieve the progressive realization of each of these rights.[52]

The main objectives of the Water Services Act are to provide for[53]:

- the right of access to basic water supply and to basic sanitation;
- the setting of national standards and norms and standards for tariffs;
- the preparation of water services development plans;
- a regulatory framework for water services institutions;
- the establishment of water boards and water services committees;
- the monitoring of water supply and sanitation services;
- intervention by the Minister or by the relevant Province;
- financial assistance to water services institutions;
- a national information system;
- the accountability of water services providers; and
- the promotion of effective water resource management and conservation.

[51]Republic of South Africa, Constitution of the Republic of South Africa No. 108 of 1996.
[52]Republic of South Africa, Constitution of the Republic of South Africa No. 108 of 1996, Section 27.
[53]Republic of South Africa, National Water Act (Act 36 of 1998), Chapter 6, Part 3, August 1998.

The National Water Act 36 of 1998

The National Water Act 36 of 1998 is enacted in terms of Section 24 of the Bill of Rights in the Constitution that states that everyone has the right to an environment that is not harmful to their health or wellbeing; and to have the environment protected for the benefit of present and future generations through reasonable legislative and other measures that[54]:

- prevent pollution and ecological degradation;
- promote conservation; and
- secure ecologically sustainable development and use of natural resources while promoting justifiable economic and social development.

The purpose of the National Water Act is to ensure that the Nation's water resources are protected, used, developed, conserved, managed and controlled in ways which take the following into account:

- meeting basic human needs;
- promoting equitable access to water;
- redressing the results of past racial and gender discrimination;
- promoting the efficient, sustainable and beneficial use of water;
- facilitating social and economic development;
- providing for growing demand for water use;
- protecting aquatic and associated ecosystems;
- reducing and preventing pollution and degradation of water resources;
- meeting international obligations;
- promoting dam safety;
- managing floods and droughts; and
- establishing suitable institutions and to ensure that they have appropriate community, racial and gender representation.

The Municipal Structures Act (Act 117 of 1998)

The Municipal Structures Act (Act 117 of 1998) sets up the basis for the establishment of municipalities in the 'A' (Metropolitan Municipalities), 'B' (Local Municipalities) and 'C' (District Municipalities) categories. It also defines how municipalities are established, the functioning, committees and mayoral options and committees, as well as the division of powers and functions between municipalities who have concurrent jurisdiction. It is through the Structures Act that district municipalities have the powers and functions necessary to perform the water services authority function as contained in the Water Services Act. The Minister of Provincial and Local Government may however authorise a local municipality, after

[54]Republic of South Africa, Constitution of the Republic of South Africa No. 108 of 1996, Section 24.

consultation, to be a water services authority. In which case the power and function for water services will shift from district municipalities to local municipalities.

The Municipal Systems Act (Act 32 of 2000)

The Municipal Systems Act (Act 32 of 2000) which focuses on the internal systems and administration of a municipality including public accountability and community involvement, in policy formulation and decision making, guidelines for making bylaws, establishing Integrated Development Plans, establishing a performance management system, delivering municipal services and credit control and debt collection. This Act introduces the differentiation between authority and provider functions of a municipality, by providing that a municipality may deliver services itself, or by way of an external mechanism by entering into a service delivery agreement. This makes it consistent with the Water Services Act which specifically defines a water services authority and a water services provider, and distinguishes the constitutional obligation to ensure services delivery, but not necessarily to provide it. It also allows for alternative mechanisms for providing municipal services as either internal or external mechanisms.[55]

The Municipal Finance Management Act (Act 56 of 2003)

The Municipal Finance Management Act (Act 56 of 2003) is meant to ensure sound and sustainable management of the financial affairs of municipalities. It covers various matters for both municipalities and municipal entities, which include managing revenue and debt, budgeting, responsibilities of mayors and municipal officials, procurement, reporting and intervention in the instance of financial problems. Access to water and sanitation is a basic human right which means a balance must be struck between duty to ensure sustainable service delivery and the provision of free basic water. Municipalities have stringent reporting obligations in terms of this Act.[56]

The annually enacted Division of Revenue Act which gives effect into Section 214(1) of the Constitution of the Republic of South Africa, 1996 requires an Act of Parliament to provide for the equitable division of nationally raised revenue among the three spheres of government with all local government infrastructure grant funding which is now consolidated into the 'Municipal Infrastructure Grant', aimed at assisting the poor to gain access to infrastructure.[57]

The Disaster Management (Act 57 of 2002)

An Integrated Disaster Management Structure is established (see Fig. 9), which focuses on preventing or reducing the risk of disasters, mitigating the severity of disasters, emergency preparedness, rapid and effective response to disasters and post-disaster recovery; the establishment of national, provincial and municipal

[55]Republic of South Africa, Municipal Systems Act (Act 32 of 2000), South Africa.
[56]Republic of South Africa, Municipal Finance Management Act (Act 56 of 2003), South Africa.
[57]Republic of South Africa, The Division of Revenue Act, South Africa.

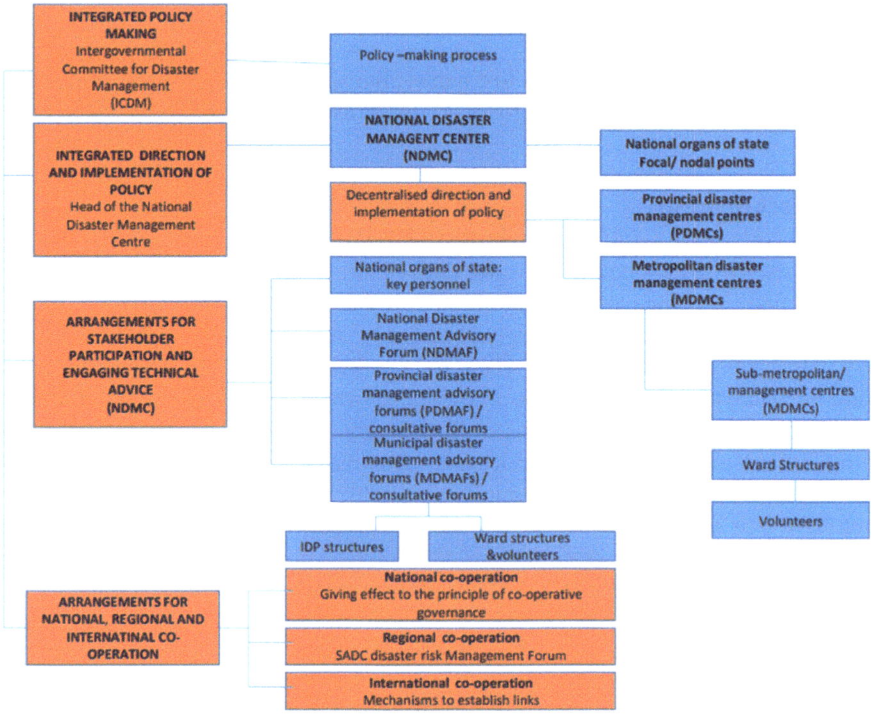

Fig. 9 Integrated disaster management structure (J Dyssel, Disaster Management overview, legislation and guidelines, Disaster management and resilience, Department of Cooperative Governance, South Africa, available at: https://www.cdm.org.za/attachments/article/1/IDDR%20SUMMIT%20-%20NDMC-%20DISASTER%20MANAGEMENT%20OVERVIEWLEGISLATIONS%20AND%20GUIDELINES%202017-2018.pdf)

disaster management centres; disaster management volunteers; and matters incidental thereto.[58]

Constitutionally the primary responsibility for disaster management in South Africa rests with the government. In terms of Section 41(l)(b) of the Constitution of the Republic of South Africa, all spheres of government are required to "secure the well-being of the people of the Republic". Disaster management is listed as a functional area in the Constitution, meaning that both the national and provincial spheres of government are competent to develop and execute laws within this area and have powers and responsibilities in relation to disaster management. Disaster management has also been 'assigned' to local government through the promulgation of the Disaster Management Act, 2002 (Act no 57 of 2002).[59]

[58]Republic of South Africa, The Disaster Management Act 57 2002, South Africa.
[59]Republic of South Africa, Constitution of the Republic of South Africa No. 108 of 1996.

7 Alien Tree Impact and Identification

This section was covered in detail in Chap. 1, but since the research was performed as part of this larger study, a summary of the process and results is included here.

7.1 Background

Trees consume large amounts of water daily. Indigenous trees usually consume an amount that is sustainable in the region, otherwise they would not survive. Colonisation that occurred throughout the last few centuries introduced many non-native i.e. alien floras to the Cape region. These include most-notably black wattles, eucalyptus trees and pine trees, all of which are present in the Western Cape area and consume many tens of litres a day. While transpiration does occur through the leaves of these trees releasing some water back into the atmosphere, it does not have the same effect a rainforest might. High wind speeds in the Cape area blow this moisture away, preventing clouds from forming due to the transpiration of the leaves.

It is estimated that these trees have a severe impact on the water supply of the region, preventing thousands of litres of water from entering the surrounding dams and reservoirs. The hypothesis here is that satellite imagery and data can be used to identify sections of alien trees and the impact thereof. The alien trees can then be removed to increase the amount of available water at catchment areas.

7.2 Example Approach

There are many different sources of satellite data, with many being free. However, for the purposes of this research, most of it was unusable in the timeframe given. The issue is that most of the data's resolution is simply not high enough to allow for tree identification by humans, meaning that complicated machine learning procedures would be necessary.

Trees however, are relatively static and take a few years to grow, and this means that older but higher resolution data is useful in this regard. Google Earth provides Digital Eye 0.5 m resolution imagery from 2017 (at the time of this research it was 8-month-old data). Since ongoing tree removal for most areas is not occurring it was determined to be valid data.

Since this is a proof of concept of a water-saving method, the entire region was not analysed. Instead a mountainous section with run-off to the main supply dam of Cape Town, Theewaterskloof Dam, was examined manually.

Using known size and colours of trees in the area, a section of forest was observed as being pine. This is shown in Fig. 10.

Once the trees were identified, the area of the forest were selected using a polygon tool, allowing the area to be calculated (Fig. 11).

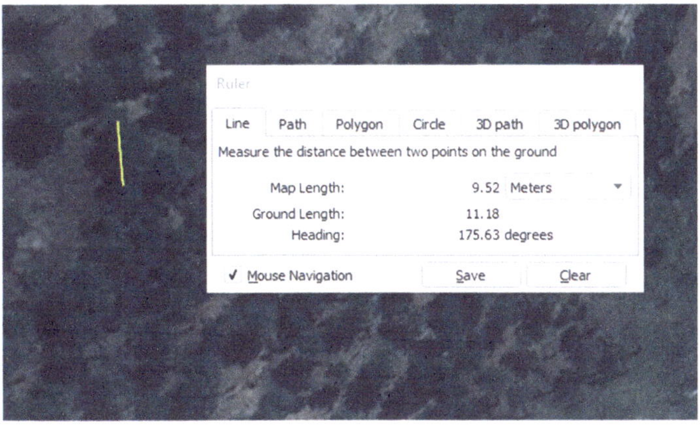

Fig. 10 Measurement of plants using google earth—pine forest near Theewaterskloof Dam

Fig. 11 Identified forests and affected river

Each tree uses 35.5 L per day[60] according to a paper by the CSIR. Using a manually counted tree density of 0.075 trees per m^2 and an area of 3,487,201 m^2, the total water use can be estimated by:

$$Water\,Use = Area \times Density\,of\,trees \times water\,use\,per\,tree$$

$$Water\,Use = 3,487,201 \times 0.075 \times 35.5 = 9,284,673\text{L/day}$$

This number represents a significant amount of water. This one section of alien forest uses almost ten times the amount of water as the current largest desalination plant can produce, and contributes a major part to Cape Town's water deficit.

7.2.1 Further Work

This method would firstly have to be extrapolated to all the catchment regions. Further in the future this could be done using machine learning. The government or other NGO's would then need to send people to clear forests, starting at the most accessible. The remote sensing tool could be enhanced to monitor regrowth in previously cleared areas as well as develop a priority list for areas requiring clearing.

Allowing the natural fynbos to replace the alien trees would ensure that a large amount of water is saved at relatively low cost, and provide other benefits to the environment such as promoting indigenous flora and fauna.

8 Rainwater Harvesting Analysis Using Earth Observation

This section discusses the other two proposed satellite applications. The first is to use rainfall prediction and modelling products to analyse the rainfall trends and patterns within the Western Cape, with the goal of (i) determining if the existing catchment areas are receiving enough rainfall to help the province with the water crisis, and (ii) to try identify areas which exhibit increased rainfall over the past few years to propose the installation of new catchments. Unfortunately, the full analysis was not possible due to the time frame of the project and instead, data from existing rain gauges throughout CCT were analysed to comment on the changes in rainfall patterns across the city. The next proposed application is then discussed, which uses satellite imagery to calculate the surface areas of roofs within CCT and its surrounds. Originally, it was hoped that an algorithm could be developed to automate this procedure, thereby allowing the Western Cape Province to calculate the total available roof surface area of the buildings within the province and to determine how much of these roofs could be equipped with rooftop rainwater harvesting

[60]Gush, M.B., Dye, P.J., Geldenhuys, C.J. and Bulcock, H.H., 2011. Volumes and efficiencies of water-use within selected indigenous and introduced tree species in South Africa: Current results and potential applications. In: Proceedings of the 5th Natural Forests and Woodlands Symposium, Richards Bay, 11-14 April.

systems. Average rainfall levels for different areas could be used in conjunction with this information to quantify how much water could be attained via rainwater harvesting methods.

One technique to overcome water shortages without the difficulty inherent to water distribution over long distances is rainwater harvesting. Rainwater harvesting involves the collection of rainwater at each property rather than just feeding it into storm drains.

Water usage is increased during summer season. However, CCT receives the majority of its rain in winter. During winter, rainwater harvesting can also be used to increase underground water levels if there is excess water. It will help rebalance the water cycle, reducing the possibility of drought. However, it is calculated that the amount of water that could be collected using rainwater harvesting techniques is too small to meet the overall water demands of the City. Rainwater harvesting is still useful in reducing the overall demand on the water supply network. Other advantages and challenges of rain water harvesting are discussed below[61]:

8.1 Benefits of Rainwater Harvesting

Increase in Underground Water Level and Demand Reduction: Groundwater is extracted by industries and residents with the help of pumps to fulfill their daily usage. Due to this, underground water stores are being depleted on a daily basis. Rainwater harvesting can help reduce the strain on underground water stores thereby aiding in recharging underground water aquifers.

Reduces Floods and Soil Erosion: Rainwater harvesting can help reduce flooding during the rainy season. Moreover, it is beneficial to reduce soil erosion. It also helps reduce accumulation of pesticides and fertilizers in lakes and ponds.

Suitable for Irrigation: Harvested rainwater is largely chemical free and ideal for use in agricultural irrigation. Large storage reservoirs can also help in fire reduction which are common during summer months. This is also becoming a major problem in CCT.

Reduction in Utility Bills: Collected water can be used in both domestic and industrial applications, thereby reducing utilities bills.

Reliance on Infrastructure: It also decreases the reliance and requirement on big water infrastructures.

8.2 Challenges of Rainwater Harvesting

Although the benefits of rainwater harvesting are plenty, it also comes with certain disadvantages which are covered below.

[61]Conserve Energy Future, What is rainwater harvesting, https://www.conserve-energy-future.com/advantages_disadvantages_rainwater_harvesting.php, [last accessed: 18 June 2019].

Unpredictable Rainfall: It can be difficult to predict rainfall in the area. Thus it is a difficult decision to install rain water harvesting system. It is not economically viable in areas with low rainfall.

Initial Cost: The initial capital outlay depends on the technology used as well as the size and complexity of the rain water harvesting system. If installed with above ground storage systems with 150 square metre collection area, it can cost in the range from R15,000 to R20,000.[62] Similar system with underground storage ranges from R30,000 to R35,000.[63] Typical payback period depends on different factors like amount of rain-fall in the area, water used and roof top area.

Residential rainwater harvesting system can take up to 15 years to pay back. Commercial systems can pay back faster than the residential system due to large area for rain water collection that can be up to three years.[64] Also, the lifespan of rain water harvesting systems can be up to fifty years with proper maintenance.[65]

Maintenance: Storage tanks as well as the roof top itself require timely maintenance as they can get polluted due to dirt etc. in periods of low rainfall. Installation of filters on inlets and their proper cleaning can reduce maintenance costs.

Storage Limits: There are some limitations in collection as well as storage, which limits the utilisation of harvested rain water. During heavy rainfall, the collected rainwater may exceed the storage capacity of the tanks which leads to wastage. Urban areas generally suffer more from storage limitations due to lower space availability. Also, storage costs a large percentage of total cost thus smaller storage is often preferred by users.

Seasonality: Rain water harvesting systems will only collect and store adequate water during the rainy season (i.e. winter in CCT). This means that the systems may not have sufficient capacity to assist with water demands in the dry season.

8.3 Rain Water Harvesting Types

The stored water can be used for a variety of applications including irrigation, cleaning and sanitation. There are two types of rainwater harvesting which are as follows[66]:

[62]Use Rain Water, Cost of systems, http://use-rainwater.com/cost-of-a-rainwater-harvesting-system.html, [last accessed: 18 June 2019].

[63]Water Solutions, Water tank sizes and prices, http://www.watersolutions.co.za/water-tanks/size-and-price/, [last accessed: 18 June 2019].

[64]H Williams, What is the payback time for a rainwater harvesting system, Green Home Construction and Home, Published on: 31-7-2013, http://thegreenhome.co.uk/heating-renewables/rainwater-harvesting/what-is-the-payback-period-for-rainwater-harvesting-system/, [last accessed: 18 June 2019].

[65]S. R. Ghimire et al., Life cycle assessment of a commercial rainwater harvesting system compared with a municipal water supply system, Journal of Cleaner Production, Volume 151, May 2017.

[66]The Constructor Civil Engineering Home, Methods of rainwater harvesting—components, transport and storage, https://theconstructor.org/water-resources/methods-of-rainwater-harvesting/5420/, [last accessed: 18 June 2019].

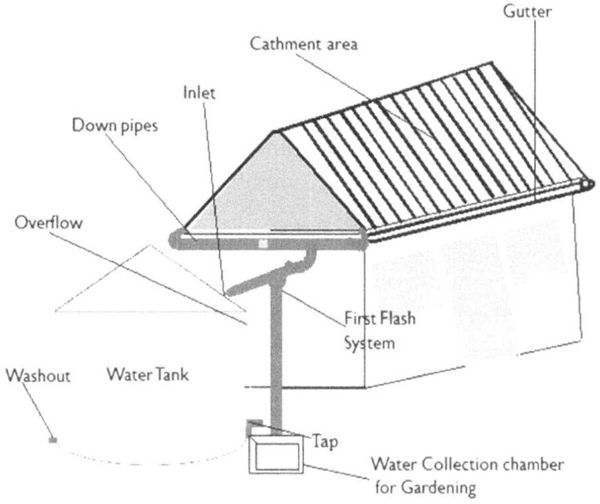

Fig. 12 Rainwater harvesting system components

Rooftop Rainwater Harvesting

With this, the roof of permanent structures are used as the rain collection area. It can be used for direct storage or underground water level recharging. This method is cost effective and efficient if installed correctly provided there is sufficient rainfall and roof area.

Surface Runoff Harvesting

In CCT, surface runoff in the city center is discharged into storm water drains which run into the sea. Aquifers can be recharged by using this rain water and this method is called surface runoff harvesting.

Irrespective of the type of harvesting system employed, rainwater harvesting systems have four key components, namely: the catchment area, the transportation of the water, the storage of the collected water and purification of the stored water. The key components of roof top rainwater harvesting systems are shown in Fig. 12.[67] This work will focus on roof top rainwater harvesting mechanisms.

8.4 Catchment Areas

Earth observation from space plays a vital role in the selection and estimation of catchment areas. The following attributes can be improved in selection of catchment area.

[67]K Mlay, Rain water harvesting technology, Rain Water Technology Blog, http://rwhtech.blogspot.com/2015/12/componentsof-rainwater-harvesting.html, [last accessed: 18 June 2019].

- Rainfall prediction and pattern analysis
- Surface area mapping.

The purpose of rainfall analysis from satellite data is to predict rainfall patterns over the CCT area to determine if the catchment areas are receiving enough rain to cope with the drought situation. This will also dictate the size of the required catchment areas.

Climate and rainfall are presumed to have a complex, non-linear relationship affected by numerous different intricate phenomena. For predicting accurate rainfall, advanced computer modeling and simulation systems are required. The next section will discuss the most common rainfall prediction algorithms and their specifications along with pattern analysis. This information can be used by government to assess the current rainfall trends in the Western Cape and to determine if the installation of new small catchment areas is required.

8.4.1 Different Satellite Prediction Techniques

There are several prediction algorithms which are useful to predict rainfall. The majority of rainfall estimation techniques use geostationary satellite infrared images and radar data. Different techniques have different temporal and spatial resolutions and incorporating different satellite imagery. Results from satellite-based algorithms need to be validated with ground data as their accuracy can be affected by topography, geographical position, climate and algorithm for rainfall predictions.[68]

Precipitation Estimation from Remotely Sensed Information

PERSIANN CSS algorithm is used for satellite-based rainfall prediction. This algorithm is mainly used for extracting cloud features from geostationary infrared satellite imagery for predicting acceptable rainfall distribution. The basic process of this algorithm is to extract cloud patches from satellite cloud images. From the cloud patches, a feature extraction is performed which categorize cloud patches based on temperature and precipitation. The angular spatial resolution to estimate rainfall in PERSIANN algorithm is available at 0.04°.

Climate Prediction Center Morphing Method

The CMORPH method provides half-hourly global precipitation estimates which are derived from passive microwave satellite scans that are propagated by motion vectors coming from geostationary satellite infrared data. Microwave sensor data obtained by scanning the shape and intensity of the precipitation properties, can be reformed additionally by the performance of a time-weighted linear interpolation. Therefore, this method provides spatially and temporal complete microwave precipitation analysis which is self-determining of IR temperature data. Cloud systems

[68]Accessed on: 22-06-2018, Comparison and validation of eight satellite rainfall products over the rugged topography of Tekeze-Atbara Basin at different spatial and temporal scales, Available at, https://www.hydrol-earth-syst-sci-discuss.net/hess-2017-504/.

and their movements can be determined using IR data, and it also gives good measurement of cloud top properties.

NRL Blended Technique

The US Naval Research Laboratory (NRL) has developed a blended technique for estimating rainfall from satellite imagery. NRL-Blend provide rainfall estimation at a minimum of every three hours at 0.1° resolution between 60°S to 60°N using data of both geostationary visible and infrared imagers and passive microwave imagers. The NRL-Blend satellite-based precipitation technique is dependent on a statistical relationship that comes from a precise, near real-time collective of passive microwave and infrared pixels from GEO and LEO orbits, as their individual sensors scan patterns uninterruptedly intersect in space and observation time.

Due to the different sensor frequencies, polarization and scanning modes, different precipitation algorithms can be applied to the datasets from the different sensors. This produces different precipitation estimation characteristics and possible outcomes. One method to merge the different datasets is by choosing one passive microwave (PMW) sensor as a reference, and to match the frequency of the satellite-derived rainfall histograms of the other satellite sensors to the reference histogram. Unfortunately this method cannot provide data on country or city level.

Tropical Rainfall Measuring Mission Multi-Satellite Precipitation Analysis

Tropical Rainfall Measuring Mission (TRMM) Multi-Satellite Precipitation is a calibration-based system in which precipitation estimation is derived from different satellite imagery and rain gauge analysis where possible, with a spatial resolution 0.25° and temporal resolution three hourly. The estimation of Tropical Rainfall Measuring Mission Multi-Satellite Precipitation Analysis (TMPA) is created in four steps:

- Calibration and combination of the microwave precipitation estimates is achieved by averaging passive microwave data of precipitation from separate frames of view with each dataset to 0.25° spatial grid from the time range of 90 min from minimal three hourly observation time
- Making of infrared (IR) precipitation estimates by calibrated microwave precipitation
- Merger of microwave and IR data to get best estimates
- Integration with rain gauge data.

The abovementioned methods use different techniques and algorithms to determine rainfall prediction and estimation (see Fig. 13). Unfortunately, it is difficult to get area specific rainfall prediction or estimation due to very large angular resolution of geosynchronous weather satellites. This also provides much larger coverage area thus evaluation of cloud movement is easier. Working methodologies of rainfall prediction algorithms are discussed above. These algorithms focus on potential rainfall capacity of the clouds which may differ from actual rain. Rainfall prediction data is easily accessible and thus not covered in this work.

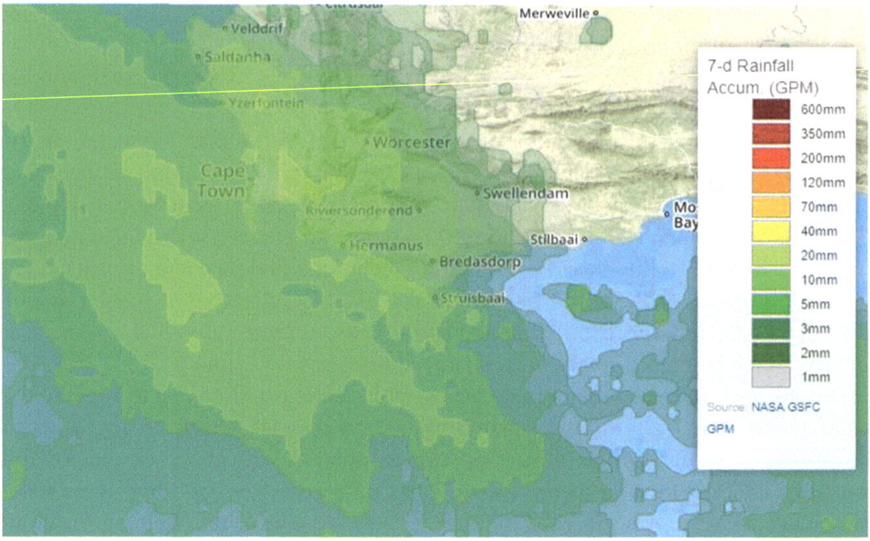

Fig. 13 Seven-day rainfall prediction data from 16-06-2018 (NASA Precipitation and Applications Viewer, Goddard Space Flight Center, https://pmm.nasa.gov/precip-apps, [last accessed: 18 June 2019])

8.4.2 Rainfall Pattern

CCT has a Mediterranean climate with an average annual rainfall between 560 and 1400 mm.[69] The rainy season is from April to September. The moisture content evaporated from land and water due to temperature results in cloud formation which in turn results in rainfall. A term used for the condensation of this moisture or water vapors, which falls down the earth due to gravity is called "precipitation".[70] This precipitation pattern can be predicted using several techniques as discussed earlier, but the only method to verify this precipitation or rainfall pattern is to physically measure the amount of rain that has fallen on a particular area by using a rain gauge. Usually this precipitation or rainfall is measured in mm or inches.

Aim of Rainfall Pattern Analysis

The aim of this section is to analyze the rainfall patterns around CCT using existing rainfall data recorded at different rain gauge stations. There are several analyses discussed for rainfall prediction which will be useful for the rainwater harvesting location and new dam construction. In this section, several rainfall stations have been included to obtain a better picture of rainfall patterns within the city thereby

[69]SA Explorer, Cape Town Climate, http://www.saexplorer.co.za/south-africa/climate/cape_town_climate.asp, [last accessed: 18 June 2019].

[70]United States Geological Survey (USGS), Precipitation and the water cycle, https://www.usgs.gov/special-topic/water-science-school/science/precipitation-and-water-cycle?qt-science_center_objects=0#qt-science_center_objects, [last accessed: 18 June 2019].

allowing identification of regions of the city which receive more rainfall during the year. Those areas will be suitable for water harvesting or potentially for new dam constructions. The details of the rainfall stations are given below.

Selected Locations

The data of 11 key stations have been selected for rainfall pattern analysis. From each location a complete record of rainfall data from 2011 to present has been obtained and analysed. These key stations are located at Brooklands, Constantiaberg, Woodhead, Newlands, Tygerberg, Blackheath Upper, Wemmershoek, SAEON Dwarsberg, Steenbras, Theewaterskloof, and Dwarsberg as shown in Fig. 14.

Data Source

Rainfall data for the 11 locations was obtained from "The Climate System Analysis Group at the University of Cape Town" website, which compiles data from several sources like the CCT water dashboard, SAEON, Agricultural Research Council (ARC), Fynbos Fire Project, and some other sources.[71] This data has been used to determine the rainfall pattern over CCT.

Analysis

Figure 15 depicts how the annual rainfall pattern in 11 locations in the CCT region varied between 2011 and 2017.

Figure 16 shows the maximum, minimum and average annual rainfall recorded in the CCT region since 2011. It shows that the minimum rainfall for the last seven years is fairly constant at approximately 400 mm while more variation can be noticed in the maximum rainfall—ranging from 1000 to 2500 mm per year. Moreover, in 2013 and 2014, an average of 1200 to 1400 mm annual rainfall was recorded in the CCT region. The overall trend indicates that since 2011, annual rainfall increased between 2011 and 2014. However, there was a sudden drop in 2015. Rainfall recorded in 2015 was approximately half of what was recorded in 2013. Post 2015, the minimum and average rainfall level stabilized at 400 mm and 800 mm respectively, while the maximum rainfall increased from 1400 mm to 2100 mm.

Figure 17 shows the quarterly rainfall pattern in the CCT region for the past 7 years. It shows that 60% of the region's annual rainfall is recorded from May to August (which are the rainy months), while rest of the year only has 40% of annual rainfall. This implies that rainwater harvesting would be very efficient in May, June, July and August. Furthermore, this also means that all water reservoirs including dams must have adequate storage to get through the other eight months of the year when rainfall is less.

---

[71]Ibid.

Fig. 14 Location of rainfall gauges (Climate System Analysis group (CSAG) at the University of Cape Town, Current season's rainfall in Cape Town, http://www.csag.uct.ac.za/current-seasons-rainfall-in-cape-town/, [last accessed: 18 June 2019])

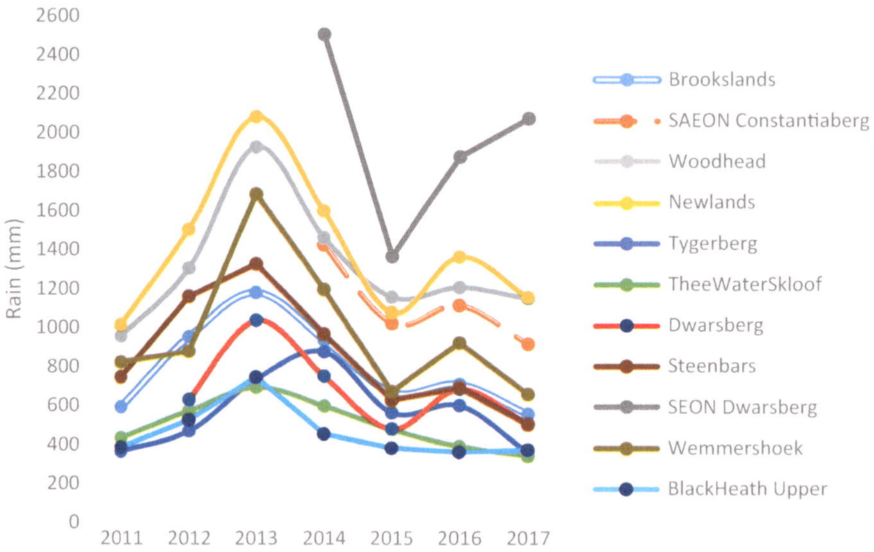

Fig. 15 Annual rainfall pattern at 11 locations in Cape Town

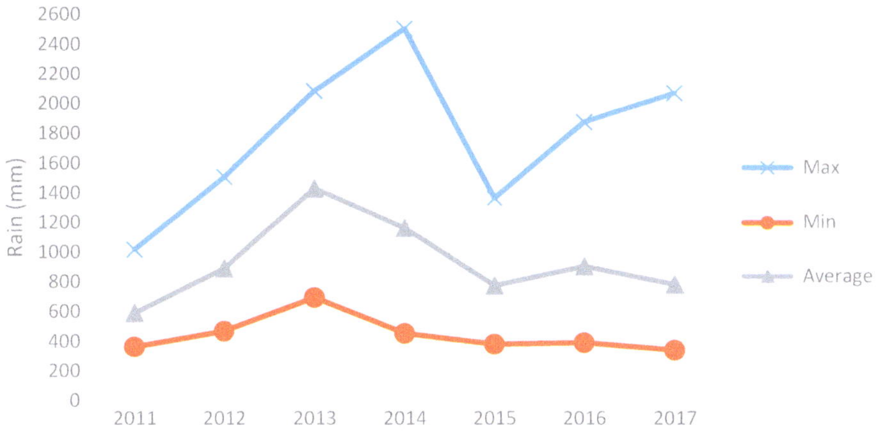

Fig. 16 Maximum, minimum and average annual rainfall for Cape Town region

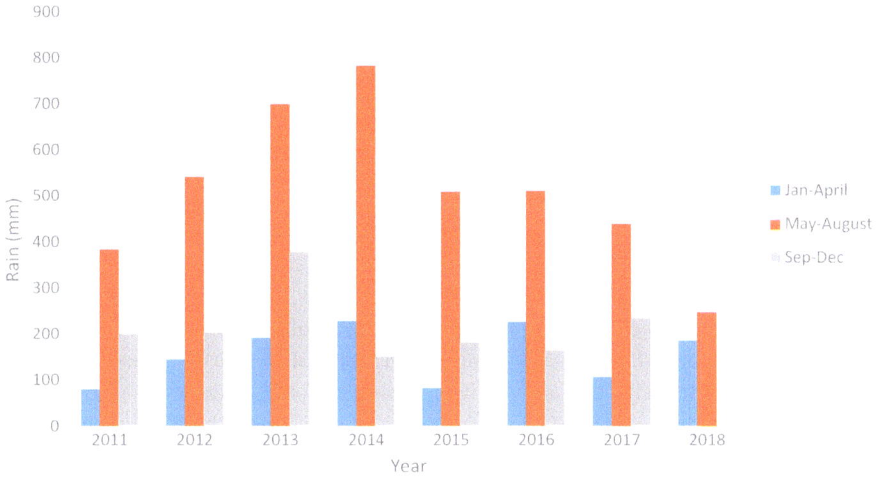

Fig. 17 Quarterly rainfall pattern in the Cape Town region

8.4.3 Surface Area Mapping

The larger the catchment area, the more rain can be collected. This is particularly important for roof top harvesting. Satellite data can be used to identify potential rooftop area availability to harvest rainfall. Two methodologies are proposed for the identification of relevant rooftop areas for rainwater harvesting.

Open Street Maps

The following procedure is used to identify rooftop areas:

- Divide the relevant CCT region into smaller areas to allow for improved building area estimation
- Estimate the rooftop area of a typical building within all smaller areas using Google Maps
- Average all rooftop areas for the smaller areas
- Count the number of buildings in CCT using the Overpass-Turbo application programming interface (API) along with the Open Street Maps (OSM) database
- Calculate the rain harvesting area by multiplying the average building area by number of buildings.

The above procedure works on the node identification of buildings in OSM. The Overpass-Turbo API is an online, read-only API used for data mining and querying of OSM map data for selected areas. In order to obtain relevant data and information from the OSM database, suitable queries and/or filters need to be applied to the data set.

Table 2 displays the average rooftop area of ten different sites around CCT calculated using OSM. This data, calculated using Google Maps, is used to estimate the rain water harvesting potential of each area (see Table 3).

Rainwater harvesting potential for nominal rainfall is 11.725 billion litres over the year. This averages to 32 million litres per day which is very low compared to the 565 million litres the city consumes each day.[72] However, the benefit of rainwater harvesting is that it will not prevent water loss in the form of pipe leakages over long distances. In addition to water loss, it will also reduce maintenance costs associated with pipelines as well as reduce the overall demand on the water supply system.

The amount of water that could be harvested is calculated for a small area around Signal Hill as depicted in Fig. 18 using actual rain data for 2017. This is done to compare the results with the above CCT analysis. Table 4 shows the amount of water which could be extracted in a typical year if all of these houses had a rainfall water harvesting systems installed. The data used is provided using the rainfall pattern monitoring shown in the previous section.

As actual rain data is similar to our assumed data, the results are comparable to actual values.

[72]J Evans, Cape Town's water consumption figures go up again, News 24, South Africa, https://www.news24.com/SouthAfrica/News/cape-towns-water-consumption-figures-go-up-again-20180320, [last accessed: 18 June 2019].

Table 2 Average area calculation over Cape Town

Serial No.	Cape Town areas	Average rooftop area (sq m)
1	Simmons town	200
2	Ocean view	215
3	Fish Hoek	140
4	Camps Bay	300
5	Rondebosch	270
6	Bellville	250
7	Good wood	280
8	Hout Bay	350
9	Muizenberg	100
10	Mitchells Plain	115
11	Kuils river	135
12	Strandfontein village	240
13	Khayelitsha	210
14	Sommerset	150
	Average Cape Town Rooftop Area	211.07

Table 3 Cape Town rain water harvesting potential

Parameter	Value	Units
Average area per house	211	sq meter
No of houses	123,491	number
Total area	26,056,601	sq meter
Rainfall in 2017	0.5	mm
Collection efficiency	0.9	na
Total water	11.7254705	million cubic meter
Water in million litres	11725.4705	million litres

8.5 Discussion

Rainfall prediction methods and recent rainfall patterns have been analysed. The rooftop surface area is estimated and based on the results of rainfall prediction and pattern, the amount of rainwater that could potentially be harvested is presented for the City of Cape Town. It has been shown that if rainfall harvesting systems were implemented on all the roofs in the City, up to 30 ML of water per day could be captured during the rainy season. Small houses and commercial buildings can install rainfall harvesting systems to partially meet their daily water needs. In order to reduce overall costs, harvested water can be used for general washing and toilet flushing, which tend to make up the majority of domestic water usages. It is not feasible to use collected rainwater for drinking water as the cost of the required water filtration and purification systems becomes prohibitive.

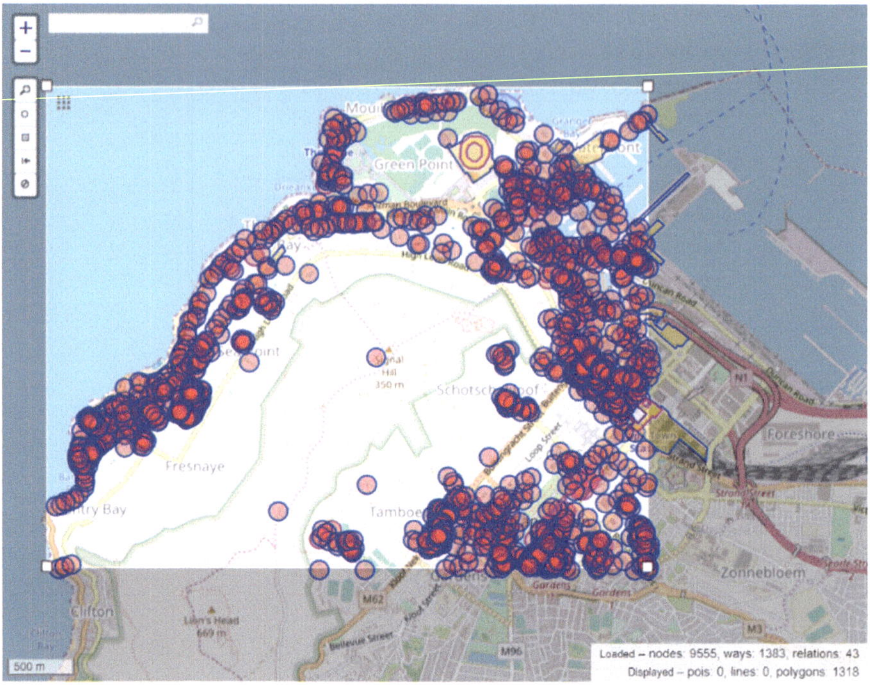

Fig. 18 Rainfall harvesting area depicting buildings from the overpass-turbo API used with OpenStreet Maps (M. Raifer, Overpass Turbo API—open source web-based data filtering tool for OpenStreetMap, https://overpass-turbo.eu/, [last accessed: 18 June 2019])

Table 4 Rainwater harvesting potential

Parameter	Value	Units
Area per house	190	sq meter
No of houses	1383	
Total area	262,770	sq meter
Rainfall in 2017	0.518	mm
Collection efficiency	0.9	
Total water	0.1225034	cubic meter
Water in million litres	122.50337	million litres

By subsidizing these rainwater harvesting systems, the City can incentivize its citizens to install these systems and reduce the overall demand on the water supply infrastructure. It is also recommended that the City mandate that all future housing developments be equipped with suitable rainwater harvesting and greywater systems.

It is recommended that the Western Cape Province perform an in-depth analysis of the rainwater patterns in the area to determine if there have been any significant changes in these patterns. This, supplemented with the installation of additional rain gauges throughout the province, will allow for the identification of potential new catchment areas. Installing multiple, small catchments in high rainfall areas can be used to help combat the water crisis. These smaller catchments can be used to provide water to local communities to help reduce the amount of capital and time required to construct the new water sources and integrate them to the main infrastructure.

9 Other Applications for Water Resource Management

This section discusses other areas where space-based RS applications can be used to optimize and monitor water resource management activities, specifically with respect to: (i) monitoring potential subsidence associated with groundwater extraction, (ii) monitoring sea-water intrusion at groundwater extraction points, (iii) detecting and monitoring leakages along water pipelines, (iv) environmental impact monitoring of desalination plants, (v) locating new groundwater sources, (vi) tracking and predicting urbanization and population growth to assist in planning new water infrastructure, and lastly, (vii) various technologies which can be used to reduce the societal and economic impacts associated with droughts.

9.1 Using Groundwater Reservoirs and Avoiding Subsidence

CCT is facing an unprecedented water shortage and as such requires careful monitoring and management of available water resources. About 98% of the city's water supply comes from six main dams, 2% comes from smaller dams, groundwater and surface runoffs. A small portion (about 2%) of the water is being obtained from groundwater sources that include the Atlantis Aquifer, the Cape Flats Aquifer (CFA) and the Table Mountain Group (TMG) Aquifer.[73] However, 17%[74] of the City's budget will be spent on expanding groundwater extraction from these aquifers, and in future it is expected that the share of groundwater supply will be increased.

The Cape Flats Aquifer has a capacity of more than 600 Mm^3, and the Table Mountain Group Aquifer has more than 1000 Mm^3. Total groundwater storage is estimated to be much larger than the total storage of water in dams (i.e.

[73]Gibson D., "Water Crisis in Cape Town: Lessons to be Learnt, Part 1: How The City's Water Supply And Management System Needs To Change" South Africa, February 2018, www.kas.de/suedafrika, Accessed [07-06-2017].
[74]City of Cape Town 2018/19—2020/21 Draft Budget, May 2018.

900 Mm^3 of water storing capacity of six major dams of Cape Town), and is also not affected by evaporation.[75] This water can be classified as 'fresh water', within the drinking water standards (with <1,000 mg/L Total Dissolved Solids).[76]

The potential problems associated with extraction of groundwater from these aquifers, especially during periods of drought, include insufficient recharge, subsidence and sea water intrusion. The groundwater reserves may not recharge at a sufficiently sustainable rate relative to the rate of withdrawal. In some conditions where lots of groundwater is extracted, the aquifer itself can collapse and land subsidence occurs.[77] Groundwater is also crucial for supporting natural ecosystems.[78] For coastal areas, like Cape Town, if a large volume is extracted there is a risk of seawater 'intruding' into the aquifers that would make this water unusable for domestic or industrial users unless prior desalination is performed. Furthermore, seawater intrusion can have catastrophic effects on the local environment and biomes. This implies that the extraction of groundwater must be carefully managed and monitored.[79]

Satellites can't solve CCT's water drought, but better imaging and better information received from satellites will allow for detailed monitoring and prevention of the aforementioned issues.

It is suggested that the Government installs continuously operating Global Navigation Satellite System (GNSS) receivers at all facilities where groundwater extraction is being (and will be) performed. Data from GNSS receivers can be efficiently used for different applications in municipal areas, for example topographical and cadastral surveys, and serving as base stations for engineering. dInSAR (Differential Interferometric Synthetic Aperture Radar) and in situ micro-gravity measurements can provide independent and continuous measurements of subsidence (in the range of millimetres) and hence should be employed to monitor potential subsidence at areas where groundwater extraction is performed. All these cases require special consideration for the instruments' monumentation, management, measurement and data processing to detect and measure subsidence.[80]

dInSAR is used to observe phase changes between two images taken in a given time and thus can be used to monitor subsidence, lateral deformation, or upliftment

[75]"WATER OUTLOOK 2018 REPORT" Revision 25—updated 20 May 2018, Department of Water and Sanitation, City of Cape Town.

[76]J Cameron, "Radical ideas to solve Cape Town water crisis—from expert who spotted Day Zero", BizNews, Feb 2018, https://www.biznews.com/thought-leaders/2018/02/09/solve-cape-town-water-crisis, Accessed [18 June 2019].

[77]S. A. Leak, "Land Subsidence From Ground-Water Pumping", U.S. Geological Survey, https://geochange.er.usgs.gov/sw/changes/anthropogenic/subside/, Accessed [18 June 2019].

[78]D Venton, "How Satellites Can Monitor California's Underground Water", Wired Magazine, July 2018 https://www.wired.com/2015/05/groundwater-insar/, Accessed [18 June 2019].

[79]N Narrandes, "Groundwater belongs to all South Africans"—WWF, Cape Town Etc Magazine, Jan 2018, http://www.capetownetc.com/news/wwf-emergency-bylaws-groudwater/ Accessed [18 June 2019].

[80]Wonnacott R., Hartnady C., Engelbrecht J., "The management of scarce water resources using GNSS, InSAR and in situ micro gravity measurements as monitoring tools, South African Journal of Geomatics", Vol. 4, No. 3, August 2015.

of the surface. In areas where groundwater extraction could cause subsidence, the subsidence can be avoided by monitoring through satellites and managing the groundwater extraction activities accordingly. The subsidence can be restricted by taking measures such as suitable locations for artificial recharge of underground sources. Satellite-based models coupled with other models like groundwater presence and flow models can be used to develop such measures.[81]

9.2 Monitoring Water Leakages

Water leakage through pipelines is contributing its part in the water crisis of CCT. Leakages cause a huge waste of the precious resource and result in large revenue losses for the City. The consequences of loss of water through the water distribution network and pipelines are obviously much worse in areas facing water shortages. According to the "Water Market Intelligence Report 2018",[82] South Africa consumes almost 4,000 Mm^3/year, including 41% non-revenue water (NRW) at national level which equates to a loss of R 6.3 billion in revenue each year. Around 21.3% of the Western Cape's fresh water is lost as NRW.

The main contributing factors for pipe leakages are the age of distribution pipelines, the material quality, maintenance and operation regimes (e.g. cathodic protection systems), local soil types and in-line water pressure. Underground leakages have to be identified and repaired to reduce the number of leaks and save this precious resource. Underground water leakages are not always visible. Hidden leaks can often go unidentified for extended periods of time, resulting in large water losses and sometimes causing sink holes and other damage to existing infrastructure. Currently CCT utilizes certain ground-based techniques to identify leaks.

Some of the ground-based leak detection methods are based on pressure monitoring, flow monitoring, acoustic surveys, geophones, correlators, thermal imaging and infrared scanning. Satellite-based leak detection[83] involves using remote sensing satellites' images to identify leaks and is more accurate[84] than ground-based systems. Satellite-based leak detection can identify more leaks, over a larger area within reduced time-frames, when compared to ground-based technologies. The main advantage of satellite-based systems over ground-based systems is the time required to perform a survey. Ground-based leak detection surveys are

[81] S. A. Leak, "Land Subsidence From Ground-Water Pumping", U.S. Geological Survey, https://geochange.er.usgs.gov/sw/changes/anthropogenic/subside/, Accessed [18 June 2019].

[82] L Basson et al., Water—2018 Market intelligence report, GreenCape, 2018, Page: 25, https://www.greencape.co.za/assets/Uploads/GreenCape-Water-MIR-FINAL-20180405.pdf, Accessed [06 June 2019].

[83] Dr. D Paudyal, Could satellites be the secret to detecting water leaks, ESRI Australia, Sep 2016, https://esriaustralia.com.au/esri-australia-blog/could-satellites-be-the-secret-to-detecting-water-leaks-blg-37, Accessed [06 June 2019].

[84] P Schumi, Advanced leak detection technology utilizing satellite imagery, Water Asset Management Conference, Hydromax USA and Utilis, Oct 2017, https://www.waterassetmgmt.com/pdfs/presentations/s3-3-advanced-leak-detection-technology.pdf, Accessed [06 June 2019].

conducted yearly or half-yearly, while satellite-based surveys can be conducted on a monthly basis (or even shorter). With this technology, leaks can be identified and repaired more rapidly than before.

In the CCT, a significant amount of water is lost through pipeline leakages. "Cape Town water losses were recorded around 16% in October 2017".[85,86] Comparing this with the recorded national water losses due to pipeline leakages, CCT has one of the best water distribution networks in South Africa. However, satellite-based technology can be employed to reduce these losses even further to overcome the water crisis. Furthermore, the current situation of dam water levels is such that CCT cannot afford such water leakages. Although the rainy season has started, and rainfall has slightly increased in comparison with the last three years,[87] the accumulative water level in major dams of Cape Town is around 38.1%[88] of total storage capacity as on 18 June 2018 i.e. 342,229 ML. If it rains significantly throughout the season, the dams will still not be filled to their maximum capacity to realize water security for its citizens.

Typical water pressures in CCT water transmission pipelines are around 20 bar and in distribution piping between 2 and 4 bar (depending on the local topography). A small leak in the transmission line and distribution network can cause hundreds of litres of water loss if leakage detection is delayed.

The CCT has more than 11,000 km[89] of water network infrastructure and employs three teams for active leak detection and maintenance.[90] The teams are responsible for field measurements, visual surveys and non-visible leak detections using electronic measurement devices. CCT also publishes guidelines[91] for finding and fixing water leakages at home or underground. Satellite-based modern leak detection technology can result in more efficient monitoring and maintenance of water infrastructure and can cut down water losses.

---

[85]P Flower (Director of Water and Sanitation, CCT), City above average on pipe leaks, News 24, South Africa, Feb 2018, www.news24.com/SouthAfrica/Local/Peoples-Post/city-above-average-on-pipe-leaks-20180219, Accessed [06 June 2019].

[86]D Olivier, Cape Town water crisis: 7 myths that must be bust, The Conversation, Nov 2017, https://theconversation.com/cape-town-water-crisis-7-myths-that-must-be-bust-86582, Accessed [18 June 2018].

[87]World Weather Online, Cape Town Monthly Climate Averages, https://www.worldweatheronline.com/cape-town-weather-averages/western-cape/za.aspx, Accessed [18 June 2019].

[88]City of Cape Town, Dam Levels Report (as of 20 June 2018), https://resource.capetown.gov.za/documentcentre/Documents/City%20research%20reports%20and%20review/damlevels.pdf, Accessed [20-JUN-18].

[89]Southern Africa Stainless Steel Development Association, Water woes, Sassda Newsletter, March 2017, https://sassda.co.za/wp-content/uploads/2016/11/1-March_To-Build_Water-Woes-p.15.pdf, Accessed [18-JUN-19].

[90]Z Basholo, Overview of water demand management initiatives: a City of Cape Town Approach, Feb 2016, South Africa, https://www.greencape.co.za/assets/Water-Sector-Desk-Content/CoCT-WCWDM-presentation-Z-Basholo-Western-Cape-Water-Forum-160204-2016.pdf, Accessed [18 June 2019].

[91]City of Cape Town, DIY Guide to finding and fixing water leaks, http://resource.capetown.gov.za/documentcentre/Documents/Procedures,%20guidelines%20and%20regulations/Water_Finding_Fixing_Water_Leaks.pdf, Page-17, Accessed [18 June 2019].

Remote sensing technology is the modern way of leak detection and is currently being used by many developed countries. Both active and passive remote sensing technologies aid in leak detection. Satellite images are acquired and processed after radiometric corrections. Based on the spectral signature, water leakages can be detected and mapped into GIS for use by the relevant repair and maintenance department. Use of this technology results in effective leak detection and repairs thereof. Utilis[92] is a private company specialised in water leak detection with the help of satellite technology and operates around the world. Utilis holds a U.S. patent[93] for a "System and Method of Underground Water Detection" to overcome water shortage problems (see Fig. 19). It has many partner companies from developed countries that use Utilis services for water leak detection and to develop effective repair implementation plans.

Spectrometers measure the electromagnetic radiation reflection of different materials. Each material/element produces a unique spectral signature. L-band synthetic aperture radar (SAR) images are employed for water leak detection. The reason for choosing L-band radar is that its electromagnetic waves penetrate almost everything on the ground (i.e. trees, shrubs and grasses) and make it useful for detection of water leakages and seepage. Also, the SAR images are weather independent and can be captured during both day and night time. By analysing the earth surface temporal changes and variation over time, leakages can be identified. This technology appears to be very effective in monitoring water transportation networks from reservoirs to water distribution networks to identify ruptures and leakages in long transmission pipelines, however it has limitations in urban distribution networks where buildings, cars and other modern infrastructure introduce noise and interference. Multi-spectral and hyperspectral images are used to resolve this issue. Hyperspectral images are able to gather information from hundreds of spectral bands,[94] thereby enabling identification of water based on its spectral signature even in noisy and congested areas.

9.3 Desalination

Desalination is the process of extracting salt and minerals from saline water in order to create fresh or potable water. Many countries use desalination to overcome their dependency on traditional water supplies. For example, Dubai use desalination plants to supplement the water they receive from groundwater stores. In parts of Australia, desalination plants have been installed due to an annual decrease in

[92]Utilis, Satellite-based infrastructure intelligence, https://utiliscorp.com, Accessed [18 June 2019].

[93]Utilis Israel Ltd, System and method of underground water detection, US9285475B1, https://patents.google.com/patent/US9285475B1/en, Accessed [18 June 2019].

[94]D Weaver, Satellite sensors: How to tap into capabilities of multi and hyper-spectral satellite data to monitor ROW encroachments, leak detection to vegetation health, Pipeline Association of Houston, Mar 2017, https://www.houstonpipeliners.net/docs/PAH_2017_03_-_How_to_Tap_into_Satelliate_Data_to_Monitor_Pipelines.pdf, Accessed [18 June 2019].

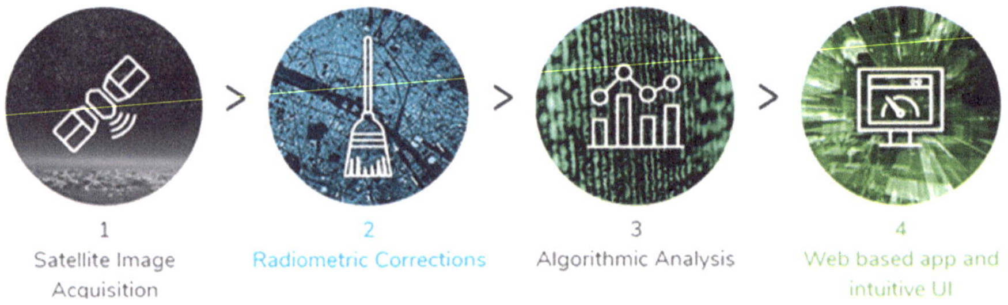

Fig. 19 Utilis satellite-based water leak detection system (Utilis, Leak detection, Utilis, USA, https://utiliscorp.com/leak-detection/, Accessed [18 June 2019])

rainfall resulting from climate change. To meet the water demand of Dubai "1574.7 million litres on average",[95] only 0.5% of this demand is supplied using groundwater and the remaining 99.5%[96] is supplied from desalinated water. The Jebel Ali desalination plant takes 10.6 billion litres[97] daily from the Persian Gulf and turns it into fresh drinking water. Despite all these efforts of desalination only 9% becomes drinkable water.[98]

After facing their worst drought from 1997 to 2010, Australia applied a number of techniques which included water recycling, water trading, grey water and storm water collection. However, the largest response was for seawater desalination and potable water.[99] In five cities of Australia, namely Perth, Sydney, Melbourne, Adelaide and South East Queensland desalination plants were constructed. The desalination plants of South East Queensland and Perth are providing an average of "145 billion litres per year"[100] for half of the population of Perth.

CCT's first desalination plant became operational in May 2018 at Strandfontein and is a temporary desalination plant. Cape Town's Deputy Mayor, Ian Neilson said that the tender for the Strandfontein plant is only for two years but they have the option to prolong it until the City builds a permanent desalination plant. The desalination plant will supplement the existing water supply in the area. CCT has two more planned desalination plants at V&A Waterfront and Monwabisi. Strandfontein and Monwabisi desalination plants should contribute 14 million litres to the daily water supply.

However, the water produced from desalination does not fulfil the demands and is very expensive—it costs approximately R40/kilolitre while water from the dams

[95]K. D. Atherton, "Dubai's massive desalination plant", Popular Science, Feb 2017, https://www.popsci.com/dubais-desalination-plant-photos#page-2, Accessed [18 June 2019].
[96]Ibid.
[97]Ibid.
[98]Ibid.
[99]International Water Association, Desalination—the Australian Experience, International Water Association, Aug 2016, http://www.iwa-network.org/news/desalination-australian-experience/, Accessed [18 June 2019].
[100]Ibid.

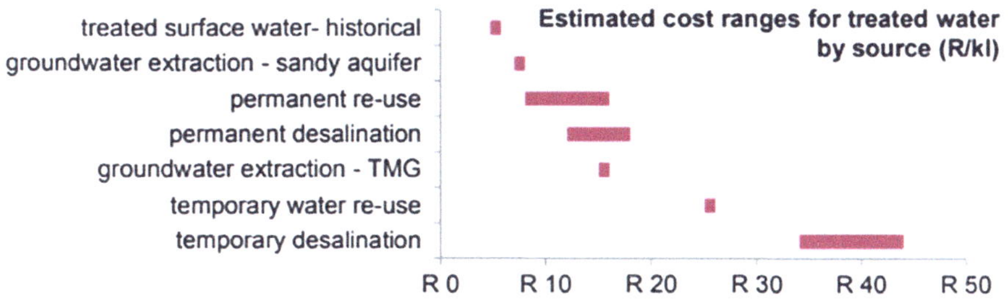

Fig. 20 Treated water cost (City of Cape Town, Cape Town Water Outlook 2018, Department of Water and Sanitation, City of Cape Town, Jul 2018, https://resource.capetown.gov.za/documentcentre/Documents/City%20research%20reports%20and%20review/Water%20Outlook%202018%20Presentation.pdf, Accessed [18 June 2019])

cost approximately R5/kilolitre.[101] Figure 20 shows the costs associated with various treated water technologies.

Brine has a high temperature, high salinity level and high concentration of suspended materials. In discharge points at desalination plants, brine causes negative effects in the local water environment if not properly managed. Due to its high salinity and high temperature, brine results in low concentrations of oxygen which influences the water quality. Discharged brine sinks to the bottom due to its higher density than sea water which causes it to be harmful for sea species and due to its high chlorine content it can kill sea organisms near the discharge point.

CCT has recently installed desalination plants with low potable water production, if it has to cut down its groundwater demand just like some of the other countries have done, South Africa may consider installing large desalination plants at different bays. In order to monitor the effect of desalination on the marine environment, some satellite-based technologies like "Sea Surface Temperature (SST)", "Structure Similarity Factor (SSIM)", "Mean Squared Error (MSE)" and "Peak Signal to Noise Ratio (PSNR)" are used. These technologies are already implemented in Dubai to monitor their desalination plants through DubaiSat-1.[102]

As shown in Fig. 21 for understanding of spatial and spectral characteristics of satellite data, the area near the discharge point is divided into a 3 by 3 matrix window. For each visible band the MSE, SSIM, PSNR and average reflectance are calculated. It is clear that the water in window 8 of Fig. 21 near the discharge point is highly polluted, window 8 area has a low PSNR and high MSE in blue channel

[101]N. Shelly "Cape Town's first desalination plant comes online", Times Live, May 2018, South Africa, https://www.timeslive.co.za/news/south-africa/2018-05-21-cape-towns-first-desalination-plant-comes-online/, Accessed [18 June 2019].

[102]"Muhairi A. et al. "Using medium and high resolution satellite images in monitoring water quality surrounding the discharges of desalination plants in the UAE", 978-1-4244-9566-5/10/$26.00 IEEE https://www.researchgate.net/publication/220823227_Using_medium_and_high_resolution_satellite_images_in_monitoring_water_quality_surrounding_the_discharges_of_desalination_plants_in_the_UAE".

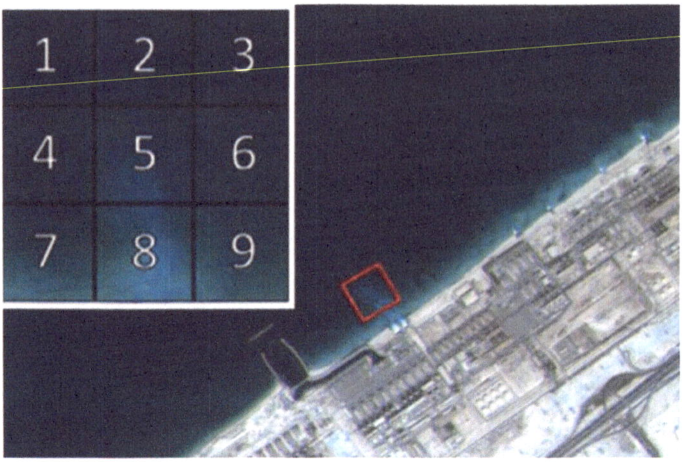

Fig. 21 Largest discharge point of Jebel Ali plant (DubaiSat-1) (Ibid.)

while it has low SSIM in green channel and a high reflectance. The most suitable method for monitoring brine levels is the reflectance ratio and SSIM because a big difference in both these means abnormality due to discharge water.[103] Table 5 shows the risks and their potential impacts and mitigation thereof associated with desalination plants.

9.4 Finding New Aquifers

To combat water crises, CCT may seek new water sources. Exploration and identification of new groundwater sources (i.e. aquifers) is very challenging and numerous studies and surveys are required. These studies include lithology, porosity, geomorphology, soil characteristics and weather conditions. These studies are time consuming and their results are only applicable to small localised areas. These studies can be done with the aid of satellite imagery which can cover a large area survey in short intervals. Satellite data does not identify the water resources directly but it can be processed and analysed for inferences of water resource exploration. Remote sensing data along with thematic maps from Geographic Information System (GIS) and ground surveys have been very useful in exploration and inferencing of groundwater resources.[104]

[103]Ibid.
[104]"Krishnamurthy, J., Venkatesesa, K.N., 1996. An approach to demarcate ground water potential maps through remote sensing and GIS. International Journal of Remote Sensing 7, 1867–1884".

Table 5 Impacts and mitigation of desalination plant (M. Dawoud and M. Al Mulla, Environmental impacts of seawater desalination: Arabian Gulf case study, International Journal of Environment and Sustainability, ISSN 1927-9566, Vol 1 No 3, 2012)

Risk	Potential impacts	Mitigation measures
Brine discharge	Salinity and temperature increase	Brine water diluting brine water with cooling water or seawater
Air pollution	Emission of SO_2, NO, and CO_2	Natural gas use Renewable energy use

Groundwater resources are found in areas with:

- High infiltration capacity
- Low surface run-off/slope
- High soil porosity and permeability
- Thicker porous and permeable zones
- Dense vegetation.

Soil infiltration capacity determines the groundwater potential and depends on soil porosity, soil permeability and the surface run-off. For the steep slope/inclined regions, the surface run-off becomes very high, thereby reducing the infiltration capacity of the soil. The areas with dense vegetation also have high infiltration capacity. Thicker and porous permeable zones are also important factors for the possibility of groundwater resources. Image processing techniques are applied on satellite images to enhance and identify lineament, slope, drainage features and for generation of Digital Elevation Models (DEM). This data in terms of thematic maps is fed into GIS containing field data results and other geological and lithological data. GIS is used for the integration of all thematic layers to identify zones that exhibit a high potential of groundwater.[105] Field data and geological survey reports are used to validate the GIS outcome of potential groundwater zones. Use of satellite images and GIS tools is a cost-effective solution that can perform groundwater aquifer explorations over large areas in a short duration of time. The identified zones from this solution will exhibit a very high likelihood of water resources and can then be targeted for drilling and additional testing.

Prospective systems can be developed and used by the Western Cape government. Although, CCT has three large and well-known aquifers namely the Cape Flat Aquifer (CFA), Table Mountain Group (TMG) and Atlantis Aquifer with abundant underground water, the above discussed GIS-based solution can aid in exploration of new aquifers for identification of new prospective zones for

[105]"Kamal Khodaei and Hamid Reza Nassery, Groundwater exploration using remote sensing and geographic information systems in a semi-arid area, Arabian Journal of Geosciences, April 2011".

groundwater extraction. This system can also be made part of the National Groundwater Information System (NGIS)[106] of South Africa for regional and national use.

9.5 Satellite-Based Population Modelling for Future Water Planning

Increase in urbanization is a global trend and therefore monitoring and mapping of urban resources and population growth is of great importance. In order to estimate water demand in the future, it is important to have correct estimates of population growth and density in different areas of the city. The number of households has increased from 1,068,573 in 2011 by 18.3% to 1,264,849 in 2016 within the greater Cape Metro area.[107] The population has increased from 2,892,243 in 2001 by 2.57% per year to 3,740,026 in 2011[108] and is expected to be 4,477,365 (approx.) by 2018.[109]

The most recent census in South Africa was done in 2011. However, the population increase and water demand can be tracked by preparing models from analysing satellite(s) data assisted by topographic maps, cadastral vector data and census data. Comparing population models of the city for different years would provide better understanding of the dynamic relationship between urban expansion and population growth[110] and thus water resources can be managed and planned for increased population requirement of the city in coming years.

9.6 Other Satellite Based Solutions for Economic and Environmental Problems

Satellite-based solutions to other problems associated with water droughts and/or scarcity are listed in Table 6.

[106]Department of Water and Sanitation, National groundwater information system (NGIS), Republic of South Africa, http://www.dwa.gov.za/Groundwater/NGIS.aspx, Accessed [18 June 2019].

[107]Western Cape Government, "Socio Economic Profile, City of Cape Town 2016", available at: https://www.westerncape.gov.za/assets/departments/treasury/Documents/Socio-economic-profiles/2016/City-of-Cape-Town/city_of_cape_town_2016_socio-economic_profile_sep-lg.pdf, Accessed [18 June 2019].

[108]Statistics South Africa, City of Cape Town, South Africa, http://www.statssa.gov.za/?page_id=1021&id=city-of-cape-town-municipality Accessed [18 June 2019].

[109]Ibid.

[110]Taylor & Francis, Thomas Alan Morton & Fei Yuan, "Analysis of population dynamics using satellite remote sensing and US census data", Geocarto International, 2009.

Table 6 Satellite based solutions to economic and environmental problems

Impact type	Problem	Satellite based solutions
Economic	Crop shortage	(a) Precision agriculture with GPS and remote sensing satellites (b) Weather monitoring
	Livestock shortage	(a) Mapping grazing areas (b) Multipurpose trees for alternate feed resources, soil mapping (c) Better water management for livestock (d) ARGOS* based livestock tracking
	Suffering agriculture based businesses	(a) Incorporating GPS and soil sensors in agriculture equipment (b) Soil mapping to dictate quality and quantity of fertilizers (c) Weed mapping
	Suffering timber industry	(a) Health mapping of forestry (b) Wild fire monitoring (c) Tree farming area definition
	Suffering fishing industry	(a) Ocean monitoring for increase in salinity due to desalination plants (b) Water quality monitoring
Environmental	Problems for marine life	(a) Ocean monitoring for oil spills (b) ARGOS monitoring (c) Water quality monitoring
	Problems for wild animals	(a) Forest monitoring (b) ARGOS monitoring
	Migration of wildlife	(a) Forest monitoring (b) ARGOS monitoring
	Extinction of endangered species	ARGOS monitoring
	More wild fires	Wild fire monitoring
	Air pollution	Environmental monitoring
	Disease outbreak	(a) Weather monitoring (b) Water Reservoir quality inspection (c) Spatial mapping and prediction (d) Tele-health
	More urbanization	Urban change mapping
	Informal settlements	Urban change mapping

*ARGOS is a satellite-based position estimation and data collection system that uses the Doppler effect for calculating the location of ARGOS transmitters and is used extensively in monitoring meteorological and oceanographic data as well as tracking wild animals. The French Space Agency (CNES), the National Aeronautics and Space Administration (NASA), and the National Oceanic and Atmospheric Administration (NOAA) started the ARGOS system in 1978 (ARGOS, International Cooperation, http://www.argos-system.org/argos/who-we-are/international-cooperation/)

10 Recommendations

This section discusses recommendations that the Western Cape Province and CCT can consider to assist their efforts in battling the water crisis and preventing "Day Zero". The major recommendations of any drought management plan are: (i) develop a comprehensive and integrated Drought Management Plan which contains policies to support short, medium and long term mitigation strategies, (ii) use EO data to continually monitor levels of surface water and groundwater stores to allow for equitable sharing of water resources among different sectors, (iii) include a broad range of stakeholders, including politicians, private entities and local representatives to achieve effective implementation of the drought management strategies, and (iv) engage with water and drought experts and scientists to formulate, optimize and improve the drought preparedness, mitigation and management strategies. Another important goal of drought management programmes is to diversify the number of water sources used to meet the demands of the province. This is currently planned, with CCT exploring groundwater extraction schemes and new desalination plants.

From a legal perspective, South Africa should aim to create new policies and restructure institutions to decentralise their water resource management schemes. In this regard, South Africa should look to the Murray-Darling River Basin Integrated Water Resource Management plan for guidance as it is a great example of a decentralised water resource management system that successfully mitigated and managed the impact of a local drought in Australia. The legal framework is structured on a "top-down" approach where power is redistributed from a central government authority down to the catchment level. This allows for effective monitoring of performance and allows for local parties to be responsible for their water resources, thereby accounting for the local differences in geography, climate, context and demands. This decentralised approach also helps overcome possible shortages of human resources and capacity.[111]

This document has shown that three, relatively small forests of alien trees (specifically pine trees) located near the Theewaterskloof Dam can consume up to 10 ML of water per day. This water is being lost to these alien trees and not adding to the water stores of CCT. It is recommended that the Western Cape Province develop a programme to perform identification and classification of alien trees located in catchment areas using remote sensing data. The programme can aim to automate the identification process based on recommendations contained within Sect. 7 of this document and in consultation with SANSA. Once complete, the newly developed programme can be used to identify dense alien tree forests and prioritise the removal thereof based on the density of the forest and its impact on the water resources within the Western Cape Province. The system can also be used to verify that removals are performed to satisfactory levels and can also monitor

[111]K Mokoena, "Decentralisation of water resource management: a comparative review of catchment management authorities in South Africa and Victoria, Australia", University of Cape Town, 2015.

regrowth in previously cleared areas. The system can be extended to also look at analyzing dam levels after removals are performed. Alien tree forests that are located in catchment areas will be specifically targeted and removed.

Section 8 of this document discussed the various rainfall pattern forecasting tools currently available, as well as how much water could be collected using rainwater harvesting techniques. It is suggested that the Western Cape perform an in-depth analysis on the current rainfall patterns and trends throughout the province to confirm that the 6 major dams and their catchment areas are still effective in collecting water. The analysis must aim to identify areas that exhibit increased rainfall over the past 5 years, with the long-term goal of constructing numerous small catchments in these areas to supplement existing water sources. It is recommended that CCT install additional rain gauges throughout the area to aid in analysis of rainwater patterns. Data from these gauges can be used to identify high rainfall areas and to target the installation of rainwater harvesting systems.

Furthermore, the government can consider subsidizing rainwater harvesting systems as a means to increase the number of private harvesting systems, thereby reducing the demand on the water supply system. Alternatively, it can consider establishing regulations that require all commercial business properties with roof areas over a certain threshold to install rainwater harvesting schemes. It is also proposed that all future Breaking New Ground (previously termed "RDP") housing schemes are equipped with rooftop rainwater harvesting schemes. The CCT should explore re-using water collected via the storm water systems within the city.

Section 9 of this document discussed other areas where satellite-based data can be used to supplement water resource management activities. These applications include: (i) monitoring potential subsidence associated with groundwater extraction, (ii) monitoring sea-water intrusion at groundwater extraction points, (iii) detecting and monitoring leakages along water pipelines, (iv) environmental impact monitoring of desalination plants, (v) locating new groundwater sources, (vi) tracking and predicting urbanization and population growth to assist in planning new water infrastructure, and lastly, (vii) various technologies which can be used to reduce the societal and economic impacts associated with droughts. Provincial government can approach SANSA to help develop these applications.

In discussions with representatives from SANSA it was revealed that SANSA host forums where representatives from all departments of government are invited to SANSA offices to showcase the numerous space-based applications and their advantages that are available to these departments. The frequency of these forums should be increased to be bi-annual and SANSA should consider distributing a monthly newsletter to all governmental departments to raise awareness of the benefits of space-based applications.

11 Conclusion

This publication has investigated the current water crisis facing the City of Cape Town and the Western Cape Province. It has given an overview of the water sources that constitute the total water supply to the City, as well as analysed the

City's response to the water crisis. It has been found that although SANSA do perform water body mapping, irrigation monitoring, flood risk analysis and water quality monitoring using EO data, there have been no new applications developed specifically to address the drought in the Western Cape. It is clear from interviews with both public and private role players that satellite applications are not being utilized to their full extent.

Moreover, this report has highlighted certain key satellite applications that the authors think can be effective in combatting the current water crisis. It is proposed that the government establish an alien tree mapping software that uses remote sensing data to automatically identify, classify and rank alien trees located in catchment areas. The negative impact of these alien trees has been highlighted in this document—three small forests located near Theewaterskloof Dam were calculated to be consuming up to 10 ML of water per day. The alien mapping software can be used by authorities to prioritise and plan removal programmes and to evaluate their efficacy.

In addition, this analysis has shown that in densely populated areas of CCT, as much as 30 million litres of water can be collected during the rainy season using appropriate rainfall harvesting systems. It is proposed that the government investigate this further, as well as perform an in-depth rainfall pattern analysis to identify potential areas for new catchments. The government can also consider subsidizing rainfall harvesting systems to encourage citizens to install these systems at their homes, thereby reducing the overall demand on the water system.

The benefits of other satellite applications in relation to water resource management have been discussed. It is clear from this discussion that there are numerous areas where satellite applications should be implemented to aid the province's response to the water crisis. These include using remote sensing data to avoid subsidence at groundwater extraction plants, identify leaks in water pipelines, monitor the environmental impacts of the planned desalination plants and help plan future water infrastructure projects based on urbanization trends.

Satellite applications are currently not being used to their full potential by the City or the province in their response to the water crisis. The City's water resource management activities can be greatly improved and assisted by the implementation of appropriate space-based technologies.

Appendix: The Big 5 of the African Sky

"The Big 5 of the African Sky are five celestial objects that represent the best specimens of each type of deep-sky class: the Southern Pleiades (an open star cluster), omega Centauri (a globular cluster), the eta Carinae Nebula (a bright nebula), the Coal Sack (a dark nebula), and the Milky Way (a galaxy). The Big 5 are visible anywhere from within the southern hemisphere. Two of the Big 5 lie in Carina, one lies in Centaurus, and one in Crux. The fifth – the Milky Way – lies in a narrow band dividing the sky in half. The brightest parts of the Milky Way are in Sagittarius, Scutum, Norma and Carina." (Astronomical Society of Southern Africa, Big 5 of the African Sky, https://assa.saao.ac.za/sections/deep-sky/big5/).

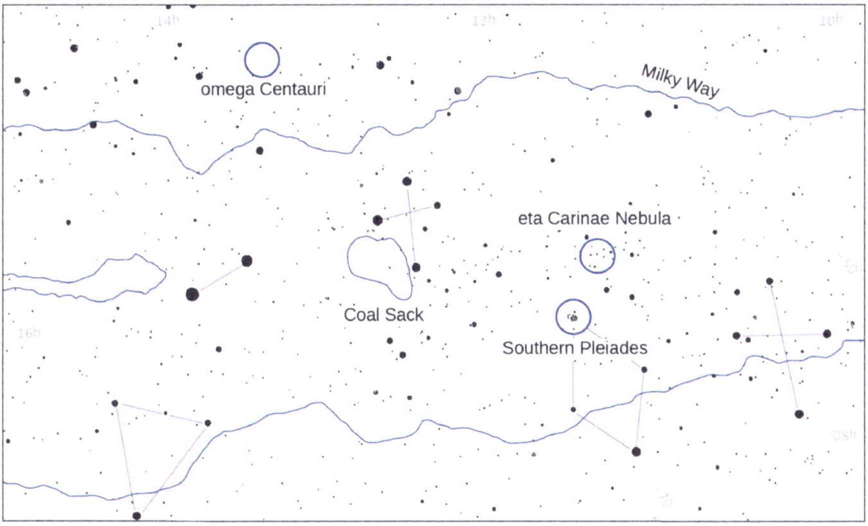

Big 5 of the African Sky, Courtesy of: Astronomical Society of Southern Africa

A. Froehlich (ed.), *Space Fostering African Societies*, Southern Space Studies,
https://doi.org/10.1007/978-3-030-32930-3

The manufacturer's authorised representative in the EU is Springer Nature Customer Service Centre GmbH, Europaplatz 3, 69115 Heidelberg, Germany. If you have any concerns regarding our products, please contact ProductSafety@springernature.com

Printed and bound by CPI Group (UK) Ltd, Croydon, CR0 4YY
30/09/2026
02225647-0011